Conducting Research in Conservation

Conducting Research in Conservation is the first textbook on social science research methods written specifically for use in the expanding and increasingly multidisciplinary field of environmental conservation.

It is a comprehensive and accessible guide to social science research methods for students of conservation-related subjects and practitioners trained in the natural sciences. Boxes provide definitions of key terms, practical tips, and brief narratives from students and practitioners describing the practical issues that they have faced in the field.

The first section on planning a research project includes chapters on the need for social science research in conservation, defining a research topic, methodology and sampling. Section II provides an in-depth focus on a range of social science methods including standard qualitative and quantitative methods such as participant observation, interviewing and questionnaires, and more advanced methods, such as ethnobiological methods for documenting local environmental knowledge and change, and participatory methods such as the 'PRA' toolbox. Section III focuses on practical issues in carrying out fieldwork with local communities, from fieldwork preparation and data collection to the relationships between the researcher and the study community. Section IV then demonstrates how to analyze social science data qualitatively and quantitatively; and the final section outlines the writing-up process and what should happen after the end of the formal research project.

Helen Newing has a first degree from Reading University in Zoology and Psychology and an interdisciplinary PhD from the Psychology and Biological Sciences departments at Stirling University. Since 1999 she has worked as a lecturer at the Durrell Institute of Conservation and Ecology (DICE) in the School of Anthropology and Conservation at the University of Kent.

C.M. Eagle has an MSc in Social Statistics and has been teaching research design and statistical analysis to conservation and anthropology students since the late 1990s at the University of Kent.

R.K. Puri is a Senior Lecturer in Environmental Anthropology and Ethnobiology at the University of Kent. He received his PhD in Anthropology from the University of Hawai'i in 1997.

C.W. Watson is Emeritus Professor of Social Anthropology and Multi-Cultural Studies at the University of Kent and a past member of the Council of the Royal Anthropological Institute.

Conducting Research in Conservation

Social science methods and practice

Helen Newing

with contributions from C.M. Eagle, R.K. Puri and C.W. Watson

LONDON AND NEW YORK

First published 2011
by Routledge
2 Park Square, Milton Park, Abingdon, Oxon OX14 4RN

Simultaneously published in the USA and Canada
by Routledge
711 Third Avenue, New York, NY 10017

Routledge is an imprint of the Taylor & Francis Group, an Informa business

Typeset in Times New Roman by Glyph International Ltd.

British Library Cataloguing in Publication Data
A catalogue record for this book is available from the British Library

Library of Congress Cataloging in Publication Data
Conducting research in conservation : social science methods and practice / by Helen Newing ... [et al.].
p. cm.
Includes bibliographical references and index.
1. Environmental sciences. 2. Environmental protection. 3. Conservation of natural resources–Government policy. I. Newing, Helen.
GE10.C67 2010
333.72072–dc22 2010004662

ISBN: 978-0-415-45791-0 (hbk)
ISBN: 978-0-415-45792-7 (pbk)
ISBN: 978-0-203-84645-2 (ebk)

Contents

List of tables

List of figures

List of boxes

Foreword

Conservation of necessity is a growing discipline because of the increasing threat to the biodiversity of the world. Every day conservation becomes a topic of greater urgency because ecosystems are being destroyed and species are becoming extinct. The United Nations target set to stop species loss by 2010 has not been met and so we need many better-trained conservationists to achieve this goal. Effective conservation is usually a very interdisciplinary subject. There is also inadequate realisation of the extent to which the social sciences are a vital part of any integrated conservation programme. Much previous work in conservation has been rather haphazard and opportunistic and so any text that seeks to bring more organisation and discipline to conservation is most welcome. I have taught many aspects of conservation to students in various different countries and I have always found that an inclusive text about conducting research is missing. One has had to draw from a variety of scattered sources.

This book will help to ensure that the novice conservationist is properly equipped before venturing into the field. I am impressed with the range of important topics that are addressed here. It is vital to plan a research project adequately and for the student to be fully equipped with knowledge of the different methods and approaches that are available. No two sites or projects are ever the same and so well-informed and adaptable researchers are needed in conservation. Fieldwork with local communities requires special skills and tact and the various topics included in the third section of this book are particularly relevant to biologists who have not had training in the social sciences. I am happy to see a chapter on the ethical issues of research as this has often been neglected in the past. Data processing has always been important, but today there are so many more facilities available and good conservation work is becoming progressively more quantitative so the chapters on quantitative analysis and statistical methods will be most useful. Any conservation project is meaningless without adequate writing up and dissemination of the results and this is well covered in part five of this book.

What I most like about this book is its comprehensive nature that brings together all the most relevant information needed to practice conservation in the field. It will certainly be used by anyone who is training ethnobotanists and conservationists. It is likely to be of use far beyond social sciences since anyone carrying out biological field work should also be made aware of the issues raised here. Most of the issues covered in this text are the same whether you are conducting conservation from a biological or social science approach. Whichever approach, conservation is so multidisciplinary that the practitioner needs knowledge from many different fields. Biological conservationists often neglect the social aspects of their work and so I thoroughly recommend this text for biologists as well as social scientists.

Ghillean Prance FRS
Scientific Director, The Eden Project

Acknowledgements

First, thanks are due to Andrew Mould at Routledge for helping to shape the initial proposal for this book, and also to Michael Jones and Faye Leerink at Routledge for their efficient guidance throughout the preparation process. Thanks also to Russ Bernard for his encouragement to go ahead with the book. Comments on the drafts by a large number of people improved the final text immeasurably; thanks especially to Callum Blaikie, Glenn Bowmann, Emily Caruso, Oliver Dreike, Christopher Eagle, Charlie Gardner, Tiffany Gilchrist, Warren Hickman, Kira Johnson, Mike Jones, Constanza Monterrubio, Gillian Newing, Michael Pearson, Adriana Puente, Laura Robson, Olivia Swinscow-Hall and David Zeitlyn. More widely, discussions and feedback on specific chapters by the MSc and PhD cohorts of 2007 and 2008 in the School of Anthropology and Conservation at the University of Kent were invaluable. Finally, Michael Fischer, Tiffany Gilchrist, Kira Johnson, Laura Robson, Lorna Slade and Olivia Swinscow-Hall have kindly permitted the use of anecdotes, unpublished correspondence and/or material from their postgraduate theses, for which I am grateful.

Above all I am indebted to my husband, Tim Fellowes, for his unfailing support and encouragement, without which the book would never have seen the light of day.

Helen Newing

Section I

Planning a research project

1 Introduction

Social science research in conservation

> The real question for debate … is not whether to integrate the social sciences into conservation but how to do so.
>
> (Mascia et al., 2003: 649)

1.1 Introduction

This book aims to provide a sound grounding in social science research methods for students and professionals working in the field of environmental conservation. Traditionally, conservation has been the domain of biologists, and an understanding of the biology of the species and habitats we wish to conserve will always be essential to successful conservation management. However, it is obvious that this alone is not enough. Most of the drivers of environmental change are social – they are to do with people – and therefore many of the biggest, most urgent challenges facing conservationists are to do with social, economic and political systems. Biological expertise is obviously of limited use in dealing with these issues, and yet around the world, practising conservationists trained in biology are called upon regularly to take management decisions regarding social issues. At a more theoretical level, our current understanding of social issues in conservation is fragmented and, in many subject areas, rudimentary. There is therefore an urgent need to develop social science expertise within the broad discipline of conservation. That is not to say that all conservationists should become social scientists, but they need to be able to work across disciplinary boundaries, whether alone or in multidisciplinary teams, and there is a strong argument that in order to do so successfully, they need a basic understanding of the principles of both the natural sciences and the social sciences. This book is designed to fill a gap in the training resources available to accomplish this goal by providing conservationists with a substantive introduction to the principles and practice of social science research.

Before going on to discuss the nature of research and explore some of the differences between the natural sciences and the social sciences, it is useful to ask why, if the social sciences are so vital to conservation, they did not become an integral part of conservation training long ago. Conservation has been a concern in policy circles and in the public imagination since the late nineteenth century (for an excellent account of its history, see Adams, 2004). However, for much of its history, the predominant approach has been to set up uninhabited protected areas where all forms of human use except research and tourism are prohibited or heavily restricted, and this has meant that there has been little need for social science perspectives. However, approaches to conservation based on untouched protected areas have been increasingly questioned since the mid-twentieth century, for several reasons. First, the establishment of a body of international law and policy on human rights following World

War II have made the often-catastrophic impacts of exclusionary protected areas on local people morally and legally unacceptable. Secondly, the break-up of the colonial empires and the emergence around the world of newly independent States in need of a means of economic development has made the set-aside of vast areas of land for the sake of Western appreciation of wilderness and wild species unjustifiable. Thirdly, and more positively, there is increasing recognition that local people are not always responsible for environmental destruction and, on the contrary, often play a vital role in environmental conservation. Studies in human ecology and ethnobiology have shown that many ecosystems and habitats around the world are maintained, rather than destroyed, by traditional forms of use, and therefore that continued use is essential to their conservation. Political ecology has developed a detailed analysis of the role of power relations in driving ecological change, and offers many detailed examples of the way in which ecological destruction has been driven not by local people but by external actors such as governments, colonial rulers, private investors and multinational companies, against the interests of local rural populations. Meanwhile the rise in international policy of the concept of sustainable development has highlighted common interests between conservationists and local people in seeking to halt large-scale environmental destruction by outsiders, and encourage forms of local use that both maintain natural ecosystems and also contribute to people's livelihoods and well-being. In the light of these changes, conservationists have started to develop new ways of working with local people and gaining their support for conservation, rather than relying on force in the form of park guards, fences and fines to exclude them altogether (for a more detailed account, see Mulder and Coppolillo, 2005 or Russell and Harshbarger, 2003).

The new 'meta-discipline' (Meffe, 1998) of conservation biology emerged in the early 1980s in response to this situation. The name of the discipline reflected its roots in field biology, but the intention was to build a much broader disciplinary base than the name suggests, so that the new generation of conservation biologists would be equipped to address contemporary conservation problems relating not only to biological aspects but also to the 'human dimensions' of conservation (Jacobson and McDuff, 1998; Buscher and Wolmer, 2007). In order to do so, it was recognised that conservationists would need training in both social and natural science aspects of conservation.

Some progress has been made to this end, but progress has been disappointingly slow. There are several reasons for this. At a practical level, there is a shortage of people with the interdisciplinary expertise to teach the new generation and a lack of suitable teaching materials for them to call upon (something this book aims to address). Some deeper challenges are related to the personal preferences of people who choose to work as conservationists (many want to spend their time working in wilderness areas as far away from people as possible) and also to deep-seated beliefs of many conservationists about the role of people as destroyers of nature. On a more theoretical level, there are some complex challenges to integrating natural science and social science perspectives, connected to very different approaches to how research should be done and what counts as 'valid' research. This chapter will set the scene for the rest of the book by introducing some of these theoretical challenges to integrating social science perspectives into the natural science-based discipline of conservation. Section 1.2 asks what is meant by research and discusses the different disciplinary traditions related to how it is done. Section 1.3 then examines the implications for the social sciences and interdisciplinarity in conservation. Section 1.4 introduces participatory research – an increasingly important element of conservation science which involves collaboration with the participants in order to support them in solving real-world problems. Finally, Section 1.5

outlines the structure of the rest of the book and makes some suggestions about how it should be used, and Section 1.6 draws some general conclusions.

1.2 What is research? Theoretical issues and disciplinary differences

> Research: 'A careful search; investigation; systematic investigation towards increasing the sum of knowledge.'
>
> (Chambers Dictionary, 2003)
>
> Social research: 'The use of controlled enquiry to find, describe, understand, explain, evaluate and change patterns or regularities in social life.'
>
> (Blaikie, 2000: 34)

Research, at its simplest, consists of finding out about things through systematic enquiry. Scientific research does so through *empirical observation* (direct observation of the 'real world' as opposed to research into ideas or texts, as in the humanities). However, beyond these broad generalizations it is not possible to give a single, definitive description of research; even the word 'systematic' can be defined in many different ways. In preparing to write this section I asked colleagues from different social science disciplines 'What is research?' The answers ranged from 'It's about proving things' to 'It's what you're doing now' (going round and talking to people). If you have been trained in the natural sciences you will probably have most sympathy with the first of these; the dominant approach in natural science research is to come up with a hypothesis and then design a way to 'prove' (or at least 'test') it. However, there are other ways to 'do' research, depending partly upon the kind of questions you are interested in, but also on different disciplinary traditions within academia. The differences between them are not superficial; they may be based on completely different premises and follow opposing principles in terms of what the different stages of the research process should involve. These include the way in which the initial questions are framed; the form of the overall research design; what is meant by 'data' and how it should be analyzed, and how rigour, objectivity and validity are defined and assessed. Unless you understand something about these differences you will not be able to communicate effectively across disciplinary boundaries.

The rest of this section summarises some features of the different approaches to research and attempts to show that they have different strengths and weaknesses – they are good at doing different things. Section 1.2.1 describes different *research strategies,* which are to do with the way the research questions are formulated and the relationship between the theory and the data. Section 1.2.2 discusses the differences between *qualitative* research (in which the data may take the form of words or pictures) and *quantitative* research (in which the data take the form of numbers, and are usually analyzed using statistics). Lastly, Section 1.2.3 introduces some deeper theoretical issues to do with our assumptions about what constitutes 'valid' knowledge and about the nature of the natural and social worlds. These assumptions have far-reaching implications in terms of the kinds of question that are regarded as legitimate in the different disciplines and the way that they should be tackled. An understanding of some of these issues is standard in undergraduate training in many social science disciplines, but it is almost unheard of in natural science training. It is not surprising, therefore, that they can be a real barrier for biologists attempting to come to grips with social science aspects of conservation.

1.2.1 *Research strategies: Which comes first, the theory or the data?*

Research strategies concern the relationship between theory and data – which comes first? – and are central to research design (see Box 2.5). The two most common research strategies are *deduction* and *induction,* which represent two ends of a range of possible approaches to research. In *deductive research*, the researcher comes up with a specific theory – a hypothesis — and then designs the data collection process in order to test it. For example, if you hypothesise that grazing of pasturelands decreases invertebrate diversity, you could select several similar areas of pastureland, subject some of them to grazing and leave others ungrazed, and measure the changes in invertebrate diversity at each site. This kind of research is usually quantitative: the data are collected in the form of numbers and the results are subjected to statistical analysis to see whether there is a 'significant difference' between the different conditions (in this case, in changes in invertebrate diversity between the grazed and ungrazed sites). 'Significant' in statistical terms is to do with the probability that any difference you find is due to the difference you have introduced — in this case, grazing — and not due to some other factor or simply to chance. By convention, a probability of above 95 per cent is taken to be significant (this will be discussed in more detail in Chapter 16). If the analysis finds a significant difference, then you can reject the *null hypothesis* that there is no difference between the two conditions and conclude that there is indeed an effect of grazing on invertebrate diversity (or more precisely, that there is a 95 per cent probability that there is such an effect). The reason for framing the research in terms of a null hypothesis – that there is no difference – rather than testing your original theory is explained in Box 1.1.

Null hypothesis testing is often referred to as *the* scientific method, which reflects its overwhelmingly dominant status in the theory of science. It offers a neat, internally coherent and intellectually satisfying way in which to test things. Experiments – in which the researcher actively makes an intervention in order to study its effects — are the most robust way to test a null hypothesis, although not the only way (see Chapter 3). Null hypothesis testing also encourages clear thinking and understanding by breaking complex phenomena down into their component parts. Indeed for scientists trained in this approach, the ability to break down complex phenomena into a set of discrete elements that can be precisely defined and 'measured' is regarded as a key skill to be cultivated.

In contrast, in *inductive research* there is no specific hypothesis. Instead, data collection is guided by a set of broad questions or issues, and the data are used to generate a theory (or hypothesis) once sufficient evidence has been collected to reveal what is going on. The most extreme form of inductive research is an approach in the social sciences known as *grounded theory* (Strauss and Corbin, 1990), in which data should be gathered on a particular issue or situation with a completely open mind so as to avoid influencing the findings according to the researcher's preconceptions and assumptions. In its 'pure' form, this is an unattainable ideal, because you need some idea of what you are looking for in order to know what to record during data collection. You can't collect data on absolutely everything. However, the grounded theory approach does highlight the fact that the more precisely the research is structured in advance of data collection, the more the results will be framed by the subjective preconceptions of the researcher. Thus, there is a balance between defining a precise focus for the research and keeping an open mind so that you don't predetermine the results.

The difference between research strategies is a frequent point of tension in cross-disciplinary communication. To many researchers who use a deductive strategy, any research that does not start with a specific hypothesis tends to seem woolly and unfocused, whereas to those who use inductive approaches, restricting data collection to a few factors according to your

Box 1.1 The theory behind null hypothesis testing: you cannot prove a positive

Never fall in love with your hypothesis – Peter Medawar

David Hume, a leading nineteenth-century philosopher, pointed out that logically, natural 'laws' about the way the world works cannot be proven by repeated observation because no matter how many individual cases you observe, you cannot *prove* that the next case will be the same. The sun has risen in the morning every day of your life, but this does not *prove* that it will continue to do so. The expectation that it will do so is based on custom or habit rather than rational proof (although it is a very sensible working assumption).

Karl Popper, one of the great philosophers of science in the twentieth century, built on this idea to develop what is widely recognized in the natural sciences as the 'scientific method'. He asserted that although you cannot prove that theories or natural laws are true by documenting repeated examples, you *can* prove them false, because it only takes one negative case to disprove the general law (provided everything else stays the same – an important proviso). If the sun failed to come up one day, we would either have to reject the hypothesis that the sun rises every morning or to modify it ('the sun rises in the morning every day except when there is an eclipse ...'). Therefore, according to Popper, scientists should proceed by coming up with hypotheses and then rather than seeking to *prove* them, they should be as critical as they can in order to see if they can *disprove* them. This is done by developing a *null hypothesis* – for example that x doesn't cause y or that there is no difference between x and y – and then setting out to see if it can be disproved. Thus science advances by means of a series of working hypotheses, each of which stands until someone comes up with the evidence to disprove or modify it.

preconceptions of what is important may seem to be prejudging the case. However, a little thought should make it obvious that different strategies are appropriate in different situations. If very little is known about the subject you are researching, then it may be most valuable to start with broad, open-ended research, whereas if you are building on substantial research that has already been done, then there may be a good foundation for generating null hypotheses that can be tested through a much more focused approach. The choice of research strategy is also related to the kinds of question you wish to address. Null hypothesis testing is good for examining cause-effect relationships, correlations, and (numerical) patterns of variation across a population. However, many kinds of research, especially in the social sciences, are not concerned with these issues but instead aim to build up an in-depth understanding of complex situations. Using null hypothesis testing may simply not be an efficient way to do this. This will be explained further in Chapters 2 and 3.

1.2.2 The quantitative/qualitative divide

Research can be categorised as either *quantitative* or *qualitative* according to two different stages in the research process: data collection and data analysis (see Table 1.1). *Quantitative data* are data in the form of numbers, whereas qualitative data are non-numerical and usually

Table 1.1 The quantitative–qualitative divide

	Quantitative	*Qualitative*
Data	Data as numbers (or, in common usage, data easily convertible to numbers)	Non-numerical data – most commonly either words or pictures
Analysis	Statistical	Critical analysis and construction of a narrative account

take the form of words or images. *Quantitative analysis* involves the use of descriptive and inferential statistics (see Chapters 15 and 16), and since statistics require numbers, it can only be carried out on quantitative data or on numbers derived from qualitative data. *Qualitative analysis* involves cross-checking, summarising and synthesising from different sources in order to build up a narrative account (see Chapter 14), and can be carried out on any kind of data, although it is most commonly associated with qualitative data. Both quantitative research and qualitative research can be either inductive or deductive, although deductive research can draw the most rigorous conclusions if it uses quantitative approaches (as in the example of null hypothesis testing in the previous section).

The quantitative - qualitative divide is probably the biggest challenge to interdisciplinary communication. From a quantitative perspective, qualitative research can seem very anecdotal and inconclusive. Qualitative scientists usually present their results in the form of a narrative presentation – a talk – and researchers not trained in qualitative methods are likely to wonder where the 'data' are (numbers) and be left with the feeling that nothing has been 'proved'. For many quantitative scientists, statistical analysis to calculate probabilities is essential for 'real' science; any other type of research is second-rate because it doesn't 'prove' (or disprove) things with the same degree of rigour. On the other hand, qualitative scientists often find that the null hypothesis testing approach and the process of breaking down a complex situation into a set of discrete factors that can be expressed as numbers can make the whole exercise simplistic and reductionist, or even downright misleading, because the world may simply not work like that (see below). Communications are further hampered by the fact that quantitative scientists present their results using statistics – graphs, tables, and the numerical results of diverse statistical tests – which may be incomprehensible to those who are not trained in quantitative methods.

Several of the above points are to do with *epistemological* questions — questions about the nature of knowledge (see Box 1.2). What counts as knowledge? What level of evidence or proof is needed before we really 'know' something? Are some forms of knowledge more valid than others? What is the relationship between knowledge and belief? Can 'knowledge' ever be completely objective? Quantitative science and qualitative science have very different positions on these questions. From a quantitative perspective, the only valid form of scientific knowledge is that which has passed the test of statistical significance; anything less is unsubstantiated conjecture. From a qualitative perspective, reducing complex phenomena to simple numbers and statistics may in its turn produce knowledge that is not regarded as valid or useful; other forms of knowledge may be more appropriate.

Biological research is almost entirely quantitative, whereas the social sciences include both quantitative and qualitative approaches. Most biologists are initially extremely resistant to the idea that anything other than numbers should be regarded as data; it goes against all their training. As a result, there has been a tendency for individual biologists who expand their research to include 'human dimensions' of conservation to favour (quantitative) questionnaire surveys

> *Box 1.2* Key terms: epistemology and ontology
>
> *Epistemology:* To do with the nature of knowledge. What counts as knowledge? How objective can it be? Are some kinds of knowledge more 'valid' than others? What level of evidence or proof is needed before we really 'know' something?
>
> *Ontology:* To do with the nature of the 'real world' and how it is organized. For example:
>
> - Are all aspects of the 'real world' organized according to general laws or is it more chaotic than that?
> - Can it be split up into discrete elements, or is the whole greater than the sum of its parts?
> - Are the natural and social worlds similar in these respects?
> - Finally, are all aspects of the 'real world' reducible to an objective truth – a single right answer – or are some aspects constructed by people's actions and interpretations?

(for example, see White et al., 2005), even when these are not the most appropriate tool. This can lead to justified criticism from social scientists, which can deepen the divide even further. This book covers the full range of basic qualitative and quantitative methods in the social sciences in order to help address this problem.

However, like the differences between research strategies, the differences between quantitative and qualitative research are not as irreconcilable as they may seem. The two approaches are good at providing different *kinds* of information and can therefore be used to complement one another. Quantitative research is good at addressing very focused questions concerning correlations or cause-effect relationships between different variables; statistically significant differences between different populations, and the prevalence of different factors within a population (for example, what proportion of people think or do a certain thing). Qualitative research is good at providing an overview of an issue or situation, disentangling its complexities, and providing an in-depth understanding of different perspectives. Studies that include both qualitative and quantitative components are known as *mixed-methods studies*, and when well-designed, can combine the best of both approaches in order to provide complementary insights into the overall topic of interest. Mixed-methods studies are increasingly common in conservation research and will be discussed in more detail in Section 3.4.2.

1.2.3 Underlying debates: differences between the natural and social worlds

The different research strategies and the quantitative and qualitative approaches to research are often perceived to be aligned along the divide between the natural and social sciences, but this is not quite accurate. Inductive and deductive strategies are used in both the natural sciences and the social sciences, and whilst the natural sciences are almost entirely quantitative, the social sciences include the full spectrum of quantitative and qualitative approaches. However, there are additional points of debate that focus specifically on the similarities and differences between the natural sciences and the social sciences. Indeed, one of the most heated debates in science theory has been about whether the social sciences can or should follow the same methodological framework as the natural sciences (a principle known as 'the unity of scientific method'), or whether there are fundamental differences between the natural and social worlds that make this impossible or inappropriate. This section gives a brief introduction to some of the debates about the similarities and differences between the natural and social sciences. It is underlain by both epistemological questions — questions concerning the nature of 'knowledge' – and *ontological* questions, which are to do with our assumptions about how the world works, and in this context, whether the natural and social worlds work in the same way (see Box 1.2). For a more in-depth account, see Blaikie (1993), on which this section is loosely based.

Natural science approaches to research are based on the premise that the natural world is made up of discrete elements which are organized according to a set of universal natural 'laws'. The law of gravity and the law of thermodynamics are two examples. Research, then, should work by defining the different elements in terms that are precise enough to allow them to be observed and measured objectively, coming up with a theory about the natural laws by which they are organized or related, and then collecting data to test the theory (or the null hypothesis associated with it). Many of these points are linked to a philosophical approach to science known as *positivism* (see Box 1.3).

However, *all* of these assumptions have been questioned by science theorists, particularly in terms of how far they apply to the social world. Can the social world be broken down into a set of discrete, tightly defined elements in the same way as the natural world? Perhaps the whole is greater than the sum of its parts, and therefore attempts to understand it by breaking

Box 1.3 Positivism

Positivism has been the subject of an extended and complex debate since the term was coined in the nineteenth century, and the term is used in many different ways (Blaikie, 1993: 16) encompassing both epistemological assumptions (about the nature of knowledge) and ontological assumptions (about the nature of the world). Some of the basic underlying assumptions are as follows:

1. The world consists of a factual reality that can be broken down into discrete elements.
2. This reality can be observed *objectively* – that is, independently of the subjective views of the observer.
3. Only *empirical* knowledge – knowledge based on direct, objective observation – is valid.
4. Through direct, objective observation of multiple examples ('cases'), it is possible to test for general 'laws' that explain the way the different elements relate to each other – for example, in terms of cause and effect.

Why is it called positivism?

The word 'positivism' was coined in the nineteenth century at a time when 'the scientific method' was being hailed as a new tool with tremendous potential to bring about positive improvements in the living conditions of the time.

it down into individual variables will not reflect its organization in the 'real world'. Are there universal laws that govern the way people behave and all the complexities of how societies and cultures work? Many social scientists would say that every social situation is unique in terms of its social, historical, political and cultural context, and therefore that looking for universal laws from specific cases is of limited value and may even be downright misleading. Lastly, is it possible to observe the social world objectively? The problem here is that you are inevitably a part of the social world you are studying. Your presence will have an effect on what people say and do and therefore on the data you collect, and your interpretation of what is going on is likely to be coloured by your own unconscious assumptions and preconceptions to a far greater extent than in studies of plants and animals.

A further complication in the social sciences is that the objects of research – the people you are studying - have their own interpretations of the social world around them, and it is debatable whether all aspects of the social world exist independently of these interpretations. For example, consider carrying out a study of collaborative management systems for protected areas. Is it meaningful to try to produce a single 'true' account of how power is shared — of who has the power to take and enforce decisions? There may be formal rules that set this out, but a study that only looked at the formal rules would give a very superficial understanding of how such institutions work and may not reflect what actually happens. It can be argued that the nature of power relations is in the eye of the beholders – they can only be described in terms of the actions of the participants and the meaning that the participants give them. Since different actors may interpret them differently, then rather than seeking to determine the 'true' nature of power relations, the research may be better targeted at seeking to determine the different ways in which power relations are 'constructed' and interpreted by the different groups of people involved. This is an important point, as it means that rather than a single objective truth, there may be multiple meanings or interpretations that are equally valid. The relevance of this issue depends partly on the questions you are interested in studying: Are you interested in an absolute answer (the 'truth'), or in people's perspectives? However at a deeper level, social scientists have questioned whether there *is* an objective truth about all aspects of the social world, or whether at least some aspects of it are *socially constructed.* This is an alternative to the positivistic position outlined in Box 1.3.

There is one more way in which issues of objectivity are of particular concern for the discipline of conservation. Conservation has been described both as a crisis discipline (Bradshaw et al., 2007; Czech, 2006; Redford and Sanjayan, 2003) and as a mission-driven discipline (Meine et al., 2006) – a discipline that attempts to promote a certain set of values and a certain practical outcome (in this case, to halt or at least slow the global rate of loss of biodiversity). The Code of Ethics of the Society for Conservation Biology emphasizes the professional responsibility of all conservationists to contribute to the 'mission' of conservation (http://www.conbio.org/aboutus/Ethics/), and most conservation researchers are passionate about this 'mission'. But how can you retain your objectivity as a scientist if you are passionate about a particular outcome? Is it acceptable to design a research project to provide evidence in support of a particular conservation policy position, or does this introduce an unacceptable bias? What should you do if your results appear to undermine a policy that you believe is vital to the conservation mission? These are difficult questions and have been the subject of much debate (for example see Mulvey and Lydeard, 2000; Gill, 2001; Meffe, 2007; Roebuck and Phifer, 1999; Scott et al., 2007; Nelson and Vucetich, 2009). There is a fine line between providing insights that are useful in achieving an agreed outcome and presenting evidence to support a particular outcome that you believe is important (and conservationists disagree on the extent to which we should be doing one or the other of these). However, recognizing that no science can be entirely objective makes this balance less difficult to negotiate. As long as you are aware of your own prejudices and are vigilant in watching for ways in which they may be influencing your research, they are manageable.

1.3 Conservation, the social sciences and interdisciplinarity

In this chapter so far, I have first outlined the importance of incorporating the social sciences into conservation research in order to develop new interdisciplinary approaches, and then shown that there are very substantial barriers to interdisciplinarity. Where does this leave us? How 'interdisciplinary' does conservation need to be and how can it strive to overcome the barriers? This section addresses these questions first by looking in more detail at the arguments for interdisciplinarity in conservation and the different social science disciplines that are relevant, and then by discussing what exactly we mean by interdisciplinarity and the implications in terms of interdisciplinarity expertise in conservation.

The argument for interdisciplinarity in conservation biology is based on the need to inform decisions about applied conservation problems that span disciplinary boundaries. In this context, interdisciplinarity is valuable as 'a means of solving problems and answering questions that cannot be addressed satisfactorily using single methods or approaches' (Marzano et al., 2006), and ideally the problem, not the discipline, should define the tools of study. However in order for this to be possible, the researcher(s) needs to know what 'tools' are available in different disciplines and how to use them properly. The trouble is that there is a very large and growing range of social science disciplines that are relevant. Of the 'traditional' disciplines, geography, anthropology, sociology, psychology, economics, and law all cover different aspects of conservation. In addition, many new environment-related social science disciplines and sub-disciplines have been created in the past 50 years, as global concern about the state of the environment has grown. Environmental education was formally established as an academic discipline in the 1960s (Stapp et al., 1969) and has grown into a major discipline in its own right. The field of environmental economics has also grown rapidly since the 1960s, largely in response to concerns about environmental sustainability (Pearce, 2002). The discipline of development studies offers insights into natural resource management and participatory approaches to community development. Ethnobiology, which has developed out of environmental anthropology, deals with 'the complex relationships, both past and present, that exist within and between human societies and their environments' (International Society for Ethnobiology, *http://ise.arts.ubc.ca/*). Environmental psychology – 'the study of transactions between individuals and their physical settings' (Gifford, 1997: 1) – includes research on attitudes to the environment, common pool resources and aspects of environmental education. Conservation psychology is a more recent sub-discipline which, like conservation biology, combines research with conservation advocacy and action (Clayton and Brook, 2005; *http://www.conservationpsychology.org/*; Saunders and Myers, 2003). This raises a series of questions for conservation as a discipline. How far can or *should* it broaden from its traditional roots in biology to cover a seemingly infinite range of topics? What subjects should be included, and how can they be integrated into a cohesive whole? And how much can any one individual be expected to learn of all the different disciplinary perspectives that are valuable in conservation research? There has been an extended debate about these issues in the leading conservation journals since the 1980s (for example, see Balmford and Cowling, 2006; Borgerhoff Mulder, 2007; Clark, 2001; Eriksson, 1999; Inouye and Dietz, 2000; Jacobson and McDuff, 1998; Kroll, 2005; Lopez et al., 2006; Martinich et al., 2006; Meffe, 1998; Newing, in press; Noss, 1997, 1999; 2000; Nordenstam and Smardon, 2000; Perez, 2005; Saberwal and Kothari, 1996; Scholte, 2003; Siebert, 2000; Takacs et al., 2006; Touval and Dietz, 1994; White et al., 2000).

One obvious step towards clarifying these issues is to define clearly what we mean by interdisciplinarity and to think about the degree of discipline-specific expertise that is required in order to integrate aspects of the different disciplines that are involved. This is the aim of the rest of this section.

In its broadest sense, interdisciplinarity has been defined as 'any form of dialogue or interaction between two or more disciplines (Moran, 2002: 16; see also Dillon, 2008, Lau and Pasquini, 2008). The most common classification scheme is shown in Box 1.4. Multidisciplinary research projects include separate components from different disciplines, working in parallel. A simple example in conservation research would be a study of a particular species of animal that focuses principally on its ecology but also includes a questionnaire survey in order to explore local people's knowledge and views. In more integrated forms of interdisciplinarity, the different disciplinary components are more closely intertwined. Data collected using one disciplinary approach may be used to inform the subsequent collection of data by a different approach, or two disciplinary strands of research may run in parallel and feed into one another repeatedly throughout the research project. For example, ethnobiological research into the same species of animal might involve an initial stage during which people are asked about its distribution and ecology, then this information could be used to inform the design of an ecological study, then in a third stage, the findings could be presented to hunters in order to explore how the distribution of the species has changed over time. Transdisciplinary studies integrate elements from the different disciplines to such an extent that the original disciplines are no longer 'visible'; instead, novel approaches are used that transcend disciplinary boundaries (and in time, may form the basis for a new discipline or subdiscipline). Participatory mapping, described in Chapter 10, could be considered as transdisciplinary, since it combines methods from geography and anthropology to such an extent that they are no longer separable.

There is variation, then, in the degree and type of interaction, in its duration, and in the emphasis on theoretical integration versus applied problem-solving (see also Section 3.4.2). In practical terms, interdisciplinary research may involve a temporary collaboration between researchers from different disciplines in order to address a specific practical problem;

Box 1.4 Key terms: interdisciplinarity

Interdisciplinarity: 'Any form of dialogue or interaction between two or more disciplines' (Moran, 2002: 16).

Classification based on the degree to which the different disciplines are integrated:

- *Multidisciplinarity:* the use of two or more disciplines in parallel
- *Interdisciplinarity:* more integrated approaches where different disciplinary perspectives inform one another
- *Transdisciplinarity:* transformative approaches involving the creation of novel perspectives that transcend disciplinary boundaries. (Based on OECD [1972] in Klein, [1990]; Barry et al. [2008].)

the adoption and eventual incorporation of a specific method from one discipline by another; a progressive convergence of different disciplines in terms of both subjects and methods; or the emergence of a new, fundamentally distinct 'interdiscipline' in its own right (Klein, 1990; Barry et al., 2008). However, whatever the degree and type of interaction, interdisciplinary research must be 'rigorous on both sides' of the disciplinary divide in order to produce 'good research' (Harrison et al., 2008). In the first example above, it is not acceptable just to tack a poorly designed questionnaire onto a natural science project. Therefore, an ecologist wishing to include a questionnaire in an otherwise natural science based study must either have a sound knowledge of the principles of questionnaire design and survey research, or else seek help from a social scientist.

In practice, very few people are comfortable working in both the natural and the social sciences, and therefore research projects that bridge the natural - social science divide often do so by means of multidisciplinary teams (teams of researchers, each of whom has expertise in a different discipline or subject area). This solves the need for each team member to have expertise in a wide range of disciplines. However, it does not mean that they no longer need to know anything about disciplinary differences. If researchers are not familiar with the different approaches to research described above and the implications for interdisciplinary collaboration, then teamwork will be hampered both by a lack of basic comprehension and also, at a deeper level, by different ideas about how research should be done and what counts as 'valid' research. Indeed, multidisciplinary teams – especially those involving both natural and social scientists — are notorious for these kinds of problems (for example see Boulton et al., 2005; Campbell, 2003, 2005; Fox et al., 2006; Jakobsen et al., 2004; Marzano et al., 2006; Strang, 2009). Natural and social scientists 'ask different kinds of questions, employ different methods, collect different kinds of data, use different analytic tools and produce different kinds of outputs (Strang, 2009). Unless the team members can appreciate each other's work, then it is inevitable that they will have problems in working together. The more integrated the different elements within an interdisciplinary project are, the greater the level of understanding of disciplinary differences that is needed in order for successful teamwork.

Bearing this in mind, the bottom line in terms of interdisciplinary expertise in conservation is not that every conservationist *must* (or even *should*) be able to work on both biological and social issues, but that they must understand enough about the different approaches to be able to communicate with one another professionally across disciplinary boundaries. As Lidicker (1998) states, what is needed is 'to educate people to understand a specific intellectual area, to appreciate the contributions of other disciplines, and to be comfortable working in teams'.

1.4 Conservation and participatory research

The term 'participatory research' is used in different ways by different people, but it is used here to refer to collaborative research in which the aim is to support local people in analyzing their situation and coming up with plans for action (for example see Little, 1994; Borrini-Feyerabend et al., 2000; 2004). A participatory approach to research has important implications for how each stage of the research process is done. How and by whom should the aim and specific objectives of the research be defined? How much control should be retained by the researcher? How involved should the participants be in different stages of the research process? Who 'owns' the results and decides how and to whom they are presented? What happens if the research uncovers something that does not support the interests of the participants? There are no hard and fast answers to these questions, but this section introduces some of the background and principles of participatory research in order to set the scene for their consideration in the rest of the book.

Participatory approaches originated in development studies, and in order to understand some of the thinking behind their use, it is useful to be aware of how they have evolved. Box 1.5 describes their evolution in the field of development. For a more detailed account, see Mikkelson (2005: 53–84) or Kumar (2002: 29–34).

The basic premise underlying the use of participatory research in conservation is that if there is really a common interest between local people and conservationists, then conservation researchers can best contribute to a solution by providing information to feed into local people's decisions. For example, local people may be concerned that wild resources such as fish-stocks or game are becoming scarce and ask for technical assistance in managing them more sustainably. Alternatively, if their lands are threatened by large-scale development projects or by incursions by outsiders, they may ask conservation researchers for assistance in documenting their land claims and natural resource use, thus supporting their efforts to lobby for a halt to the destruction.

A word of warning: The term 'participatory research' and associated terms are not always used consistently. Specifically, 'participatory research' is sometimes used to refer to a specific set of research methods, and sometimes (many academics would say more accurately) to the collaborative approach described above. Box 1.6 defines some key terms in the ways that I use them in this book. I use the phrases 'PRA methods' or 'participatory methods' to refer to a set of methods (some of which are described in Chapters 9 and 10), and the term Participatory Action Research (PAR) to refer to a collaborative approach. It is important to recognize that use of participatory methods does not mean that you are 'doing' PAR, because PAR is defined not by the methods but by the relationship between the researcher(s) and the participants – specifically, the reason why the participants are participating and the level and forms of their participation. Are they just going along with what the researcher asks of them, or do they want the research done in order to address their own concerns? What role and level of involvement do they have in each step of the research process? It follows that you cannot always *choose* to take a PAR approach in a particular project; it depends on the level of interest and motivation amongst the participants.

1.5 The structure of this book

The aim of this book is to provide an orientation in the principles of social science research and an introduction to a selection of relevant social science methods for conservationists who are receiving training elsewhere (or have already been trained) in the natural sciences. The book is organized in four sections. The rest of Section I is to do with the process of designing a research project, and should be read from the start of Chapter 2 to the introduction of Chapter 4 in its entirety. Chapter 2 discusses some possible sources of ideas for a research topic and then describes how you should work from the broad topic to a specific focus for the research, formulated as a set of aims and objectives. This is done by searching the literature, talking to friends and colleagues and making time to think. Chapter 3 describes the different types of *research design structure* and discusses their strengths and weaknesses. It then introduces the different methods that are discussed later on in the book and discusses how you should choose between them, and how they can be combined in *multiple-methods* or *mixed-methods* studies. Finally, Chapter 4 discusses the issue of sampling. Sampling is about who you choose to talk to, where you go to collect your data and what

Box 1.5 Evolution of participatory approaches to development

Participatory approaches to research originated in development studies. From the 1960s, development practice shifted from an initial emphasis on state-to-state financing and technical assistance to a more grassroots approach that involved working directly with local communities. It was reasoned that the goal of development – ultimately, to improve economic well-being and relieve poverty and suffering – would be achieved most effectively by working directly with the intended beneficiaries and listening to what they defined as their needs. Standard research techniques used up to that point, such as questionnaire surveys, were slow, faced problems of inaccuracy, and could not give in-depth insights. Therefore a methodological approach was developed in the 1970s known as *Rapid Rural Appraisal* (*RRA*), which attempted to adapt conventional social science techniques using a mixture of methods to allow researchers to carry out a quick appraisal of a particular situation incorporating local people's views, while minimising some common biases in short-term qualitative studies.

Typically, RRA involved a field visit of two to six weeks by a multidisciplinary team of researchers, who would carry out archival searches (searches of locally available written documents), talk to officials and local 'experts' (such as locally based researchers and staff of non-governmental organizations), and travel from community to community carrying out informal and formal interviews and holding community meetings or workshops in order to gather data on people's needs and views. However, RRA was increasingly criticized, partly on the grounds that it wasn't appropriate for outsiders to make decisions that would have major effects on local people's lives based on a visit of only a few weeks.

In order to address this criticism, a more participatory approach was developed in the 1980s known as *Participatory Rural Appraisal* (*PRA*) (Chambers, 1992; Pretty et al., 1995). PRA emphasizes the *empowerment* of local people to take decisions and act for themselves, and participatory research is often contrasted with *extractive research*, where researchers simply extract information and make recommendations according to their own analyses and interpretations.

The term 'PRA' has become pervasive within the field of development; it is no longer used only in connection to rural studies or appraisals. The term has stuck in spite of the creation of many new acronyms, some of which are defined in Box 1.6. However it is used in different ways by different people. In addition to referring to a philosophical approach based on empowerment, it is also often used simply to refer to the use of a set of interactive methods involving structured group tasks that are frequently used in community workshops (see Chapter 9). Many of these methods were first developed as part of RRA techniques, but they have come to be known as PRA methods.

PRA has been criticized in its turn for the level of control that is exerted by researchers – especially in a community workshop setting – and the lack of true empowerment of local people (for example, Cooke and Kothari, 2001; Cornwall and Hickey and Mohan, 2004; Mosse, 2005; Pratt, 2003; Twyman, 2000). However, PRA methods have become part of the standard range of social science methods in development, and the ideal of empowerment remains strong, even though many projects fail to put it into practice.

Box 1.6 Key terms: participatory approaches to research

Rapid Rural Appraisal (*RRA*)*:* Typically, RRA involves a two- to six-week field visit to a rural location by a multidisciplinary team, who *appraise* the local situation by carrying out archival searches, informal and formal interviews, and community meetings or workshops.

Participatory Rural Appraisal (*PRA*)*:* PRA evolved from RRA but puts the emphasis on empowerment of local people to take control of their own lives. The term 'PRA' is also used to refer to a set of interactive methods that are commonly used with groups of people (often in community workshops) to assist in analyzing their own situation and planning actions. Many of the methods consist of structured tasks involving the use of visual representations such as diagrams and maps, or practical exercises that are carried out on community lands.

Participatory Action Research (*PAR*)*:* An alternative term for research that aims to support local people in analysing their own situation and taking actions to solve their own problems. The beneficiaries may actually identify the need for the research and approach a researcher, or the research project may be suggested by a researcher and the beneficiaries may see its potential value for their own purposes. They should be involved in defining the aims and focus of the study, and may also be involved in developing a research design and in gathering and analyzing data. They also decide whether and how to act on the findings. PAR is commonly contrasted with *extractive research.*

Extractive research: Research which aims to collect information from the participants for use by others. This may be either for the purposes of 'pure' theoretical research, or to inform decisions by outsiders such as the staff of non-governmental organisations, state agencies, policy institutions or private investors.

examples you choose to collect data on. Many people assume that sampling is only relevant for quantitative research, but this is not so: these are important decisions in any research, although they are approached rather differently in quantitative research and in qualitative research. Once you have read the introduction to Chapter 4, you should be able to identify the other chapter sections that are most relevant to your proposed study.

Section II consists of a run-down of a range of methods in social science research, each of which is the subject of a separate chapter. This section is not intended to be read in its entirety; once you have read Chapter 3 you should be able to select the methods that seem to be most appropriate for your study, refer to the relevant chapters to check that your selections are appropriate, and then read those chapters in full. The methods that have been selected include the full range of basic techniques that involve recording what people say and do, from the most qualitative (participant observation: Chapter 5) through different kinds of qualitative interview (Chapter 6) to the most quantitative (questionnaires: Chapter 7). A few more specialist techniques have also been included that are particularly relevant to conservation. Chapter 8 describes a set of techniques that are used in ethnobiology to analyze local environmental knowledge systems and local perceptions of social and environmental change. Chapter 9 deals with community workshops, and describes a range of 'PRA' techniques that are commonly used in a workshop setting. Lastly, Chapter 10 describes one specific participatory technique – participatory mapping – which has experienced a boom in use over the past ten years, particularly in projects that aim to support local communities in securing their rights to land and resources. The emphasis throughout the book is on conservation fieldwork rather than analysis of documents or policy, and therefore methods for the latter, such as content analysis, are not discussed in detail (although see Chapter 14 for some further reading).

Section III concerns a set of issues that are rarely addressed in methods books, which are to do with issues and dilemmas that are likely to arise in connection to fieldwork. All the chapters in this section call on the field experiences and suggestions of young researchers carrying out fieldwork for the first time. Chapter 11 is concerned with practical aspects of fieldwork, including equipment, travel and health concerns, and also the process of collecting and managing the data in the field. Chapter 12 moves on to your role as a researcher in terms of your relationship with the local people – the different expectations they may have of you, the effect of your presence among them, and steps that you should take in order to become attuned to local circumstances and make the most of your time with them. Finally, Chapter 13 discusses the ethical dimensions of research, not only during fieldwork but also during preparation, writing up and publishing. It introduces professional guidelines on ethics and responsibility, discusses some specific types of responsibility towards the study community, and then sets out some general principles to bear in mind in order to ensure that you act appropriately.

Section IV moves on to data processing and analysis, and is divided into a single chapter on the analysis of qualitative data (Chapter 14) and two chapters on quantitative analysis (Chapters 15 and 16). There is a big advantage to reading the relevant chapter(s) on analysis *before* starting to collect your data; if you know what will be involved in the analysis, then you have a better chance of collecting data in a form and quantity that will allow you to answer your research questions. Qualitative analysis often starts during data collection anyway, and therefore if you are using qualitative methods it is particularly important that you read Chapter 14 before you leave for the field.

Chapter 14 focuses on analysis of qualitative field data in the form of words, but also gives some suggestions for further reading on analysis of visual data such as pictures or videos, and also on content analysis (a more formal technique for analysing texts in detail). Chapters 15 and 16 deal with quantitative analysis and give a run-down of different kinds of *descriptive* and *inferential statistics* respectively. Descriptive statistics are used simply to describe and present the data – for example in the form of frequency tables, percentages, bar charts and pie charts, or by calculating the average and some measure of the spread of the data. Inferential statistics go further than this and calculate the probability that patterns in the data collected from one or more samples reflect something more than chance variation and tell us something meaningful about the wider world. They can be used to estimate the characteristics of a population from data collected on a sample, or to compare two or more samples (or sub-sets within a sample), or to see if there is a significant relationship between two or more different variables. It is not possible in a single chapter to include step-by-step instructions on exactly how to do each test (but suggestions are made for further reading); rather, the chapters set out some of the underlying principles and then give a systematic run-down of the most commonly used tests in social science research. The aim is to give clear guidance on which test is appropriate in a given set of circumstances. It is assumed that conservation biology students will have (or be developing) an understanding of statistics from their natural science training, in which case the information provided should be enough to allow you to identify the appropriate statistical tests for use with social science data and apply the simpler tests. For readers with little knowledge of statistics, Chapters 15 and 16 should be enough to give you an overview of how statistics works and what it can do, and point you towards further reading. However, you would be wise to consult a professional statistician before carrying out statistical analysis, and preferably before finalizing the research design, since the design determines what tests you can use and therefore what questions you can attempt to answer. This is especially important if you are considering using some of the more complex multivariate techniques.

Finally, Section V is concerned with writing up the results of your research and disseminating them to different audiences. Chapter 17 gives a detailed account of what you need to do when you have finished data collection and analysis in order to write a report, whether it is an academic thesis or a non-academic report such as a consultancy report. Chapter 18 discusses some of the different tasks that may be necessary after the thesis or report is finished, including reporting back to the participants, preparing additional reports or other kinds of material for different audiences, and, if you intend to continue in research, developing further research proposals based on what you have found.

1.6 Conclusion

Social science research in conservation is an expanding field. However, whilst there are now many highly experienced social scientists working in conservation, the majority of conservationists continue to be trained predominantly (often, entirely) in the natural sciences. There are very few researchers who understand the principles of both natural science and social science approaches to research, and therefore it is not surprising that progress in developing integrated approaches to conservation problem solving has been slow.

Clearly, it is not realistic to expect conservation professionals to be experts in everything from beetle ecology to indigenous knowledge systems to trade networks to international policy negotiations, but the crucial point is that students should get some basic training in different methodological approaches to conservation research, both so that as established professionals, they can work successfully with colleagues who have specialized in different areas, and also so that they are equipped to specialize themselves in any area they choose, whether it is natural science based or social science based or bridging the divide between the two. Understanding the differences between disciplines in terms of how research is structured and evaluated, and being aware of the different methodological tools that are available in the social sciences and the relative merits of each of them, is a first step in this process.

This book aims to provide sufficient guidance on social science methods to complement the training in biology and ecology that is widely available to people who are entering the field of conservation research. In this way we hope it will contribute to the emergence of a new generation of conservation professionals who are comfortable working collaboratively across disciplinary boundaries, whatever the specific subject area in which they choose to develop their own expertise.

Summary

1 This book aims to provide a grounding in basic social science research methods for students and professionals working in the field of environmental conservation. The underlying premise is not that every conservationist *must* (or even *should*) work on both ecological and social issues, but that they must understand enough about the different approaches to research to be able to communicate with one another professionally, at least at a basic level.

2. Different disciplines have very different ways of organizing and doing research, and this creates a major barrier to understanding, especially between the natural and social sciences.
3. Typically, natural science research works through the formulation of a hypothesis, collection of numerical (*quantitative*) data on a set of specific, predefined *variables*, and *statistical testing* to calculate the probability that any patterns in the data are 'meaningful' (due to something other than chance variation). The hypothesis is then accepted or rejected.
4. Social science research is much more varied in terms of the research strategy (*inductive* to *deductive*), the kind of data collected (*qualitative* or *quantitative*), the kind of analysis carried out (*qualitative* or *quantitative*) and the tools and standards by which validity is assessed.
5. Different disciplines and traditions of research are also based on different underlying *epistemological* and *ontological* assumptions (assumptions about the nature of knowledge and the nature of the world respectively).
6. The relationship between pure and applied research is also controversial. Conservation is a 'mission' discipline that attempts to promote a particular set of values. However this raises issues to do with *objectivity*.
7. Interdisciplinarity can be defined as 'any form of dialogue or interaction between two or more disciplines'. The greater the level of interaction, the greater the challenges in terms of reconciling fundamental differences in approaches to research.
8. No one individual can be trained in all the different disciplinary perspectives that are relevant to conservation. However, in order to make interdisciplinary research possible, individual researchers need enough understanding of different disciplinary approaches to have an overview of what tools are available to address a given research problem, and to be comfortable working in teams with practitioners of other disciplines.
9. Participatory action research (PAR) is research that is based on a philosophical approach that emphasises *collaboration* and *empowerment* of the participants. The researcher and the intended beneficiaries of the research work together to address practical problems.
10. In PAR, local people are involved in identifying the need for research, defining the aims and objectives, and applying the results. They may also play a role in the actual data collection and analysis.
11. PAR often makes use of PRA methods – methods that involve structured group tasks, often based on the production of visual representations such as maps and diagrams. However, PAR is defined by the philosophical approach rather than the methods. Any methods can be used in PAR, and PRA methods can be used either in PAR or in *extractive* research.

Further reading

Adams, W. (2004) *Against Extinction: The Story of Conservation*, London: Earthscan. [A very readable account of the origins and development of the conservation movement, and of the evolution of perspectives on wilderness, people and society.]

Klein, J.T. (1990) *Interdisciplinarity: History, Theory and Practice*, Detroit: Wayne State University Press. [An authoritative book exploring historical and theoretical aspects of interdisciplinarity, with illustrative examples.]

Mulder, M.B. and Coppolillo, P. (2005) *Conservation: Linking Ecology, Economics and Culture*, New Jersey: Princetown University Press. [An introduction to the 'human dimensions' of conservation.]

Russell, D. and Harshbarger, C. (2003) *Groundwork for Community-based Conservation: Strategies for Social Research*, Walnut Creek, CA; Oxford: Altamira Press. [An excellent introduction to community aspects of conservation, including both theoretical issues and many aspects of practice.]

Websites

http://www.conbio.org/ The home page of the Society for Conservation Biology (SCB) – the main international professional organization for conservationists.

http://www.conbio.org/workinggroups/SSWG/ The home page of the Social Science Working Group (SSWG) of the SCB. The resources section includes a selection of methodological tools for use in social science conservation projects.

2 Defining the research topic

> The difficulty in most scientific work lies in framing the questions rather than in finding the answers.
>
> (A.E. Boycott, 1928: 60)

2.1 Introduction

The full research process can be divided into several stages, which are listed in Box 2.1. This is the first of three chapters on *research design*, which is the beginning of the process. Different people use the term 'research design' in slightly different ways, but I will use it to mean steps one to three in Box 2.1 – the process of identifying a research topic, defining what you wish to find out about it, and planning what you will actually do (the *methodology*) in order to address your aims.

This chapter examines the first two stages and Chapters 3 and 4 will describe how to develop a methodology. Section 2.2 describes some possible sources of ideas for a research topic and the issues that should be considered when choosing from them. It is aimed mainly at university students who are looking for a subject for their thesis. Section 2.3 is the core of the chapter, and describes some common steps in developing a specific focus for the research. Developing a focus is probably *the* most important skill in the research design process; you need to end up with a framework that is coherent, targeted, informative and doable. This involves narrowing down the scope of the project both in theoretical or conceptual terms (what *exactly* you are interested in) and also in practical terms (the geographical area, population and/or time period the project will cover). The final product should take the form of a framework defining the research at different levels – an overall aim or aims, a set of objectives and, usually, some more detailed research questions or topics connected to each objective. The level of detail that is needed varies greatly, but the overall rule is that it must be specific enough to inform your plan for what you will actually do – at least during the first stages of data collection.

The whole of the above process should be informed by a thorough, systematic literature search. In this electronic age the literature search can be a daunting task, simply because of the overwhelming amount of information available. Therefore it is vital to develop strategies to search efficiently and effectively. Section 2.4 describes in some detail how this can be done, using both restricted academic sources and also the expanding range of open-access public sources.

2.2 Identifying a topic or problem

If you need to choose a research topic from scratch for a thesis, there are several possible sources of ideas. There may be something suitable based on your own experience outside

Box 2.1 Stages in the research process

1 Identify a general topic or problem.
2 Develop the specific focus of the research. This involves defining what you will try to find out about the topic by developing a set of specific aims, objectives, and research questions and defining the *scope* of the study.
3 Develop a *methodology*: a plan for what you will actually do to address your aims, objectives and research questions.
4 Carry out the plan (data collection and analysis).
5 Write up the results.

study – something you have been interested in since childhood or involved in as a volunteer or in paid employment. You may be aware of a particular problem facing conservation managers or local people at a site you know and be interested to frame your research to inform their choices. Alternatively, you might be interested in following up an issue that you have come across during classes or reading. Lastly, in most university departments the staff will have some suggestions for suitable student projects that fit in with their own research agendas.

There are advantages and disadvantages to each of these different sources and the best option depends upon your own interests, confidence and skills and on what you wish to get out of the research process. Following up questions you have identified yourself in class or from reading is a skill that needs practice; the dangers are that either you will choose something too broad or theoretical to be practical, or that you will develop your questions only to find half way through that they have been extensively studied before. The same applies to interests that originate in your own personal experiences. However, if you develop a project successfully from scratch in this way it will stand you in good stead for an academic career – and at the same time allow you to look in depth at an issue you are passionate about. If this appeals to you, check with your supervisor whether the topic is suitable before developing your ideas in depth, and if they agree to the topic, start a broad literature search to see what other people have done before.

An alternative is to choose a research problem that has been identified by conservation practitioners or others affected by conservation practice, such as local communities, private landowners or state agencies. In participatory action research, the need for the research may have been identified by the people in the study community itself. In this case, you should listen carefully to exactly what they need and make sure they are involved in the process of defining the specific aims and objectives of your study. Working with practitioners or potential beneficiaries on a current practical problem can be especially rewarding because you know the results will be of practical interest to your collaborators, and if all goes well, there is a good chance that your findings will be acted upon. If you do a good job you may even be offered further employment. On the other hand, their ideas may be different from yours about what the most important issues are, how they should be framed, and how you should study them (and – most problematically of all – possibly also in terms of what they expect the results to show). You may have to compromise both on the specific focus of the study and also on some of the theoretical aspects of the research design in order to meet their needs. It is vital to make absolutely sure before you commit to a project that there is a mutual understanding of what the project will aim to do and what it will involve, and if you are a student, your supervisor should be closely consulted in this process. You should also ask for confirmation (in writing) from your proposed partners that you will have any infrastructural support and cooperation you need once the project is under way.

Perhaps the simplest option is to take up a project suggested by your teachers. The research problem should already be well-defined and well-targeted in relation to the existing literature; your research is likely to be part of a wider programme that in many cases will provide you with background information, methodological guidance, contacts and logistical support for your study; and in addition, your supervisor will have a special interest in making sure that you have strong support so that you design and carry out the research appropriately. On the other hand, you have less scope for following your own interests and will gain less experience in research design.

In practice, the process of selecting a subject area often involves a combination of some or all of the above sources. Perhaps the most important thing to think about when choosing between different options is the balance between your interests and the practicalities of 'doing' the research. If you end up with a project that is well-supported but does not really interest you, you will not do as well as you should. On the other hand if you decide to fulfil your life's dream and go and work at a remote, exotic location on the other side of the world you may have the experience of a lifetime – but you may also end up with a piece of research that does not reflect your full potential because of logistical difficulties and the lack of close supervision. What you need, then, is a subject that is practical within the limits of the time and resources you have available, at a field site where you can be sure of sufficient cooperation and logistical support, but that also stimulates your curiosity. Wherever the original idea comes from, you should discuss your ideas with your supervisor before you develop a detailed outline.

2.3 Defining the specific focus of the research

Once the broad topic is decided upon, the next step is to define more specifically what you want to accomplish. In order to do this you must draft a title and overall aim or aims; select a field site; identify the different factors or issues or questions within the overall topic that will be covered and decide which ones to focus on; define your terms, and construct a set of concise, coherent objectives and/or research questions. This is the beginning of an ongoing process of refining your ideas and narrowing down the focus of the research, which will continue at least until you have a complete research design, and quite possibly through data collection, analysis and even the report-writing.

Box 2.2 lists some of the different stages in this process, set out in the standard order in which they are done. If the subject is new to you, start with some introductory background reading – perhaps a section from an undergraduate textbook or a review paper – so that you do not waste time coming up with ideas that have been considered before. Reading should aim to find out what has already been done, how it has been done and what can be learned from the results. Make notes as you read and build up your own overview of the issues (see Section 2.4 for more details).

Once you feel you know enough to develop some initial ideas for your own research, you should try to write down a specific title and an overall aim or set of aims for the project. In collaborative research, the collaborators should have the opportunity to work with you in this process. The level of involvement that collaborators want to have is extremely variable – some may simply confirm that they are happy with the general topic and leave the details up to you, whereas others will want to get thoroughly involved. In participatory action research, ideally the beneficiaries should be involved in defining a specific focus, and they

Box 2.2 Instructions: how to refine and focus the research topic

- If the topic is new to you, read some introductory books or articles.
- Once you know enough to generate ideas for your own research, draft a title (or a few alternative titles) and aim(s) to discuss with your supervisor or manager.
- Once the topic is confirmed, contact managers at one or more potential field sites to see whether they would be supportive of the study and if so, ask for their comments – both on the subject matter and on whether what you propose sounds practicable.
- Meanwhile, continue reading in order to:
    - Define concepts and key terms as precisely as possible.
    - Develop a list of subtopics or issues that you will focus on.
    - Develop a list of research objectives or questions (and for quantitative research, variables: see Section 7.2.2).
    - Start to consider other aspects of the methodology, including the research design structure (Chapter 3), sampling (Chapter 4) and the methods you will use.
- Once the field site is decided upon, find out in more detail about the logistics of working there (see Chapter 11) and define the practical scope of the project.
- Keep reviewing the draft research design throughout, checking back for consistency between the different levels (aims, objectives, research questions and so on), for a single, well-defined focus, and for clarity of thinking and expression.

may also have some input into designing the methodology and even collecting and analysing data. This is only really possible if you can visit the field site before finalizing the research design and spend time with people discussing what you will aim to do. It may also be constrained by the participants' limited knowledge of technical aspects of the research process – after all, if they knew enough to design and carry out a research project for themselves, then they would not need your input. So you need to be realistic about how participatory the research can really be as you work through the different steps that are described below.

The title and aims should be as concise as possible. If you find that you are writing long, complex sentences in an attempt to include all the different aspects you have thought about, then you almost certainly need to narrow down the scope of the project further. If you really cannot decide which direction to take, write down several possible titles (or aims), then in the following steps, try to assess which one will work best. If possible, bounce your ideas off interested friends and colleagues; the more times you try to explain your project to others (and have to fend off their questions and criticisms) the more clearly you will develop your own ideas.

When you are reasonably happy with what you have written, discuss it with your supervisor or line manager. Ask them if it makes sense and sounds both interesting and practicable. If so, and if you have not yet done so, get in touch with managers or decision makers at one or more potential field sites in order to sound them out and check the practicalities of what you propose to do. If the project will need their support (and if this has not been agreed in advance), the most important aim at this stage of the correspondence is simply to check whether that they are willing to give such support before you commit to any one particular site.

This makes it sounds like a highly systematic and organized process, but in fact it tends to be chaotic and very variable. The field site may be fixed before you begin to think about the research design, or it may remain uncertain until late on in the process. You may draft what you think is an excellent title at the start, but then as you go into the issues in more detail it may seem irrelevant or poorly formulated and therefore you may need to adjust the wording. Then you are ready to go on to the next level of detail and define the aims and objectives – after which you must check them back once more against the title to make sure they match one another. Then you can progress to yet another level of detail – perhaps the specific research questions – and once more check back against the broader levels to make sure everything is consistent. Thus it is normal to move back and forth repeatedly between the different activities and levels until they form a coherent whole. This probably sounds more complicated than it really is, because it is difficult to write about it in general terms; every project develops differently. The final structure should be simple and logical – something that you can easily keep in mind as you proceed with the research and that will lend itself to a coherent narrative in your final report.

2.3.1 *Drafting a title*

In order to draft a title, you need to work from the original, broad research topic to a more specific subtopic or question. If you have no particular question in mind, a good way to start the whole process is to brainstorm about what factors might be relevant (see Box 2.3). You should end up with a small number of groups of factors, each containing a list of more specific aspects. Box 2.4 gives a list that might result from an exercise of this kind on the broad topic of 'what determines people's levels of support for environmental conservation'.

The next stage is to choose how to apply this list. No single study could explore every item in the list in detail, so you have a choice: either you can keep a broad perspective and look at influences on people's levels of support in general terms, or you must pick a particular factor or group of factors to investigate in more detail. The former might involve documenting people's levels and types of support for conservation and then interviewing them to explore the reasons they give for their support or lack of it (without prejudicing their answers

Box 2.3 Instructions: how to brainstorm

1. Write down everything that occurs to you uncritically until you run out of ideas.
2. Review the list and cut out any items that, on reflection, seem irrelevant (or just plain silly).
3. Sort the remainder into groups of similar or related factors. Within each group, sort further into subgroups as appropriate.
4. Within each group and subgroup, cut out any duplicates and perhaps rephrase some others so that the differences between them that you had in mind are clearly expressed.

Box 2.4 Brainstorming example: what determines people's levels of support for environmental conservation?

- Family background: class, parents' professions, parents' attitudes to the environment, place of residence (rural/urban), etc.
- Personal characteristics: age, sex, level of education, profession, ethnicity, current place of residence (rural/urban).
- Economic interests: reliance on natural resources (e.g. farming, hunting, fishing, gathering of wild products); income from nature-based tourism, etc.
- Recreational use of natural areas: walking, cycling, riding, bird-watching, etc.
- Personal contacts: friends who work in conservation.
- Cultural characteristics: attitudes to the natural world, including spiritual beliefs related to nature.
- Historical factors: past experience of 'environmental conservation' activities; conflicts such as restrictions on access and resource use; past benefits, and past experiences of crises in nature – destruction of natural sites, shortages of natural resources, etc.

by mentioning any of the above factors). In contrast, the latter could involve a questionnaire in which you ask people detailed questions about, say, their income from tourism, document their levels and types of support for conservation, and then use statistical analysis to test whether there is a relationship between tourism-related income and conservation support.

These two approaches reflect the two most commonly recognized *research strategies* – fundamentally different ways of framing a research project (Box 2.5; see also Section 1.2.1). We will return to the concepts of *inductive* and *deductive* research throughout this book, because they have implications for almost every aspect of the way research is designed, carried out and analyzed.

The difference between deduction and induction is at the heart of a lot of debate about the 'best' way to do research, and is often closely linked in this debate to the relative merits of quantitative and qualitative approaches (although either strategy can be used with either approach). In fact, however, as long as they are used appropriately, inductive and deductive strategies are equally valid; they simply give different *kinds* of information that address different *types* of research questions. In the first example in Box 2.5, if all goes well you will be able to conclude that tourism-related income either does or does not influence levels of support for conservation. However, you will learn nothing about the relative *importance* of income in influencing levels of support compared to other factors at that particular study site. It may be that something else – for example, spiritual beliefs or a history of conflicts over access to natural resources – is far more important in determining support, even if income does have an effect. In the second example, if all goes well you should gain a much more balanced picture of the *main* factors influencing levels of support at that particular study site, but you probably will not be able to quantify their relative impacts and you will learn a lot less about the specific effects of tourism-related income. Longer studies sometimes combine different research strategies to find out different things. You could interview people with open-ended questions about what influences their support for conservation and build up a picture from what they tell you, and at the same time collect more specific data on income levels in order to find out whether income has an effect.

Box 2.5 Key terms: research strategies

Research strategies are about the relationship between theory and data collection. The two most widely recognized research strategies are as follows:

1 Deduction: Start with a specific theory or hypothesis, collect data only on the factors it involves, and then evaluate the data to see whether they support or oppose the theory. Deductive research is typically (but not always) quantitative, and the theory often takes the form of a null hypothesis about the relationship between specific, pre-defined variables.

Example:
Title: The effects of income from tourism on support for conservation: a case study from [the study site].
Aim: To test whether income from tourism has an effect on support for conservation.
Null hypothesis: There is no difference in support for conservation between those who receive income from tourism and those who do not.
Type of data collection: questionnaire with detailed questions on:

- level of income from tourism;
- level of support for conservation.

2 Induction: Collect information about the research topic starting from a broad perspective, without imposing your own preconceived ideas and theories. Then build theory based on what the data tell you. Inductive research is typically (but not always) qualitative and emphasises in-depth description and understanding of complex situations.

Example:
Title: Factors affecting support for conservation in [study site].
Aim: To explore what factors affect people's levels of support for conservation.
Type of data collection: Participant observation and qualitative interviews to explore types and levels of support and motivations affecting support.

(For a fuller account of different research strategies, see Blaikie [2000: 85–127].)

In summary, which strategy you use and what you choose to focus on is connected to your own interests and to the original reason for doing the research. If you are interested primarily in economics, then you may choose to focus on tourism-related income from the start. If you are more interested in a broad understanding of what determines support for conservation, then the inductive approach may appeal more. If the study aims to inform management of one particular site, then a broad approach may be safer, at least initially, to make sure you gain an understanding of the overall picture and the relative importance of different factors.

The title should reflect not only the focused-down topic of research and the types of questions you want to ask, but also the practical scope of the project in terms of the geographical area, population and timescale. Box 2.6 gives a few of an enormous number of possible titles

Box 2.6 Possible titles for research of 'what determines people's levels of support for environmental conservation?'

1 Factors affecting support for environmental conservation among residents of London, UK.
2 What makes people into 'conservation volunteers'? Motivational factors among conservation volunteers in Sweden.
3 The effects of community involvement in tourism on local attitudes towards nature conservation: a case study from Peru.
4 Harnessing financial incentives to improve landscape conservation: the UK Countryside Stewardship scheme.
5 Cultural attitudes towards elephants in Africa and India: implications for community involvement in elephant conservation.
6 Trends in international development assistance for conservation from 1992 to 2010.
7 Indigenous perceptions of nature: what does 'conservation' mean to the Huaorani?
8 Declining fish-stocks and changing levels of support for conservation among fishermen in Vanuatu.

related to research on 'what determines people's levels of support for environmental conservation', and these will be used in the following text to illustrate the different aspects of defining a research project.

The first thing to notice about this list is that some of the concepts have been modified. In several titles the term 'environmental conservation' has been replaced by related terms ('nature', 'landscape conservation', 'elephant conservation' or just 'conservation'). Each of these gives a different focus to the research. Similarly, 'support' has been replaced in some titles by 'attitudes', 'involvement' or 'perceptions' – all different subtopics within 'support'. Title 2 focuses on a specific type of support – conservation volunteering. Defining and narrowing down concepts will be discussed in more detail below.

Second, different titles concern different types of research questions. Several are concerned with *explaining* the effects on support of specific factors (tourism, financial incentives, declines in fish-stocks), whereas others are concerned with more open-ended *description* – of a population, a situation, or (in title 5) a comparison. Explanation tends to be associated with deductive research, and description with inductive research. A third type of question is found in title 4; the wording suggests that the main concern is *how* to make good use of financial incentives. Blaikie (2000: 23) divides questions into three main types – those concerned with description, explanation and prescription (how to do things in the future). It is crucial to be clear about what type of question you wish to ask, because this is important when deciding on a particular type of research design. This will be discussed further in Chapter 3.

Third, each title puts limits on who will be involved – the 'population' that will be studied. In most cases the study population is related to a geographical area – a country, region, city or National Park – but there are other types of population. Title 4 is concerned with beneficiaries of the Countryside Stewardship scheme (a UK government programme that paid farmers to manage their land in ways that were beneficial for biodiversity conservation) and title 7 with a specific ethnic group (the Huaorani in Amazonian Ecuador). Other types of study population could include networks (members of a conservation organization; subscribers to a special-interest bird-watching magazine; participants in an international policy forum) or people who attend a particular event such as an international conference. Alternatively (commonly), the study population may be a particular subset of a geographical population – for example, conservation volunteers in Sweden (title 2 in Box 2.6) or fishermen in Vanuatu (title 8). Obviously, you cannot define the geographical area or population until you know where and with whom you will do the fieldwork, so there is a limit in how far you can get with this aspects of the research design until you have decided on a field site.

Finally, another dimension to be considered in developing a specific title is time. Are you interested in a one-off snapshot of the situation (a 'synchronic' study), or in comparing the same factors at two times (a 'diachronic' study), or in understanding the processes of change over a continuous time period? If the latter, over what period of time? You could choose to assess changing levels of support for conservation before and after a particular event, such as an educational programme, or simply track how support has changed over a number of years and analyze what events or circumstances may have contributed to this. Titles 6 and 8 in Box 2.6 have an explicit focus on change over time.

Trying to draft a title – or several possible titles – is a very good way to focus your mind on exactly what it is you wish to do. However, do not agonize too much over the *exact* wording of the title at this stage – typically, it will change several times as you develop the full research design (unless you are tied down by the terms of a contract).

2.3.2 *Drafting the aim or aims*

The title should state the specific subject of the research, whereas the aim should say what you hope to do. Sometimes the difference between them is simply a matter of wording (as in the examples in Box 2.5). A project can have a single, overarching aim (taking the form of either a statement or a question) or perhaps two aims. More than this, and you probably need to narrow down the overall scope of the research or it will be difficult to work it into a coherent, focused narrative line when you write it up. Where there are two aims, often one is theoretical and the other more applied. Table 2.1 gives an example.

There are two types of statement that are frequently put forward as research aims but are not appropriate. First, there are statements of intent that are purely to do with implementation – for example, 'to increase local participation ... '. This statement says nothing about research – increasing participation means changing the ways things are done, not finding things out. However it is easy to formulate a research aim that is closely related to it – for example 'to identify factors that affect levels of local participation ... ' or 'to inform management

Table 2.1 An example of a title and aims

Title: Declining fish-stocks and changing support for conservation among fishermen in Vanuatu

Aims:

1 To explore the relationship between declining fish-stocks and changing support for conservation among fishermen in Vanuatu between 1995 and 2005
2 To inform policy concerning local participation in fisheries management

practices related to local participation ... '. Identifying factors or informing management practices involve finding things out and providing information, which are appropriate activities for research.

The second type of statement that is not appropriate is one that makes it clear what result you expect to come up with – for example 'to prove that local participation improves the effectiveness of fisheries management'. You're jumping the gun here; any statement starting 'to prove that' or 'to show that' suggests you 'know' the result before you do the research. Conservationists seem to be particularly vulnerable to pre-judging their own research results, presumably because they are passionate about their subject and often have strong opinions on the best ways to approach conservation problems. However, your research will not carry much weight if it is visibly influenced by your own prejudices. Again, it is easy to re-formulate the statement into a more acceptable format. An example could be 'to explore whether ... '. If you are really sure that local participation improves effectiveness, then rather than setting it up as a false target, it is better to formulate your aim in terms of exploring *the ways* in which it does so.

2.3.3 Defining your terms

Part of the process of refining the research topic is defining your terms. Look again at the title about fish-stocks and support for conservation in Vanuatu in Table 2.1. What is meant by 'support for conservation'? This may seem a silly question – we each 'understand' what support is – but are our understandings exactly the same? Are they precise enough for the purpose of the research? Actually, support includes two main types of things: things you *do* for conservation (making donations; becoming a member of a conservation organization; organizing conservation-related events; working as a conservation volunteer; farming organically or managing wild resources sustainably) and also something to do with what you *think* of conservation in terms of value judgements (if you are 'supportive', you are positive about conservation). Moreover, the forms that support can take – especially in terms of what people *do* for conservation – are different in different places and social contexts. If you do not define your terms prior to data collection you may confuse these different aspects, or even if *you* know what you mean, your reader will not know unless you state your definition explicitly.

Definitions become even more important when dealing with more abstract concepts such as 'trust' or 'community cohesion' or 'sustainability' (or even 'nature' or 'culture'). In any research you must define your terms clearly enough to make sure that you know what you need to collect data on and that what you say and write is comprehensible to others. However, there is enormous variation in how precise the initial definitions of concepts should be. In quantitative research, each definition must identify one or more quantifiable *variables* associated with the concept on which data will be collected. This kind of definition, known as an *operational definition,* is particularly important in questionnaire-based studies and will be discussed further in Section 7.2.2. In qualitative research a *conceptual* definition is sufficient – a definition in words of the kind given in dictionaries.

Developing a conceptual definition is not always straightforward. Many abstract concepts are extremely hard to define precisely, even though people within a particular culture have a common understanding of what they mean. Moreover, many technical terms have a 'popular' meaning that may not be the same as their technical meaning, and there may also be several different versions of the technical definition. No one definition is 'right'; some are simply more commonly used or more appropriate for a particular context than others.

Sometimes conceptual definitions differ only on a point of detail, but sometimes they are partly or wholly contradictory. In this case you need to decide how *you* will use the term, which involves looking at other people's definitions and either selecting the one that best suits your research interests, or else developing your own working definition. Box 2.7 gives instructions on how to do this. In your report you should discuss the different definitions in the literature and state explicitly what definition you are using, with sufficient clarity so that both you and your readers know what you mean.

For an example, look back at the beginning of Section 1.2, which discusses definitions of the concept 'research'. Two definitions are cited at the beginning of the section. I then give my own definition ('finding out about things through systematic enquiry') and also quote two contrasting responses from colleagues to the question 'what is research?' in order to give an indication of the range of meanings people attach to the term. In fact I collected over two pages of definitions of 'research' from written sources and asked many more people than the two who I have quoted. My own definition is based on factors that were common to the majority of definitions and relevant to the subject of this book; if I were writing a book on, say, philosophical research, a slightly different definition might be appropriate. In conclusion, it is not a question of a 'right' or 'wrong' definition; after all, 'research' is just a label that we use for the purposes of communication. It is simply a question of making clear what you are trying to communicate, taking into account how the word has been used before.

Whichever type of definition you end up with, you then need to review it carefully. Is it clear and unambiguous? Does it really reflect the original concept in the way you want, or has it changed in some way you did not intend? Does it focus too heavily on certain aspects? Does it cover only a part of the original concept and leave out other parts? Have you gone completely off subject? These questions are concerned with the *face validity* of the new definition – on the face of it (that is, subjectively), does it appear to focus on the right thing? Validity of this kind will be discussed further in relation to questionnaire research in Section 7.6.

Box 2.7 Instructions: how to develop a working conceptual definition

1 Collect together all the different definitions and usages you can find, starting with dictionaries and non-specialist sources if necessary, and then looking at the academic literature. Make sure you note down the full reference details of the source for every case (see Section 17.4).
2 Identify the different factors that are included in different definitions, and note down which ones seem to be widely accepted and which ones are included only by one or two sources.
3 Now construct your own working definition, based both on what seems to be commonly accepted and also on which factors are most relevant for your own research purposes.
4 Once you have written a definition, check it back against the definitions and usages you listed in (1), and think carefully about whether it really communicates what you wish to say.

There are, however, certain kinds of research in which definitions are purposely left quite broad until the data collection is under way. For example, in a study of fishermen's support for conservation, rather than developing *definitive* definitions of 'support' and 'conservation' yourself, another approach would be to try to find out what the fishermen *themselves* understand by conservation and what *they* would regard as support for conservation. In order to do this you simply need a working definition (known as a '*sensitizing definition*': Blumer [1954], cited in Bryman [2004: 271]) in order to be able to plan your research and communicate your ideas. Once you start talking to people you will probably come across several different definitions. Rather than being a problem, this might be an interesting finding in itself, revealing underlying differences in people's perspectives. This approach would give a far deeper insight into what the fishermen think and why they act as they do in terms of support for conservation, although it would make it harder to compare levels of support across different sites since the emerging definitions would probably be slightly different at each site. This kind of approach is common in research into people's different perspectives on a particular issue, and is also suitable for participatory approaches.

2.3.4 Developing a set of objectives

Once you have a draft aim or aims and have made progress in defining and narrowing down your concepts, you need to work on some more specific objectives. Each objective should contribute directly to some aspect of the aim and together they should address the aim as a whole. You can develop the list of objectives through another brainstorm, writing down every possible objective that comes to mind. Once you have a tentative list you should put them in a logical order, merge similar ones and be ruthless in cutting out ones that represent an interesting but distracting sideline, or that are simply impractical. There is no general rule about how many objectives you should have; they are simply part of the framework that helps to structure the research project. Four to eight is probably the norm; if there are more, it may reflect a lack of focus and it will be difficult to keep sight of the whole picture later on.

Objectives tend to be far more specific for deductive research than for inductive research. Table 2.2 gives some possible objectives for the two examples that were given in Box 2.5 to illustrate different research strategies. In each case, some of the objectives are to do with collecting *descriptive* information on individual factors of interest (in this case, different aspects of support for conservation and factors that may affect support). Other objectives are *analytical* – they are to do with the way in which different factors relate to one another. Some of these may be tackled during data collection, whereas others will be addressed only during data analysis. In addition, the last objective in each case is to do with the *applied purpose* of the research. Together the objectives represent the essential building blocks you need in order to address the research aims.

2.3.5 Further narrowing down

I have used the terms 'aims' and 'objectives' to represent the broad conceptual framework within which the research project will take shape. The terminology for these steps is not standardized: some people use 'goal' instead of aim, or state an overarching question instead. The term 'purpose' or 'rationale' may also be used, which is usually a more applied aim ('the purpose of this study is to inform policy ... '). However, most research does include something equivalent to these two layers. Beyond that, different kinds of research are structured

Table 2.2 Examples of the title, aims and objectives using deductive and inductive research strategies

Deductive study:
Title: The effects of income from tourism on support for conservation: a case study from [the study site].
Aim: To test whether income from tourism at [the study site] has an effect on support for conservation.
Null hypothesis: there is no difference in support for conservation between those who receive income from tourism and those who do not.
Objectives:

1 To identify and define the different ways in which local people in [the study site] support conservation.
2 To develop an index of 'support for conservation' based on the above.
3 To identify different sources of income for local people related to tourism.
4 To collect data on levels of support and tourism-related income from a representative sample of local people.
5 To analyze the patterns of variation in income and support within the study sample and test for statistically significant relationships between them.
6 Based on the results, to make recommendations for tourism management at [the study site].

Inductive study:
Title: Factors affecting support for conservation in [study site].
Aim: To explore what factors affect people's levels of support for conservation.
Objectives:

1 To explore local people's perspectives on 'conservation' – what they understand by the term, their experiences of it in practice, what value they place upon it, and what they regard as 'support'.
2 To document different forms of 'support for conservation'.
3 To explore people's explanations of why they do or do not give support.
4 Through focusing in on particular situations that called for support, to trace back the events and motivations that led people to give support (or not).
5 From the above, to analyze the different factors that influence support (or lack of it).
6 To make recommendations based on the results.

in different ways: in inductive, qualitative research, the objectives may provide a sufficient framework from which to develop the methodology, whereas in quantitative projects, it may be necessary to define a third or even fourth level of specific research questions in order to develop a list of variables. The important thing is to define your research at several different levels – from the broad to the specific – in the way that best allows you to think, plan and communicate clearly. You should end up with a set of points (objectives or questions or hypotheses) that are:

- clear and intelligible;
- linked to each other;
- researchable (concrete and specific enough that you can gather data on them);
- realistic, given the time and resources available;
- connected in some way to established theory and research;
- interesting – they should address something new. (Adapted from Bryman, 2004: 523)

The ultimate aim is to narrow down to a sufficient level of detail to frame the rest of the research design, including the structure, the sampling strategy, and the specific methods to be used. That may mean that you need to take some aspects of the research down to a fine level of detail and leave others quite broad. These issues will be addressed in the following chapters.

2.4 The literature search

The rest of this chapter describes how to carry out a literature search, which should be an integral part of the process described above. The literature search can be a daunting prospect, simply because of the immense amount of information available. For most subjects it is impossible to read everything, so unless your subject is very specific and little researched, do not be tempted to try it. Instead, you must develop a targeted strategy to find key publications quickly and efficiently and then expand from there, prioritizing and keeping records as you go.

2.4.1 Information sources: libraries, databases, online search engines ... and people

It is always worth asking your supervisor, line manager or colleagues who know something about the research topic for their suggestions for sources of information. If your research is in their specialist area they may rattle off several key references from memory without warning (so take a pen and notebook). However aside from this, the main sources of information are academic libraries and online sources. Even if you have access to a well-stocked academic library, the most efficient way to begin a literature search is probably through online search engines. Box 2.8 gives details of some of the principal electronic databases and search engines that cover a wide spread of subject areas. The 'google' family of search engines are publicly available with no fee, whereas access to Web of Knowledge (WoK) is restricted.

2.4.2 Searching on keywords

Before starting a search, the first crucial step is to identify key words and phrases that can be used as search terms. These may include concepts, geographical areas, taxonomic groups of animals and plants, and so on. If the subject is new to you, start by looking for an introductory textbook. If it is more familiar to you, look for a recently published review of the current issues and directions in research.

Box 2.9 summarizes a series of searches I have recently carried out on tourism certification – the process of certifying tourism operations through an independent evaluation of their performance on environmental, social and sometimes additional criteria. The first search was of my university library catalogue of paper holdings. I then searched in the (public-access) 'google' family of search engines, and finally in WoK (I would not normally search *all* these sources in a broad initial search; I did so purely for the purposes of demonstration). In Box 2.9 and the following text, the search terms that were used are marked in [square brackets].

One important point to note is that different search engines treat the words typed into the search box slightly differently. The WoK treats [tourism certification] as a single phrase and therefore would only identify sites where they occur together as a phrase. If you wanted the two words to be treated separately you would need to add a joining word in the search box like this: [tourism AND certification]. In contrast, the google search engines automatically treat [tourism certification] as two independent search terms and will list all sites that contain both words even if they are found in different parts of the same site. In order to search google sites for the phrase ["tourism certification"] it must be enclosed in double quotation marks. Other search engines have still different systems; they may treat the two words as alternative terms and list all sites that mention EITHER tourism OR certification, which would give you

Box 2.8 Electronic databases and search engines

The Web of Knowledge (WoK: http://wok.mimas.ac.uk/) is the principal international database for searching academic publications and covers the natural and social sciences, the humanities and the arts. The database uses keyword indices for articles in some 22,000 academic journals and over 60,000 proceedings of conferences, symposia and other meetings. Access is password-restricted; well-resourced universities and research institutions should have full access.

In WoK it is possible to search on an author or keywords and then to 'mark' the items that are most relevant in the list that appears. You can then print the results, cut and paste them to a Word document, email them to yourself or export them to a specialist reference database such as EndNote or RefWorks. You can also click on the link that appears for each article. This usually brings up an abstract, it may also bring up the full text of the article. There will then be further links to additional articles. Click on the name of the author to see a list of their other articles; click on 'cited references' to see links to articles cited, where these are available online; click on 'times cited' to see links to articles that have cited this one; or click on 'related records'.

A word of warning: WoK has a tendency to 'crash', so save or export your findings frequently or you will lose the lot.

Google scholar (http://scholar.google.co.uk/schhp?tab=ws) offers a very acceptable alternative for those who do not have access to the WoK, and a useful complement to WoK for those who do. It covers a mix of book references, academic journal papers and other research publications such as those by policy institutions. It also includes unpublished reports that will not feature in the WoK. Each entry includes at least three further links: to other articles that have cited the current one, to 'related articles' and to a 'web search'.

One disadvantage of google scholar is that it includes ALL academic articles that have been posted on the web – including early drafts – so it is best, as far as possible, to use google scholar to identify relevant material but then download the material directly from the site of origin.

Google books (http://books.google.com/) searches not only titles and keywords of books but also the full text of those books for which it has access. Often, therefore, it comes up with 'hits' that are not picked up by google scholar. It is particularly useful to check what books have been published and to identify individual chapters in edited volumes (which are not included in most of the large academic search engines). For many books, you can read a sample of the text online before deciding whether to obtain a copy. Google books also gives online access to full text of many books that are out of copyright.

Google (www.google.com) is widely acknowledged as the most comprehensive public-access search engine. However a big disadvantage in terms of academic research is that it includes *all* types of source indiscriminately. The sites that appear first may be shopping sites, holiday advertisements and so on, which are irrelevant to most academic searches. Also the sites that do provide information are of very variable quality. However google can be a useful additional source if you are searching for non-academic documents or for background information about a potential field site.

Box 2.9 Example of first stages of a literature search: tourism certification

1 Catalogue of the University of Kent Templeman library
A search of my university library's online catalogue on [tourism certification] produced only one record – a book of edited papers published in 2001. In order to find more sources, I used two techniques to broaden the search:

i Change of search terms. A search on [ecotourism certification] produced three books (the same one plus two more edited volumes – one from 2002 and one from 2007).
ii Searching on the class-mark by which the book is classified. This produced a long list of books on tourism, including one additional book that included the phrase 'responsible tourism' in the title, and several more books on 'sustainable tourism'. (Note that nearly the same result could be obtained at this point by going to the library and looking along the shelf holding books with that class-mark, but you'd miss books currently out on loan, awaiting re-shelving, shelved separately with outsize books, or on order).

2 Google books
A search of google books on [ecotourism certification] produced 466 hits – too many to scan through in full. Unlike the library search, therefore, I needed to narrow down the search rather than broaden it out. I did this as follows:

i Put the search terms in inverted commas so that they were treated as a single phrase rather than as two separate keywords. A search on ["ecotourism certification"] produced 76 hits – better, but still too many for an initial search for broad introductory information.
ii Specifying publication dates (to do this or specify other criteria click on 'advanced search'). A search on ["ecotourism certification"] between 2003 and 2007 produced only 25 hits – a reasonable number to assess. The most relevant, other than those already found in previous searches, was a book published in 2007 on ethics in tourism.

3 Google scholar
A search of google scholar on [ecotourism certification] produced about 3,130 hits. A search on ["ecotourism certification"] produced about 145 hits; limiting it to 2003 – 2007 reduced it to 49 hits. Of these, 14 were journal articles; six were books; nine citations; 12 unreadable characters; and the rest were reports by NGOs, government websites, international policy bodies (WTO) and university websites that would not be found through the other search engines described here.

4 Web of Knowledge
A search on [tourism certification] on WoK produced only one journal article. However a search on [tourism AND certification] produced 16 articles.

an enormous number of irrelevant sites. All sites should have information on how they treat search terms (often under 'advanced search' or 'search tips') and it is well worth spending a few minutes checking this information if you are not sure.

Two things to note from this example are as follows:

1. You can improve the search terms as you go along. In the above example I started with [tourism certification] but got few results, so I tried [ecotourism certification]. I also came across the terms 'responsible tourism' and 'sustainable tourism' and could have used them as search terms too if I was still short of material.
2. A general rule: If your initial search produces too many hits, narrow it down; if it produces too few, broaden it out. There are many ways to do this – adjusting the search terms, restricting or broadening the dates of publication, and so on. Many people deal with very long lists of hits – for example in google – by scanning through the first few pages of results until they feel they have enough. However, since results are not necessarily ordered according to your priorities, you may miss the most useful items in this way.

Every researcher develops their own approach to carrying out a literature search and has their favourite sources. I usually start with searches on google scholar and the WoK. I use google books (and the university library catalogue) to check what recent books have been published. If I need non-academic sources such as policy documents, I may also use google or I may search national or international policy sites directly. For international conservation policy, the most important sites are probably those of the relevant United Nations bodies – especially the IUCN (http://www.iucn.org) and the Convention of Biological Diversity (http://www.cbd.int) – and of the big international conservation organizations. At the national level, many governments have extensive material on their websites, arranged by ministries and subdivisions within ministries, which is especially useful for information on policy or on particular sites such as protected areas.

A note of warning: it is very easy to get lost in online searching and to waste a lot of time. You can go on searching long past the point where you will be able to process the information you have gathered. Alternatively, you may go off at a tangent or bury yourself in the first articles you find and waste several days reading documents that seem highly relevant when you first come across them but turn out to be less so once you search further. In order to make the most of your time, it is essential that you have a clear focus before you start and that you know what level of detail you are aiming at, so that you know when to stop. In an initial search, for example, you may only wish to find one or two recent review articles. In this case it may be useful to set yourself a rough time limit, after which you stop and review whether you have enough. An initial search to find one or two recent reviews should take less than an hour; if your search terms are well targeted it may only take 5 or 10 minutes. Once you have read and digested them you will be better able to target further searches to focus on more specific issues and subtopics.

2.4.3 Information management: keeping records of what searches you have done

The literature search, then, is not done in a single massive trawl of sources but in a series of successive steps at more and more specific levels, all interspersed with reading, reflection and identification of key issues for further consideration. Each stage informs the next, and

Table 2.3 Record of searches

Sources	*Search terms*
Kent University library online search	Tourism certification
Kent University library online search	On class-mark of above book
Kent University library online search	Ecotourism certification
Google books	Ecotourism certification 2003–2007
Google scholar	Ecotourism certification 2003–2007
Web of Knowledge	Tourism AND certification

therefore it is essential that you keep a record of where you have searched, what search terms you have used and what you have found. Proper records will save you from repeating searches that you have already done or spending hours trying to find again some fascinating title that you noted down without recording where you got it. Table 2.3 provides a minimal record of the above search.

In addition to keeping records of searches, it is essential to keep a record of the documents that you find. Reference details can initially be cut and paste into a Word or Excel document to create a list, or if you are using the WoK, they can be exported directly into a specialist referencing database programme such as EndNote or RefWorks. As the list expands, you should start to organize by subtopics so that you can find the references on each subtopic easily. Again, you can do this either in a Word document or in a referencing database, where you can enter keywords. An added bonus of categorizing by subtopics like this is that the categories can feed into further brainstorming on relevant issues and research questions; you simply need to decide which ones to pursue and which ones to leave out. Finally, once you start to download the full text of articles it is useful to store them separate folders according to sub-topics. For tourism certification this might include processes for the establishment of a certification system; marketing advantages; criteria and indicators; auditing processes, and wider aspects of ethical practice. Then when you are ready to start reading on a particular subtopic you can scan the titles of the documents in the appropriate folder and prioritize.

2.4.4 Starting reading

Once you have a list of document titles and perhaps some articles stored on your computer, the fun begins: you can start reading and find out what other people have done and said. If possible, start with a recent review. Read it right through and make notes summarizing key concepts, arguments and examples; otherwise a few months later it will all be a big blur. If you are working from an electronic source, it may be useful to copy and paste key sections into your notes – but *be very careful to mark these sections as direct quotes* or you may inadvertently commit one of the great sins in research: plagiarism (see Section 17.4). Plagiarism includes the unacknowledged use not only of other people's words, but also of their ideas, so make sure that it is clear which parts of your notes summarize material in the source papers and which parts are expressing your own comments and ideas. Read through your notes when you have finished and make sure they make sense and include everything you want to record. Also, make sure that you have noted down the full bibliographic details of the source (see Section 17.4) so that you can reference it properly when you come to write up your report. Then back up your notes and move on to the next article.

All the time that you are reading you should be analyzing what you read with a critical eye (see Box 2.10). Do not take everything at face value but ask yourself whether there are flaws in the argument or the research design, whether there are any unsupported underlying assumptions and whether the interpretation is really justified by the evidence. When you have finished an article, give yourself time to reflect on what it was trying to say and how relevant it is to your own research interests. Note down any new references that you want to follow up, either as you read or by looking through the reference list at the end of the article (or probably a bit of both). The whole process of reflection, critical analysis and note taking is vital in order to process what you are reading properly and put it in perspective; otherwise you run the risk of losing the big picture in a mass of detailed information.

As you read more articles, reflect on how they relate to each other – especially where they seem contradictory – and make notes on this, too. In this way you can build up your own overview of the issues. Collect different definitions of key concepts, look for 'gaps' in information and explanation, note down the details of interesting case studies and also note down suggestions for areas that need further research. Students sometimes ask me how many papers they should aim to read on a particular subject area, but of course it is not the number of papers that is important, but the amount you learn from them. You should stop reading on a specific topic (or subtopic) once further reading does not give you important new insights or information. At this point you should have a good understanding of the main theoretical issues and conceptual and methodological approaches with which to inform your own research design. Before you leave the topic altogether, write a summary of what you have found out. The summary should *describe* and *synthesize* the material you have read; include your own *commentary* and *critical analysis* (see Box 2.10) of what other people have said, and identify *gaps* in what has already been done.

The final product from the literature search is the *literature review* – a document setting out what the literature has to say about your research topic. In a thesis, the literature review forms a substantial part of the introductory chapter (see Section 17.3.1) and should form a concise, coherent narrative that introduces the theoretical background to your study. Each time you make a point that is based on a particular article, you should cite the source (see Section 17.4 on how to do this). You may not write the formal literature review until you are writing up the report after data collection and analysis, but it is useful to write a draft before data collection. It will help to focus your ideas and the more that you write now, the less you will have to go back and re-read the original sources later on.

Box 2.10 Example: critical analysis

Critical analysis in relation to the literature review involves picking apart the strengths and weaknesses of what other people have said, for example, missing steps in their argument, unwarranted assumptions, methodological flaws or interpretations that are not entirely supported by the data. An (imaginary) example might look like this:

'Smith et al. (1999) report a positive effect of the conservation education programme on attitudes towards predators among sheep farmers. However their sample sizes were small, and a lack of pre-testing or rigorous sampling makes it impossible to rule out the possibility that other factors were responsible for the difference between those who participated in the programme and those who did not.'

There is an art to efficient use of time during reading. The first few articles you read on a given topic tend to take the most time, because a lot of information is new to you. However once you are familiar with the background to the topic, you will start to come across similar material again and again, and reading and note taking will become much quicker. You only need to make notes on points that are relevant and new to you. When you pick up a new book or article, do not assume you must read it cover to cover; try to develop the technique of scanning the text to pick out the parts that are useful, and then move on to something else. In the case of a book, look first at the contents and see whether there are certain chapters that are particularly relevant to you and others less so. Within a chapter or journal article, if you start reading and find that you are already familiar with the material being presented, or conversely that it is not really relevant to your research topic, then it makes no sense to read it all the way through to the end. But if you abandon it, do still record the fact that you have looked at it, so that you will not waste time repeating yourself.

2.5 Conclusions

The initial process of defining and narrowing down the research topic is worth giving some time and effort. If done well, it should form a foundation for a strong research design. However it cannot all be done in one step; it will evolve as you develop the rest of the research design, especially the methodology. Therefore you should read Chapters 3 and 4 before you attempt to go very far in narrowing down.

Summary

1. The research process can be divided into five stages. This chapter discusses the first two stages: identifying a general topic or problem, and developing the specific focus of the research.
2. For students looking for a topic for a thesis, sources of ideas include your own experience, issues that you have come across in class, and suggestions from the academic staff where you are studying. In choosing a topic you should balance your specific interests against the practicalities of 'doing' the research.
3. Once a topic is chosen, the specific title, aims and objectives must be defined by a process of brainstorming, reading, thinking, and consulting with your supervisor or line manager and with potential collaborators.
4. Research *strategies* are fundamentally different ways of framing a research project. In *deductive* research, a theory or hypothesis is developed and data are collected to test it. In *inductive* research, data are collected starting from a broader perspective and theory is built on what the data reveal.
5. In addition to narrowing down the subject of the research and deciding on a research strategy, it is important to define the practical scope of the project in terms of the geographical area, population and timescale.
6. Conceptual or operational definitions should be developed for key terms.
7. You should end up with a set of aims, objectives questions or hypotheses that are clear and intelligible, linked to each other, researchable and realistic, connected to established theory and research, and interesting.
8. The literature search is an important part of this process and should be carried out systematically and with a well-defined focus.

Further reading

Blaikie, N. (2000) *Designing Social Research*, Cambridge: Polity Press. [A more advanced text on the whole of the social science research design process, particularly relevant for qualitative, inductive research.]

Creswell, J.W. (2009) *Research Design: Qualitative, Quantitative and Mixed Methods Approaches*, 3rd edn, Thousand Oaks, CA: SAGE Publications. [A more advanced text containing substantive chapters on the literature review, on the use of social science theory, and on the development of a 'purpose statement' for a research project.]

Pratt, B. and Loizos, P. (1992) *Choosing Research Methods: Data Collection for Development Workers*, Oxford: Oxfam. [Written for staff of development NGOs, this book provides an accessible outline of practical approaches to applied research.]

Russell, D. and Harshbarger, C. (2003) *Groundwork for Community-based Conservation: Strategies for Social Research*, Walnut Creek, CA: Altamira Press. [Essential reading for all conservationists, this book contains excellent chapters on the conceptual issues involved in community aspects of conservation as well as on practical approaches to field-based applied research. See especially Chapter 10.]

Wallace, M. and Wray, A. (2006) *Critical Reading and Writing for Postgraduates*, London: SAGE Publications. [A very clearly written and informative book that gives excellent guidance on searching the literature and on critical reading and writing.]

3 Developing the methodology

Make things as simple as possible – but not simpler.
(Albert Einstein in R.V.G Menon, 2010: ix)

3.1 Introduction

This chapter describes the next step in the research process, which is the development of a *methodology*. The methodology is concerned with what you will actually *do* in order to address the specific objectives and research questions you have developed. People often equate 'methodology' with the list of individual methods that will be used – questionnaires, semi-structured interviews and so on. However the methodology must also include an overall strategy that fits all the different methods together into a coherent design. This involves deciding on a research design structure, choosing the specific methods and developing a sampling strategy. It often also involves defining what analyses you will carry out. This chapter discusses the first two of these factors; sampling is discussed in Chapter 4 and analysis is discussed in Chapters 14 to 16.

Confusingly, the term 'research design' is used both for the overall process described above and also, more specifically, for what I am here calling the research design *structure*. The latter is to do with how the data collection is structured. There are many different types of research design structure, from laboratory experiments to surveys to case studies to longitudinal studies. Which structure you should choose depends on several factors, including the nature of the research questions you wish to address; the scope of the study in terms of geography, population and time span; the time and money available; and the current state of knowledge of your chosen subject. The following section introduces each type of research design structure in turn, and Section 3.3 discusses some factors to take into account in choosing between them. Section 3.4 gives an overview of the different methods that are described in this book, discusses some general principles in deciding which methods to use, and introduces multiple-methods and mixed-methods approaches. Finally Section 3.5 attempts to draw some general conclusions bringing together the different aspects of research design that have been discussed in this and the previous chapter.

3.2 Types of research design structure

Research designs are often divided into three basic types according to the level of control and intervention by the researcher: true experiments, quasi-experiments and observational research. In true experiments, the researcher makes some kind of active intervention – subjecting one or more groups of participants to some kind of change – and tests for its effects under tightly controlled conditions (in order to minimize the effects of other factors). Quasi-experiments mimic true experiments in that they also involve an active intervention and use comparisons to examine its effects, but they do not fulfil all the conditions of 'true' experiments (see below). Finally, observational studies involve no intervention by the researcher but, as the name suggests, are based purely on observation of events as they occur naturally. Observational studies can be structured in several different ways. The following section describes each kind of research design structure in more detail and then Box 3.6 presents a summary of the different kinds.

3.2.1 Experimental design

In experimental design, you actually *do* something or *change* something (this is known as the *intervention* or *treatment*) and then test for its effect. There are several strict principles concerning how this should be done, as follows:

- Tests are carried out both before and after the 'treatment', in order to test for its effects.
- The treatment is administered to only half the participants, known as the *experimental group*. A second group – the *control group* – is also tested at the same time intervals, but undergoes no treatment.
- The participants – usually known as subjects – are assigned randomly to one of the two groups.
- Throughout the experiment every effort is made to minimize changes of any kind other than the treatment itself.

All of these are concerned with maximizing the degree of certainty that any change observed in the experimental group after treatment is due to the treatment itself, and not to other factors (this is related to the *internal validity* of the research design: see Box 3.7). The procedure is summarized in Box 3.1.

Experimental design is most common in laboratory research, where it is possible to control conditions very tightly and thus rule out *confounding variables* – factors other than the ones you are interested in that might muddy the results. You have far less control of the conditions in field research and therefore true experiments are much rarer, especially when dealing with people, because of ethical issues about 'doing something' to them just for the sake of research.

However, certain types of intervention are acceptable. For example, you could introduce a new activity related to environmental issues to a group of school children and test the

Box 3.1 Experimental design

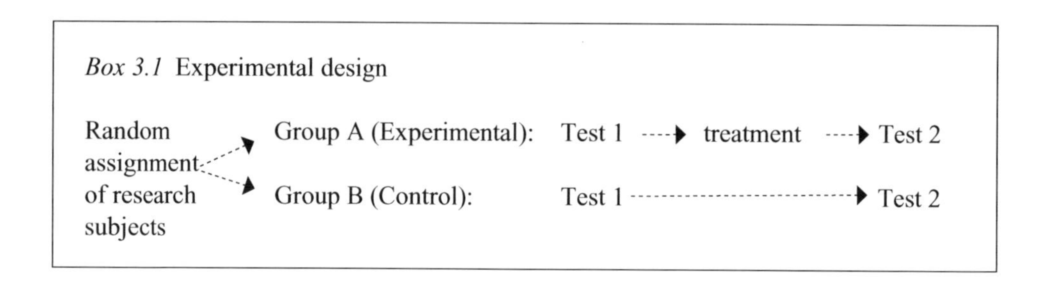

children's knowledge and attitudes before and after they participate. However, if there *were* a change, could you really be sure it was due to the new activity and not to something else that had happened in the meantime? An additional check would be to run parallel tests on a second group of children who *do not* participate in the new activity. If the two groups demonstrate a similar change in knowledge and attitudes then it is probably not due to the new activity, whereas if the experimental group shows a greater change, then it is more likely that the activity is having an effect. The two groups can be compared statistically in order to see whether knowledge and attitudes have changed significantly more in one group than in the other.

A further check in experimental design is to allocate each participant randomly either to the experimental group or to the control group (keeping equal numbers in the two groups) in order to minimize the chance of any systematic differences between the groups. In the above example, an easy way to do the research would be to work with two classes of children. You could test knowledge and attitudes to the environment in both classes, then introduce the new activity to one class, and then test both classes again. But what if one of the classes had already done a lot of work on the environment, or has a particularly pro-environmental teacher, or has 'brighter' pupils? The pre-testing should pick up any initial differences but if these are large it would then be difficult to assess the effects of your new activity on top of this. Also, you could not assume that the activity would have similar effects on children in other classes that were different in some other way. The sample would not be *representative* of a wider population of children (see Chapter 4), and therefore you could not generalize the results. A further complication is that some other change could happen to one class and not the other between the two tests; they might have a visit from an environmental campaigner, or participate in a second classroom activity on environmental issues, unknown to you. Another principle of experimental design, then, is that the participants should be randomly assigned to one of the two groups rather than bunched together in naturally occurring groups such as classes. Ideally, you would first select a representative sample of children to participate in the study (see Chapter 4), and then allocate half of them randomly to each group.

3.2.2 Quasi-experimental design

There are often practical or ethical reasons in fieldwork that make it impossible to fulfil all the conditions of true experimental design. For example, it may not be possible to collect data before the intervention is made (known as 'baseline data'), especially if the need for intervention is urgent. Alternatively, it may be unethical or politically unacceptable to introduce management changes that benefit only some people and leave others as they were before. A third possible limitation, even if it is acceptable to apply the change to only some people at first (perhaps as a trial run or, in technical terms, a 'pilot'), is that there may be practical problems in allocating people randomly to either the experimental or control groups. Many forms of management are applied at the level of a defined geographical area, which means that all the people affected live in the same place and all the people in the control group live somewhere else. In this case there are likely to be underlying differences between the two groups of people connected to where they live, and it is possible that these will 'confound' the results. Finally, and hardest of all to deal with, in field research it is impossible to keep all external factors constant.

I define a quasi-experimental design as a design that, like true experimental design, aims to evaluate the effects of a specific intervention by using comparisons, but lacks all the checks and balances of true experimental design. It may be that there is no control group, or no pre-testing, or (probably the most common case) no random allocation of participants to the two groups. This may sound like a serious failing, and in purely theoretical terms, it is. However, in practical terms, quasi-experiments may be the best option available and are a very useful tool in the armoury of field researchers as long as their limitations are recognized.

3.2.3 Observational research

Observational research is defined as research that involves observation alone rather than active intervention. Some observational research examines possible cause-effect relationships by focusing on a specific, naturally occurring change, whereas other observational research aims at in-depth description and understanding rather than explanation, and may not look at change at all.

There are many ways to structure observational research, and different structures are good at addressing different types of aims and questions: *case study design* is good at building in-depth description and understanding of a specific situation or example; *comparative case study design* compares two or more cases to draw inferences and generate theories about the differences between them; *cross-sectional design* is used to draw inferences about the characteristics of a large population and the patterns of variation within the population by collecting data from a cross-section of its members; and *longitudinal design* tracks changes over time by collecting data on several successive occasions, often years apart. The following text describes these different structures in more detail.

3.2.3.1 Case study design

Case study design involves detailed data collection about a single 'case' or situation. Examples of 'cases' in conservation could include Bwindi Forest Reserve in Uganda (for a study of the effects of forcible resettlement on local people), London zoo (educational impact of zoos) or the 2002 World Summit on Sustainable Development (processes of negotiation in international environmental policy fora). Case studies aim at a detailed understanding of the case that has been selected, both for its own sake and in order to add to broader theoretical understanding and generate theories about underlying issues.

SELECTING A PARTICULAR 'CASE' TO STUDY

Case study research in conservation often originates in a request for advice from managers of a particular conservation site, and therefore the specific case is pre-selected. But if this is not so, how do you go about choosing a case study?

The answer includes both practical and theoretical concerns. On the practical side, is the site easily accessible? Do you have contacts there who will introduce you to people? What permits do you need and how easily will you be able to get them? Is there the necessary infrastructure to support you during your visit? These questions are particularly important for students embarking on research for the first time, and many university departments (including mine) will need some evidence that these matters have been taken into consideration before they will agree to student projects at sites that have no direct contact with the university staff.

However, there are also more theoretical aspects that should be considered. What makes one particular case study more interesting than another? Will it contribute anything new to

Box 3.2 Choosing a case study: theatre as an educational tool in zoos (Penn, 2005)

For a PhD study of the use of theatre as an educational tool in zoos, Laura Penn chose Central Park Zoo in New York as a case study. At that time Central Park Zoo had the most varied and advanced zoo theatre programme in the USA and possibly in the world. However, there were also strong practical reasons for choosing Central Park Zoo: before starting the PhD, Dr Penn had worked as a member of the educational team at the zoo. Her prior involvement ensured that she would have excellent access, and indeed it was through working there that she became interested in the use of zoo theatre as an educational tool. However this would not have been sufficient reason in itself to use Central Park Zoo as a case study. Its extensive theatre programme made it a very good choice in theoretical terms.

current understanding, or just provide yet another example illustrating what we already know? The appropriate type of case study depends partly on the current state of knowledge about the research topic. If the project is addressing a new, poorly developed area of research, almost any case study should add something to the current state of knowledge. It is most useful if it is reasonably typical of the wider situation – although of course what counts as 'typical' is at this stage a matter of subjective judgement. In contrast, if the research field is already well-developed, there are benefits to choosing an extreme or unusual case that appears to contradict current theories. Some obvious questions would be *why* it has become so extreme; exactly *how* it contradicts current theories and what may have brought about this situation here and not elsewhere; and what else has developed differently as a result of the extreme or unusual circumstances.

A third approach is to choose a case study where the factors you are interested in are particularly prominent, so that you can explore them in depth. Box 3.2 gives an example in which the case study was chosen because it was a highly developed case of the subject of research – zoo theatre – that might only be a very minor feature at other zoos. Studying zoo theatre at a different zoo might have provided very limited opportunities for data collection, simply because not much zoo theatre was going on. However, note that in addition to this reason, there were also strong practical reasons for choosing the case study (the researcher's prior involvement there). Practical reasons of this kind are often unacknowledged in formal descriptions of the reasons for the choice of case study but they are almost universally an important factor – and quite rightly so. You cannot plan a research project with any confidence if you have never been to the proposed field site and have no contacts there.

3.2.3.2 Comparative case study design

Comparative case study design involves carrying out studies of two or more cases in parallel and then comparing them. Building a simple comparison into case study research is a very valuable way to structure the research; it means that rather than simply *describing* results for a single case, you can *compare* the results for different cases, which gives plenty of scope for analysis and interpretation and can generate interesting questions for further research.

The most common way to set up a comparison is to choose two cases that differ in one factor that you think may be interesting to investigate, but are as alike as possible in every other way. For example, you could choose one indigenous and one non-indigenous community in order to compare them in terms of their natural resource use. In this case you should look for two communities that are as similar as possible in every other respect – size, demographic make-up, accessibility, natural resource base available, and so on. One general rule is that it makes sense to choose two communities that are close to each other geographically. Clearly, if you were to choose one indigenous community in Peru and one non-indigenous community in Australia, any differences are just as likely to be to do with their location and broad context as with indigenousness. Even within one country, it is easiest to find similar cases in the same state or region. For a study looking at natural resource use, they should also be located within the same ecosystem.

You will never find two communities that are identical in all respects other than the one you are interested in, but you should do the best you can, and discuss the implications of any differences in your report. If you find a difference in natural resource use between the two cases in the above example, you cannot *prove* that it is because of a difference in indigenousness because there will inevitably be other differences too, but you might generate some valuable insights and theories about differences between indigenous and non-indigenous peoples, and your study will add to the growing body of 'cases' that are described in the literature. In this way, rather than attempting to test hypotheses, comparative case study design is concerned with describing similarities and differences in different examples and discussing possible interpretations. It is a powerful option in field research.

3.2.3.3 Cross sectional design

Cross-sectional design involves collecting data from a large number of individual *cases* that form a cross-sectional *sample* of the population of interest. The use of the term 'cases' here must not be confused with the way it is used in case study design, although the difference is actually just a matter of scale; in both types of usage a case is a particular example (see Box 3.3). The strategy by which the individual cases are selected is known as the *sampling strategy* and will be discussed in detail in Section 4.2, but the principle is that the sample

Box 3.3 Key term: 'cases'

A 'case' is a particular example. The term is used at different levels.

A 'case study' is a study of one particular instance. This may be a particular location (one zoo), a particular event (one international conference), an organization or network (Friends of the Earth UK) and so on.

A *'case'* in terms of sampling means one particular item in a sample of many such items. In social science research each 'case' is often a person (for example, in a sample of Friends of the Earth UK members, each case is one member), although other kinds of unit can be sampled, such as households, communities, places, events, organizations and so on. In a sample of eighty zoos, each zoo would be a 'case' (but with such a large number, would not be referred to as a 'case study'), and data would be collected for each zoo as a whole (for example, by distributing questionnaires or carrying out semi-structured interviews with the zoo directors). (See also Section 4.1).

Box 3.4 Cross-sectional survey of Essex members of Friends of the Earth

This study aimed to gather information on the social background of members of the environmental group Friends of the Earth (FoE). The scope was limited to FoE members in the county of Essex in the UK. The sample was defined as all members living in Colchester (the county town) plus two out of every three members in the rest of Essex. A postal questionnaire survey was distributed to each member of the sample. The number of questionnaires completed and returned was 126.

Note that the sample in this study is not completely representative of the overall population; it includes all members in Colchester and only some of the members elsewhere in Essex. Presumably this was for practical reasons – it was easier to reach people living in Colchester. Whilst this is a weakness of the study, it does not make it invalid. Even though the sample is not entirely representative it can still give a useful indication of the characteristics of FoE members in Essex.

Source: Greenway (1979)

should be representative of the population as a whole so that you can use the information you gather about the sample to make inferences about the population. Therefore probability sampling should be used if possible.

Cross-sectional design is good at inferring the characteristics of a large population; comparing two or more different populations or subsections within a population (for example, men and women or different age groups); and looking for relationships between different characteristics of the cases (for example, income and attitudes). It is the commonest type of design in national surveys and other very large-scale studies, and data are usually collected by means of a questionnaire survey, partly because of the logistical challenge in using other methods across a large, geographically dispersed sample and partly because the quantitative data provided by questionnaires are appropriate for making statistical inferences. However, it can also be used across smaller areas. Box 3.4 gives an example of a cross-sectional survey that was carried out for an undergraduate thesis on members of the environmental organization Friends of the Earth in the county of Essex in the UK.

3.2.3.4 Longitudinal design

Longitudinal studies are extremely valuable in tracking change over time. The defining characteristic is that data are collected repeatedly on the same thing at intervals – either from a new sample of cases each time or (preferably) from the same sample each time (a 'panel study'). Conservationists will be familiar with longitudinal research in the natural sciences in the form of ecological monitoring, which involves gathering data on ecological variables at repeated intervals in order to track ecological change – often over a period of years or even decades.

Longitudinal studies are very common in some areas of the social sciences, such as health studies and education. In relation to social aspects of conservation, they have probably been used most often to track changes in environmental attitudes and behaviour. Within each individual data collection exercise, data collection may be structured by any of the other types of research design structure described here. Box 3.5 gives an example of a longitudinal study

Box 3.5 A longitudinal survey of public attitudes and behaviours toward the environment

The UK government's Department for Environment, Food and Rural Affairs (DEFRA) and its predecessors have run a nationwide questionnaire survey in England every three to four years since 1986 in order to track changes in environmental attitudes and behaviours. Over this time some questions have been kept constant while others have been altered or added in order to cover new issues or changes in perspective. The results are used to inform government policy on environmental issues and, by looking at the attitudes and behaviours of different subgroups, to target different groups in terms of publicity and awareness-raising activities.

The 2007 questionnaire was administered face to face to 3,618 people spread across a total of 378 different locations in England. It included sections on household and individual attributes; well-being; travel; energy and water efficiency; recycling, composting and reducing waste; purchasing behaviour; and environmental awareness and attitudes.

Source: DEFRA (2007)

using a *cross-sectional* design to monitor changes in public attitudes towards the environment in England. Many other types of longitudinal study use a *case study design* to track changes at a particular site, and thereby to inform ongoing management.

3.3 Choosing a research design structure

Box 3.6 summarizes the different types of research design structure that have been described above. Often it is obvious which structure is appropriate for a particular study as soon as you have defined the questions you want to answer and the scope of the study. If you need to cover a very large population, cross-sectional design is probably the best option on both theoretical and practical grounds. If you want to investigate a particular incident or situation, a case study design is appropriate. If your questions are to do with explanations – what effect a particular factor might have, or what causes a particular phenomenon – then the 'tightest' option is experimental design, but you have to weigh this against practicalities. Is it ethical to intervene just for the sake of research? Even if so, can you fulfil all the conditions of true experimental design and still keep a reasonably naturalistic situation that reflects the 'real world'? If not, then you could choose between a quasi-experimental design, comparative case study design, or perhaps a longitudinal study using a repeated cross-sectional design.

It should be clear by now that different types of structure have different strengths and weaknesses. One factor to consider when choosing from them is that of *validity*. In broad terms, validity is to do with 'whether the evidence which the research offers can bear the weight of the interpretation that is put on it' (Sapsford and Jupp 1996: 1, in Bell, 2005: 117–18). In other words, do the data address what you want them to address? This depends not only on precisely how specific items of data are collected (validity at this level is known as *measurement validity* – see Section 7.6) but also on the overall research design and how tightly it addresses the original research questions. Three of the most important types of research design validity are summarized in Box 3.7.

Box 3.6 Classification of research design structures

Experimental design: Random assignment of participants to experimental and control groups in order to test for an effect of the 'treatment' on the experimental group. Data are collected pre-and post-treatment for both groups.

Quasi-experimental design: Also involves treatment (active intervention by the researcher) but may lack random assignment of participants to groups, inclusion of a control group, or pre-treatment testing.

Observational design: Just observation – no treatment.

Case study: In-depth examination of a single case.

Comparative case study: Comparison of two or more contrasting cases.

Cross-sectional design: A representative sample is selected from a large population.

Longitudinal design: Data are collected repeatedly at different times, in order to track change over time.

Internal validity is most relevant to research that aims at explicit explanation rather than description and understanding. It is strongest in experimental design, because there is a strong internal logic that allows you to rule out 'confounding variables' and test for causes and effects with a high degree of confidence. Cross-sectional studies may also be used to test for relationships between variables, but usually it is only possible to test for correlations, not causes (see Chapter 16). Of the other types of research design structure that use comparisons in order to explore cause and effect relationships, the weakest in terms of internal validity is probably comparative case study design. You can identify differences between the case studies and discuss possible reasons for them, but you cannot test rigorously for a causal link because you cannot control for all the potential confounding variables.

External validity is more to do with sampling than with the research design structure, since generalization beyond the sample is only valid in statistical terms if probability sampling is used so that the sample is *representative* of the study population as a whole

Box 3.7 Key terms: different types of validity related to the research design

Internal validity: A study has high internal validity if its internal design allows you to draw conclusions with a high level of theoretical rigour. Internal validity is highest in experimental design.

External validity: External validity is to do with the extent to which the results can be generalized from the sample to a larger population. External validity is highest in cross-sectional design.

Context ('ecological') validity: Context validity is to do with how far the situation under which the research is carried out represents 'real life'. Context validity is probably highest in case study design using unobtrusive methods such as participant observation.

(see Chapter 4). Cross-sectional research is the only type of research that treats external validity as a primary concern; it lays great stress on probability sampling. At the other extreme, case studies make no pretensions to external validity – they do not claim to produce generalizations past the particular case (although the results can be compared with other cases reported in the literature in order to build theoretical understanding). Experimental studies – particularly those in the laboratory – hardly ever consider external validity; they often rely on volunteers (or the student population that is immediately to hand) rather than attempting to find a representative sample. A vast amount of lab research in experimental psychology has been carried out on students and in strictly theoretical terms, therefore, much of the knowledge built up in psychology is 'known' only with respect to students, not to people in general. However some common sense is needed here; if there is no reason to assume that students are different from everyone else in terms of the factors being studied, then generalizations to the population as a whole tend to be accepted as working theories until and unless someone comes up with a reason for not doing so.

Finally, the concern underlying the concept of *context validity* is that if the situation is artificial, then people's behaviour is likely be different from the way in which they would behave in 'real' life, and therefore it is hard to know what the results tell you. Laboratory studies are the weakest in terms of context validity and case studies are the strongest, because they can best take into account the complexities of the specific social context. Observational studies are generally stronger than studies that involve direct intervention by the researcher, and unobtrusive methods such as participant observation have higher context validity than pre-arranged interviews or questionnaires.

3.4 Choosing your methods

3.4.1 The range of methods described in this book

Obviously, the research design cannot be finalized until you have chosen the methods that you will use to collect the data. Section II of this book consists of a detailed description of each of a range of social science methods in turn; this section provides an introduction to each of the different methods that is included and presents some general principles to bear in mind as you choose between them.

Probably the most important factor to consider in choosing one or more specific methods is whether a quantitative, qualitative or *mixed-methods* approach (see Section 3.4.2) is appropriate for a particular project. Quantitative methods provide data in a format that is easy to express in numbers and the results are therefore easy to analyze using statistics, which means that inferences can be made from a sample to a larger population, and hypotheses about cause-effect relationships or correlations between different variables can be tested rigorously. Qualitative methods are less precise but more flexible. They are better at exploring issues that cannot be clearly defined at the start, and at providing in-depth description and understanding. They are also better at taking the social and cultural context into consideration. However, they are less appropriate for highly targeted studies that aim to test specific hypotheses or make statistical inferences. Whilst qualitative data *can* be analyzed quantitatively, it takes more effort to process the results for this purpose, and there is more imprecision in doing so than would be the case with quantitative data.

Figure 3.1 places the different methods described in this book on a continuum from the most qualitative to the most quantitative. *Participant observation* (Chapter 5) is the standard method of social anthropologists. It involves spending time with the people you wish to

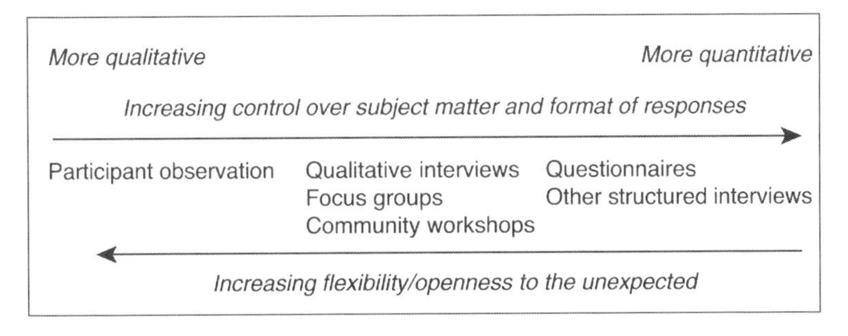

Figure 3.1 The qualitative–quantitative continuum in social methods.

learn about – getting to know them, joining in their activities, chatting with them – and making systematic observations about what they do and say. It is therefore almost always associated with a case study design. The principle is that only by joining in their lives can you develop a really in-depth understanding of what is going on. Participant observation is a time-consuming process – traditionally, anthropologists aimed to spend at least a year in the field – and this is often a barrier to its full use in applied conservation projects. However, it is sometimes possible to use participant observation over much shorter timescales, especially if you already know the people involved. Relevant examples in conservation would include participant observation at a single protected area or in a specific event such as an international conservation policy conference.

Qualitative interviews are conversations about the subject of research in which the researcher imposes varying degrees of structure and direction (see Chapter 6). Most interviews are conducted with a single respondent, although they can also involve groups of people. They are categorized according to the amount of structure that is imposed; thus there are informal interviews (chats with people you meet as you go about your daily business – a central component of participant observation), unstructured interviews (free-ranging conversations that you arrange in advance in order to explore some aspect of the research topic) and semi-structured interviews (where you use a written '*interview guide*' to make sure you cover a pre-defined list of topics). All forms of qualitative interview involve discussion back and forth between interviewer and interviewee and therefore, to a greater or lesser extent, can take their own direction according to the issues that come up. They can be used either in exploratory, inductive case studies, where each interview may have a slightly different interview guide according to who you are talking to and what you have learnt up to that point, or in more structured studies where a set of semi-structured interviews is carried out with a sample of people, using a fixed interview guide. The strength of qualitative interviews is in providing background information and context, generating ideas, discovering the unexpected, and providing in-depth information on each participant's views, perspectives and motivations.

Focus groups (Section 6.2.4) represent a form of qualitative group interview. A typical focus group involves between six and ten people, lasts from one to three hours, and covers from four to twelve questions or topics for discussion. Their great strength – and weakness – is that they are based upon discussion among a group of people rather than independent statements by each person alone. A group discussion is often better than a set of individual interviews at bringing out contrasting views, encouraging reflection and producing in-depth explanations of the reasoning behind the views that are expressed. On the negative side, what each person says will be influenced by what others have said and therefore focus groups are poor at uncovering patterns of variation between individuals or finding out what proportion of people hold a particular view. For some direct comparisons between focus groups and individual interviews see Kaplowitz and Hoehn (2001) and MacMillan et al. (2002).

In contrast to qualitative interviews, *structured interviews* consist of a pre-defined set of questions or tasks that are presented to all respondents in precisely the same way. There is no 'conversation' in the sense of discussion back and forth between the interviewer and the interviewee. They are often referred to as *quantitative interviews*, because the fact that they are standardized means that it is easy to express and analyse the results quantitatively (as numbers). *Questionnaires* (Chapter 7) are the best-known form of structured interview, and involve asking a fixed set of short-answer questions in the same way and in the same order of each of a set of respondents. They give the researcher the highest level of control of any method over the form and content of the data collected, and therefore provided that you know exactly what you want to ask – what variables you need to collect data on and what values they might take – you can do it most efficiently and precisely by using questionnaires. This means that you need to know a lot about the subject and sometimes also about the study site in advance; if you design a questionnaire based simply on your personal hunches, then it may turn out to be completely off-target. You cannot change the questions part way through data collection in order to follow up interesting leads.

Context validity is lower with questionnaires than with more qualitative methods, because the questionnaire creates an artificial situation. People's stated views may differ from those that they express in everyday life, either because they want to appear in a good light or simply because questionnaires do not give people much time to think before giving an answer (see MacMillan et al., 2002). Questionnaire surveys cannot compete with qualitative methods in terms of providing an in-depth, contextually relevant understanding of a particular situation, but, on the other hand they are excellent for generating statistics about the characteristics of a sample of people and inferring the characteristics of the wider population from which the sample is taken; external validity is high (provided that probability sampling is used; see Section 4.2.1).

Chapter 8 discusses a range of other kinds of structured interviews that are common in studies of people's perspectives on natural resources. First, it describes a set of techniques that are widely used in a branch of ethnobiology known as *cultural domain analysis*. These techniques aim at uncovering local knowledge of plants, animals or other features of the environment and finding out how this knowledge is structured and how it varies between individuals. Second, it describes a range of techniques that are used for documenting changes over time, from seasonal cycles to longer-term trends to historical events. Applications in conservation research include finding out about local people's knowledge and understanding of natural cycles; of trends and variability in environmental conditions; about the history of land use or conflicts over natural resources, or about the particular circumstances and events that have led to the current situation.

Chapter 9 describes the toolbox of PRA techniques and their use in community workshops. Workshops are regarded by many researchers as a tool for feeding back results rather than as a research method in their own right, but they can provide a wealth of information on local people's perspectives, and in participatory approaches they provide a powerful way in which to think things through jointly with groups of the intended beneficiaries. They do have weaknesses as data collection tools from a theoretical perspective. Because they involve large groups of people, there are serious problems in terms of group response effects and sampling. There are also more practical challenges in terms of how to record what is going on, especially when several subgroups are run in parallel in a single workshop. They are therefore not usually appropriate as *the* major method in a theoretical or academic study. However, if workshops are used sensibly and with full recognition of their methodological weaknesses, then they can be a valuable component of a multiple-methods research design. Moreover some of the structured PRA techniques described are valid as stand-alone sources of data.

Participatory mapping (Chapter 10) is a more advanced technique that is widely used in applied conservation research. It involves mapping natural resources in collaboration with local people based on their knowledge of their surroundings, and of all the methods described in this book, it is the one that most successfully spans the natural and social sciences. It can be a basic exercise that involves drawing a simple diagrammatic map from memory, or you can use a conventionally produced map or aerial and satellite photos to provide an accurate base map of the main geographical features, then work with local people to fill in further details, and finally check the results through on-the-ground verification using a GPS. Participatory mapping has several advantages compared to mapping based purely on ecological techniques. Rather than having to start from scratch you can build on local knowledge, which is particularly valuable for rare or hard-to-detect species. Furthermore, it can tell you a lot about what local people think is important – what they consider is worth mapping. In applied conservation work, a third significant advantage is that the process of participating in the mapping process can help to make people think about the limits of the natural resources available to them, or convince them that they have the power to act to conserve and protect them. The map itself can be an invaluable tool to support applications for land and resource rights, especially for indigenous peoples.

Table 3.1 summarizes the different social science methods and their common uses, and also notes some of the most important links with different research design structures and research strategies.

Traditionally, different social science disciplines have favoured different methods. Thus social anthropology is strongly associated with participant observation, sociology with questionnaires and focus groups, and development studies with community workshops. In general, the methodological barriers between disciplines are becoming less acute and many field studies now use a mix of different methods to complement one another (see Section 3.4.2). The bottom line is that the methods should be chosen to fit the research questions rather than being constrained by disciplinary or other prejudices – especially in an interdisciplinary subject such as conservation. Questionnaire surveys have been particularly prominent in many areas of conservation social science, partly because natural scientists tend to be most comfortable with quantitative methods, but qualitative methods are becoming more widespread.

The choice of methods should also take practical considerations into account; in fact the whole of the research design involves constant compromise between theoretical ideals and practical considerations. If you have only three months to carry out research on a complex issue at a site that is new to you, then participant observation will almost certainly be impractical – at the end of three months you may only just be beginning to find your bearings and get to know people. A more structured technique may be more realistic, even though it will not give you such an in-depth understanding. In contrast, if you want to cover a large population that is widely dispersed (and have enough background information to target your questions well), then a questionnaire survey is likely to be the best option, as long as there are no cultural barriers to the successful use of questionnaires (see Section 7.1). In contrast, if you intend to target a small number of individuals – for example residents in a community

Table 3.1 Range of social science methods

Method	*What is involved*	*Uses*	*Chapter or Section*
Participant observation	Participating in the life of the study community and making systematic observations.	Inductive, in-depth case study of a specific situation. Not practicable for short studies unless the researcher already knows the study community well.	5
Qualitative interviews	Conversations about the research topic in which the researcher imposes varying degrees of structure and direction. May be used with individuals or with groups.	Common in case studies and comparative case studies, and in longitudinal research, either alone or as part of a mixed methods study. Can also be used in cross-sectional design. Good at exploring people's perceptions of an issue in depth. May use an inductive or a deductive research strategy.	6
Focus groups	Formal discussion groups (group qualitative interviews) involving about six to ten people.	Good at encouraging reflection and producing in-depth explanations of the reasoning behind the views that are expressed. Most often used as part of a mixed-methods study, either to 'scope' the issues early on (using an inductive strategy) or to test out initial findings from other methods towards the end.	6.2.4
Questionnaires	Standardized question and answer sessions, asked of all respondents in the same way and in the same order. Either administered by a researcher (face-to-face or by phone), or else 'self-administered' by each respondent by filling in a written form on paper or electronically.	Used to collect data on a set of pre-defined variables from a large number of people (typically over 100, and in large studies, several thousands). Questionnaire surveys usually use a cross-sectional design with cluster sampling. However they can also be used in case studies, comparative case studies or longitudinal studies, and they can use other forms of probability sampling, or quota sampling. Questionnaires are almost always analyzed quantitatively, which allows statistically valid inferences to be drawn from a sample to a parent population. They can be used alone or in *mixed-methods studies* (see Section 3.4.2).	7
Other kinds of structured interview	Various	There is a massive variety of different kinds of structured interview. Freelisting, pilesorts, ranking exercises and timelines or seasonal calendars (described in Chapter 8) can be used either in 'pure' (*extractive*) research or in *participatory action research* (see Section 1.4), and either with individual respondents or with groups of people.	8
Community workshops	Large, interactive meetings involving both *plenary* sessions (see Chapter 9) and small group discussions and exercises.	Not appropriate as the sole method in an academic study but very valuable in *mixed-methods* studies (see Section 3.4.2). Associated particularly with consultation processes and *participatory action research* (see Section 1.4).	9

Table 3.1 (*Cont'd*)

Method	*What is involved*	*Uses*	*Chapter or Section*
Participatory mapping	Mapping that incorporates local people's knowledge, with various degrees of technical sophistication.	Associated with case study research. Participatory mapping can be used in extractive interdisciplinary research – for example, to map the distribution of hard-to-find species or habitats, or to target the collection of ecological data. It can also provide detailed information on how local people perceive their surroundings and what they think is important. It is an invaluable tool in participatory action research aiming at gaining recognition for people's land and resource rights.	10

of less than about 80 people, or executives in government offices or non-governmental organizations – then qualitative interviews are usually appropriate. Any interview carried out in a busy office should be well-planned in order to make efficient use of time, which means using a semi-structured interview rather than more unstructured methods.

Finally, it is important to consider the requirements and preferences of whoever you are doing the research for – your supervisor, manager, collaborators or funders. In spite of what was said above about not being constrained by disciplinary prejudices, it is still the case that a purely quantitative study may not go down well in a social anthropology department and a purely qualitative study may not be acceptable in a natural science-dominated conservation biology department, so if you are working towards a thesis you need to check with your supervisor that your proposed methodology will be acceptable in disciplinary terms. If you are writing for policymakers, bear in mind that statistics carry a lot of weight in demonstrating that action is needed, whereas in-depth exploration of the issues may be more useful in coming up with recommendations for the type of action that is appropriate. Therefore the best option of all, if the time and resources are available, may be to use a combination of qualitative and quantitative methods – a *mixed methods* approach.

3.4.2 Multiple methods and mixed-methods research

Many undergraduate studies use only a single method, but in larger studies two or more methods can be used together. Any study that combines several methods can be referred to as a *multiple-methods study*. Two or more methods can be used to collect data on a single issue for the purposes of *cross-methods triangulation* (see Section 6.6), or they can be used to collect information on different issues. Each method may be used more or less in isolation to investigate a single aspect of the overall research project (if they are from different academic disciplines this would be a *multidisciplinary* study; see Section 1.3). Alternatively, the different methods may be more integrated in some way (as in *interdisciplinary* studies).

Box 3.8 summarizes some of the most common ways in which different methods can be combined in a single research design. In *sequential* design the different methods are used in a particular sequence, and the results from the first inform the use of the next. In contrast, in

Box 3.8 Common structures for multiple-methods research

1 Sequential (one method informs the next)
2 Concurrent

   a. For the purposes of triangulation (different methods are used to collect data on the same thing). Both approaches address the same research questions. Usually, the data are analyzed separately and the results are then compared, but sometimes the results are combined in some way and treated together in analysis.
   b. For the purposes of complementarity (different methods are used to collect data on different aspects of the overall project). The different methods may be given equal importance, or, in *embedded design,* the majority of the research may use a single method, and a second method may be used just to address one minor aspect.

concurrent design they are used in parallel – either for the purposes of triangulation or to gather data on different things, of different kinds, or at different scales, in order to address different aspects of the overall aims and objectives of the research.

Box 3.9 gives an example of a multiple-methods case study from my own research and summarizes the methods used, which included archival searches, informal, unstructured and semi-structured interviews, and community workshops. There are two different scales of study: the first looks at the case as a whole and the second zooms in on three communities to look at specific aspects in more detail. In effect, the three communities are 'sub-case-studies' within the overall 'case' of the Communal Reserve, selected because of their high level of involvement in the reserve's creation. The results from more informal methods were used to inform the design of the subsequent semi-structured interviews and of the workshops, and the discussions in the workshops were used as the subject for a further round of informal and unstructured interviews with community members. Thus there are elements that are sequential and others – particularly between the two scales – that are concurrent.

The term *mixed-methods research* refers specifically to research that combines quantitative and qualitative social science approaches. It should be clear by now that each approach has its own strengths and weaknesses; they can therefore complement one another in a mixed-methods study. However this is particularly challenging because of the fundamental epistemological differences between quantitative and qualitative approaches (see Section 1.2.2). Nonetheless, over the past 20 years much progress has been made in developing our understanding of how they can be combined successfully, and mixed methods studies are increasingly common. Boxes 3.10 and 3.11 summarize the most common ways in which quantitative and qualitative methods can be combined, in sequential and concurrent design respectively, drawing on Cresswell (2009) and Tashakkori and Teddlie (1998).

Many mixed-methods case studies combine targeted qualitative interviews with a small number of people who may have specialist knowledge (such as the managers and NGO staff working at a particular protected area, or community leaders and members of management institutions) with a questionnaire survey of a larger population, such as local residents or tourists. Qualitative interviews with 'specialists' allow the researcher to explore the issues in depth with each individual, and to be flexible to explore their different areas of knowledge.

Box 3.9 Example of a multiple-methods case study design: collaborative wildlife management and changing social contexts

This research project was designed to investigate the implications of processes of social change for the collaborative management of protected areas. The central part of the project was a case study of the Tamshiyacu Tahuayo Communal Reserve in Amazonian Peru. The case study was chosen on both theoretical and practical grounds. It was one of the best-documented cases of collaborative wildlife management in Latin America from a natural science perspective, but little detailed research had been done from a social science perspective. In addition, I already had experience of working in Amazonian Peru, and I was invited to this site by biologists working there.

The methods used were as follows.

For information about the reserve as a whole:

- Review of published literature and archival materials.
- Informal, unstructured and semi-structured interviews with key actors (community leaders; researchers; staff of non-governmental organizations; government officials).

For more detailed information about local people's perspectives on the reserve, the following methods were used in the three communities most involved in the reserve's creation:

- Semi-structured interviews with every adult resident in three communities (about 100 people in total).
- Unstructured interviews with a targeted sample of community leaders and elderly individuals in an additional two communities.
- Additional informal and unstructured interviews as appropriate and as opportunities arose during field visits.
- Community workshops in each of the three communities, informed by information gathered in individual interviews.
- A further workshop bringing together representatives of each of the three communities, to build on the issues that were raised in the three separate community workshops.
- Further informal and unstructured interviews exploring the issues that were raised in the workshops.

There would be no advantage in using questionnaires with these people; much depth and breadth of information would be lost, there would be no flexibility, and the sample size would not be large enough to justify statistical analysis of the results. On the other hand, local residents (or tourists) are likely to number hundreds or even thousands, and therefore it is impractical to carry out enough qualitative interviews to tell you something about the group as a whole. Questionnaire surveys using a cross-sectional design can be used to gather data from a representative sample of individuals (see Chapter 4) in order to draw inferences about the population as a whole.

In addition to these methods, focus groups or community workshops can be used to gain information on the kinds of issues that different groups of people are concerned about and

Box 3.10 Sequential design in mixed methods research

1 **Quantitative** (Data collection and analysis) ⇒ **Qualitative** (Data collection and analysis)

- Quantitative methods may be used at a broad scale to test a theory or concept.
- The results may help to identify issues and case studies to be investigated in more depth using qualitative methods

2 **Qualitative** (Data collection and analysis) ⇒ **Quantitative** (Data collection and analysis)

- Qualitative methods may be used to 'scope' the issue and generate hypotheses.
- The results may inform the design of quantitative surveys to test the hypotheses across a larger population.

In sequential mixed methods design, whichever approach is employed at the start of the project tends to dominate the conceptual framework and methodology.

Box 3.11 Concurrent design in mixed methods research

1 For the purposes of triangulation
Both quantitative and qualitative approaches are used to address the same research questions. The results may be analysed separately and then compared during the interpretation stage, or more rarely, they are combined and analyzed together. For the latter, either qualitative data must be converted into numbers and analyzed quantitatively or quantitative data must be analyzed qualitatively:

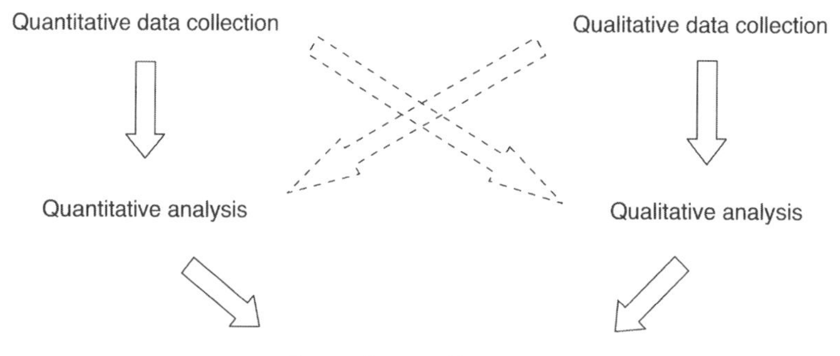

2 For the purposes of complementarity
Quantitative and qualitative approaches are used to address different questions within the overall research design, or to investigate an issue at different scales, or to collect data from different groups of people (for example, a questionnaire survey could be used for local residents and unstructured or semi-structured interviews could be used for managers or NGO staff).

their ideas about what actions are needed in the future. Many mixed-methods studies start with a set of qualitative interviews or focus groups and then use the findings to inform the design of a questionnaire. Alternatively, the results of a questionnaire survey can be presented in focus groups or workshops, both in order to collect further data in the form of discussion, comments and feedback, and also simply to fulfil the obligation of the researcher to report back to the participants (see Chapter 18). For example, Appleton and Booth (2005) discuss mixed-methods approaches combining questionnaire surveys and participatory workshops in a development setting.

However, the appropriate combination of methods depends not only upon theoretical issues, but also on practical considerations and the social and cultural context. In the study described in Box 3.9, a questionnaire survey with local people was ruled out not only because the study sought an in-depth understanding of local people's perspectives rather than data on a specific set of quantifiable variables, but also because questionnaires were unlikely to provide reliable information on the politically loaded subject of local people's relationship with the Reserve; people were more likely to talk freely in a more informal interview, once a degree of trust had been built up with the researcher. In addition, there would have been serious logistical difficulties in reaching all the communities surrounding the reserve, and therefore a cross-sectional design, which would be needed in order to provide a representative sample (see Chapter 4), would not have been possible. It was therefore felt to be more valuable to explore the issues in depth, through qualitative interviews with people in only a small number of communities, even though this was at the expense of wider coverage.

Of course, multiple-methods studies in conservation can combine not only different social science methods, but also social science and natural science methods. It is beyond the scope of this book to discuss this in detail, but the principles are the same as those described above. Many interdisciplinary studies in conservation use a concurrent design, in which natural and social science components provide information on different aspects of the whole topic. In studies of land use, archival searches and geographical or ecological mapping techniques can provide information at a broad scale, whereas participant observation, interviews or workshops can fill in the details at the local level. Participatory mapping, described in Chapter 9, combines natural science and social science sources of information in a single, highly integrated methodology.

The main constraints in combining natural and social science methods are the lack of researchers who are competent in both, and the lack of time in the field to cover them both in depth. As Harrison et al. (2008) point out, it is not sufficient simply to tack a small, poorly designed social science component onto a natural science project (or vice versa); both the natural and social science components must be done properly. For this reason, such studies are often carried out by multidisciplinary teams rather than by lone researchers. However, in a substantial single-researcher study such as a PhD it is possible to combine them successfully.

3.5 Conclusion

Table 3.2 summarizes the different aspects of research design that you may read about in published articles, and that you may need to consider in your own research, and indicates where in this book they are explained. I say '*may* need to consider' because for many types of research, some of these aspects do not need to be made explicit. Null hypothesis-based studies do not usually state explicitly that they are using a deductive research strategy, because it is obvious. At the other end of the spectrum, participant observation studies do not

Table 3.2 Summary of the different aspects of research design

		Where to look in this book
Research strategy	Inductive/deductive	1 and 2.1; 2.3.1
Research design structure	Experimental, quasi-experimental, observational, case study, comparative case study, cross-sectional, longitudinal	3.2 and 3.3
Methods	Quantitative, qualitative or mixed methods?	1 and 2.2; 3.4
	Specific method(s) to be used?	3.4.1
	Single method or multiple methods?	3.4.2
For each method	Sampling strategy and sample size	Chapter 4
	Detailed design of methodological tools (questionnaire questions, interview guide, workshop exercise, etc.)	Chapters 5–10
Analysis	Quantitative or qualitative? Specific analytical procedures?	Chapters 14–16

usually state explicitly that they are using a qualitative, inductive approach, because this is fundamental to participant observation. Similarly, they may not specify the sampling strategy or sample size; non-probability sampling is almost invariably used, and sample size may not be important provided that the principle of *saturation* is applied (see Section 4.3.1). The best way to learn which aspects you need to consider and describe in your own research is to look out for how they are treated in reports of similar kinds of study by other people. However, having an understanding of the full spectrum of approaches will allow you to draw from different disciplinary perspectives on conservation, and will serve you well if you take part in multidisciplinary teamwork.

Research designs vary greatly in their level of complexity. Most undergraduate field projects have a single, reasonably simple objective, are structured either as a case study or by cross-sectional design, and use *either* questionnaires *or* qualitative interviews as the sole method. Masters projects tend to have a wider scope with a set of related objectives; they may involve a single method or multiple methods, and may include quite complex analyses. Case study design, comparative case study design and cross-sectional design are all common. PhD projects always include multiple objectives and methods and may have quite complex research designs. However, as a general rule, the research design should be kept as simple as possible in order to address the research questions; making it more complex than necessary will just make it difficult to conceptualize and communicate the overall findings clearly.

There is also enormous variation in the degree to which the research design can or should be defined before data collection begins. Defining the research design precisely at the start gives a sense of security – you know what you should be doing during data collection – but it may only be possible to finalize the details once you are at the study site, because of the need for knowledge both of the practicalities and also of the social context. If the design is applied too rigidly, it can be unnecessarily constraining. In inductive studies there are strong theoretical reasons against defining every detail in advance; by definition, inductive studies involve the development of theory based on what the data reveals, and therefore there must be flexibility to follow up leads as they appear. However, if too much is left open-ended – especially when limited time is available – the end result may be a highly dispersed data set with a lot of gaps. Finally, in participatory action research it would be inappropriate to finalize the research design before going to the study site and consulting with the beneficiaries.

In traditional approaches to social anthropology, PhD students would be expected simply to go to the field and start participant observation with very little planning beforehand at all. Nowadays, however, students in any discipline are expected to develop a research design before embarking on fieldwork, even if it is likely that some of the details will change later on. The reasons for doing this are to further define the focus of your study, have a clear plan for how to begin the fieldwork, and become familiar with the different aspects of research design and think through their implications for your particular study while you are still able to access reference materials and supervisors. At the least, prior to fieldwork there should be a set of well-defined aims and objectives, consideration of the overall scope of the project, a selected research design structure and proposed method or set of methods, and an indication of the sampling strategy. In studies where considerable background information is already available it may also be possible to draft the specific methodological tools (such as questionnaires or interview guides or workshop exercises) and to decide on the types of data analysis to be carried out.

Once data collection has started, in deductive research, major changes to the research design should not be made unless things go disastrously wrong, but minor changes are permissible and indeed almost always necessary. In inductive research there is greater flexibility. Overall, it is quite rare to carry out a research project exactly as originally planned; very often some aspect will turn out to be impractical or irrelevant once data collection is under way. Therefore the original research design should not be treated as rigid and inflexible; it should be reviewed periodically and, if necessary, adjusted. Obviously the longer the data collection period, the more important this is; in an undergraduate project there is not usually time to make major changes, whereas in a three-year PhD the final research design is often quite distant from the original plan.

Summary

1 The *methodology* is a plan for carrying out a particular piece of research in order to address the overall aims.
2 The methodology relates to what you will actually *do*. It includes not only the specific methods that will be used but also the overall strategy by which they fit together.
3 Research designs are often classified by the overall structure of data collection. These include *experimental design, quasi-experimental design* and observational research.
4 Observational research can be further divided into *case study design*, *comparative case study design*, *cross sectional design* and *longitudinal design*.
5 Different types of design structure have different strengths and weaknesses in terms of *internal*, *external* and *context* validity.
6 In choosing specific methods, probably the most important factors to consider are the *scope* of the project, the balance between *quantitative* and *qualitative* approaches, and the time and resources available.
7 Research designs vary greatly in complexity. Undergraduate projects often have a single objective and use only one method, whereas postgraduate projects are broader in scope and methodology.
8 In larger studies multiple methods can be used. *Multiple-methods studies* in conservation can combine not only different social science methods, but also social science and natural science methods. Methods may be used sequentially or concurrently. Each method may be used more or less in isolation to investigate a single subtopic within the overall research design, or the different methods may feed into each other in a more integrated design.

9. *Mixed-methods studies* are studies that combine quantitative and qualitative social science approaches.
10. The research design is rarely defined in every last detail prior to the start of data collection, especially in inductive studies. However for any research project, prior to fieldwork there should be well-defined aims and objectives, consideration of the overall scope of the project, a selected research design structure and method or set of methods, and an indication of the sampling strategy.

Further reading

Blaikie, N. (2000) *Designing Social Research*, Cambridge: Polity Press. [A more advanced text on the whole of the social science research design process, presenting an excellent explanation and discussion of the different research strategies. Particularly relevant for qualitative, inductive research.]

Creswell, J.W. (2009) *Research Design: Qualitative, Quantitative and Mixed Methods Approaches*, 3rd edn, London: SAGE Publications. [Includes substantive sections on how to use mixed methods approaches.]

Denscombe, M. (2007) *The Good Research Guide for Small-scale Social Research Projects*, 3rd edn, Maidenhead: Open University Press. [Part I goes into more detail than is possible here on different research design structures and theoretical approaches.]

Holland, J. and Campbell J. (eds) (2005) *Methods in Development Research: Combining Qualitative and Quantitative Approaches*, Swansea: Practical Action Publishing. [Consists of reflective articles and practical case studies from the field of development that demonstrate how qualitative and quantitative approaches to research can be combined. The emphasis is very much on practical applied research, and the last section includes several examples of participatory approaches.]

Tashakkori, A. and Teddlie, C. (1998) *Mixed Methodology: Combining Qualitative and Quantitative Approaches*, Applied Social Research Methods Series Volume 46, Thousand Oaks, CA: SAGE Publications. [Gives a useful account of the history and theory of mixed methods approaches.]

4 Sampling

I sometimes wonder if two thirds of the globe is covered in red carpet.
(attributed to Prince Charles)

4.1 Introduction

Sampling is about choosing exactly what sources you will collect data from – what places you will visit, what events you will attend, which people or organizations you will talk to and so on. There is a common assumption that sampling is relevant only for quantitative research, but this is not correct. It is certainly true that it needs to be more precisely defined in quantitative research, but sampling is a key aspect of research design in any research project – you need to develop a strategy by which you decide where to spend time, who to spend time with and when you have done enough. Sampling theory provides a useful framework to think clearly about decisions of this kind.

This section introduces some of the basic concepts in sampling design, which are summarized in Box 4.1. The following two sections look in more detail at sampling strategies and sample sizes, respectively. Sampling is treated very differently in quantitative research and in qualitative research, and if you are reading this chapter with a particular research project in mind, then you probably will not need to read all the sections. You should read the introduction to each main section and then move on to the following subsections that are most relevant.

In order to develop a sampling design, an important first step is to decide on the *sampling unit* – the type of item you wish to sample. The commonest sampling unit is an individual person, in which case the sampling design is concerned with how you choose which people to talk to and how many people you should include in the study. Specific examples of the sampling unit are called *cases*. Thus in a survey of individual people, Jane Smith might be one case and Tom Baker might be another case. However, sampling units can be defined at different levels and for different kinds of item. In studies of people's attitudes to the environment the sampling units are normally individual people, so that you can examine how attitudes vary between individuals and look for patterns of variation (for example, in terms of differences between men and women or people of different age groups). In contrast, studies concerned with pressure on natural resources from subsistence use often use households as the sampling units, because in most cultures the household is the fundamental economic unit. It would not be relevant to ask the same questions of different individuals within a household, because most decisions about natural resource use are taken at the household level rather than by each individual independently. Still other kinds of study may use larger

Box 4.1 Key terms: sampling

Sampling units (= units of study, units of analysis): The 'units' about which you gather information. The most common units are individual people, but sampling units can also be groups of people (such as households or communities) or other kinds of item such as events, documents, organizations, and so on.
Case: One specific sampling unit – one named person, one specific household or event, and so on.
Study population: The set of cases you are interested in – for example, the set of all the people within a specific geographical area, or all the fishermen who use a particular coastline, or conservation volunteers in Sweden, or tourists who visit a specific National Park.
Census: A study that collects data on every single case in the study population.
Sampling frame: A complete list of all cases within the study population, for example, a list of all the tourists who have visited a specific National Park within the past three years.
Sample: The subset of cases for which data are collected.
Survey: A study that collects data from a sample of the study population – usually because the total population is too big to allow you to include every single case.

groups of people as the sampling units, such as neighbourhoods or communities. In a survey of communities around a national park, each named community would constitute one case. Alternatively, sampling units may not be to do with people at all; they could also be protected areas, non-governmental organizations, schools, events, documents and so on. The terms 'sampling unit' and 'case' are often not used explicitly in qualitative research, but decisions about the units of study still have to be made, and an understanding of these terms is therefore useful in developing a clear research design.

Once you have defined the sampling units, the next step is to define the *study population*. The study population is the complete set of cases you're interested in – perhaps all households in a particular community or all non-governmental organizations attending an international meeting. In qualitative research a definition of this kind may be adequate to commence the study, whereas in quantitative research the study population must be defined much more precisely – ideally precisely enough so that a list can be made of all the cases that make up the study population (known as a *sampling frame* – see Box 4.1 and Section 4.2.1).

In any study, the ideal would be to collect data from every case in the study population (this is known as a *census),* but except for very small populations this is not usually practicable. Instead, data are collected from a *sample* of the population – a subset of cases. Quantitative studies based on a sample rather than the total population are known as *surveys.* The sampling design in any kind of research is to do with choosing *which* cases (people or communities or events or organizations) to include, and deciding *how many* of them you need to include. These are known as the *sampling strategy* and the *sample size*, respectively, and are the subjects of the following two sections of this chapter. In multiple methods studies you may need to define a separate sampling strategy and sample size for each method that is used.

Some of the commonest pitfalls in applying sampling theory arise because several of the terms have a meaning in everyday speech that is slightly different from their technical meaning. A study population is not necessarily made up of people; it is made up of whatever the sampling units may be. A 'case' in sampling theory is different from a 'case' in case study design (see Box 3.3 in Section 3.2.3.3), although the difference is simply one of scale; in both uses of the word a 'case' is an example. Finally, the words 'census' and 'survey' are often used interchangeably in everyday speech, but there is a fundamental difference in the way they are used in sampling.

4.2 Sampling strategies

The most important division between the different types of sampling strategy is between *probability sampling* and *non-probability sampling.* The underlying principal of probability sampling is that every case – every member of the study population – has a known probability of being included in the sample, and therefore statistically valid inferences can be made from the sample to the overall population. In the simplest approach, every case has an *equal* probability of being included, in which case the sample will be *representative* of the population. No sample has *exactly* the same characteristics as the total population, but there are statistical techniques to allow for this (see Chapter 16). Sections 4.2.1.3 and 4.2.1.4 introduce some sampling methods where probabilities are known but not equal, in which case valid inferences can still be made about the population by making small adjustments during data collection or analysis.

In *non-probability sampling,* the probabilities that any one case will be included in the sample are not known, and there is no attempt to make statistically valid inferences to the whole population. Indeed, the total population need not be defined precisely, and no sampling frame is required. Non-probability sampling is appropriate for research where the main purpose is to explore people's views in depth on a particular issue or to document specialist knowledge (for example, of experienced hunters or herbalists) rather than to determine the characteristics of the whole population. It is also used in exploratory studies, and where full probability sampling is simply not possible with the time and resources available. The rest of this section will describe the most common strategies for probability sampling (Section 4.2.1) and non-probability sampling (Section 4.2.2) and discuss the relative merits of each of them.

4.2.1 Probability sampling

Table 4.1 summarizes the most common strategies in probability sampling. Of the four methods shown, simple random sampling is the most straightforward to use, at least for small populations. Systematic sampling is an acceptable alternative where it is not possible to construct a sampling frame, and if systematic sampling is not possible either, quota sampling is an additional option (see Section 4.2.2.2 – although the results should be interpreted with caution since quota sampling is not a probability sampling method). Cluster sampling is an excellent strategy for large, dispersed populations for which simple random sampling would be too demanding in terms of time and resources. Finally, stratified random sampling allows you to control the proportions of different subgroups within the sample, usually in order to be able to make direct comparisons between them during data analysis.

All the methods listed except systematic sampling require a sampling frame – a complete list of the cases that are included in the study population. In order to create a sampling frame, the study population and the sampling units must be defined absolutely precisely. What is the *exact* geographical area you wish to cover? Do you really want to include *all* residents, or only those over the age of 18? Do you count someone who lives there for six months of

Table 4.1 Probability sampling strategies

Sampling strategy	*Comments*
Simple random sampling: Pick out individual cases from a sampling frame, using a random numbers table.	Best option in theoretical terms in order to attain a representative sample. Needs a sampling frame.
Systematic sampling: Pick one case using a random numbers table, and then pick every nth case.	Good alternative when a sampling frame is not available. The only requirement is that the cases can be considered in a set order. If this is not possible, consider *quota sampling*.
Cluster sampling: Divide the population into 'clusters' (often, geographical areas), take a sample of clusters, and then take a sample of cases from each selected cluster.	Useful for a large, dispersed population. To achieve probability sampling, a sampling frame is needed for each cluster that is sampled.
Stratified random sampling: Divide the population into 'strata' (groups of cases with certain characteristics). Then take a simple random sample of cases from each stratum. The sample sizes for the different strata may mirror their proportions in the parent population (*proportionate sampling*) or may boost minority groups to ensure adequate sample sizes (*disproportionate sampling*).	Used to control the proportions of different subgroups within the sample – often in order to allow direct comparison between them in analysis. Needs a sampling frame. Proportionate sampling also needs prior information on the composition of the study population.

the year as 'resident'? These kinds of question should be decided as far as possible during the processing of narrowing down your research design (see Section 2.3), although they may need to be reviewed once you are at the field site as further issues arise. Once you have defined the population there may be an existing list you can use as a sampling frame (for example an electoral role or government census, a list of members of a conservation organization or an Internet discussion group, or the mailing list of a tour operator), or if not, it may be possible to construct your own list by conducting a quick census at the beginning of the research or by combining existing lists from different sources. However the advantages of doing so must be balanced against the time it will take.

4.2.1.1 Simple random sampling

See Box 4.2

4.2.1.2 Systematic sampling

Systematic sampling offers a good second-best option to simple random sampling when it is not possible to construct a sampling frame – for example, if you wanted to sample households in a large community for which no census data were available. Starting on a specific street and counting the houses as you walk along, pick one house using a random numbers generator and then pick successive houses at a fixed interval (for example, every tenth house you come across). Keep going until you have walked down every street sampling the houses in this way. As long as you start with a random number, every house in the community has an equal probability of being picked, and therefore systematic sampling fulfils the basic premise of probability sampling.

The big advantage of this method is that very little information is needed about the total 'population' before you start collecting data. You do not need a sampling frame, and you do

Box 4.2 Instructions: how to take a simple random sample

1 Define the study population and decide on a target sample size.
2 Construct a sampling frame.
3 Number the cases sequentially, starting with number one.
4 Use a random numbers generator to pick the cases to be included in the sample.
5 If for any reason one of the cases you have picked cannot be included (for example, if the person has moved away or refuses to take part), do not just take the next case on the list but use the random numbers generator to pick a replacement case.

not even need to know the exact population size (in this case, the total number of houses in the village). You do, however, need a *rough* idea of population size so that you can set the sampling interval at a level that will give you an appropriate sample size. For example, if you think that there are about 3,000 houses in total, you could select every tenth house, which would give you a sample of 300. If there turn out to be more than 3,000 houses you must keep going until you have systematically sampled them all.

Systematic sampling is only possible if the cases can be counted in order. If the houses in a village are arranged in straight lines along clearly defined streets then it is straightforward, but if they are scattered haphazardly it may be difficult to keep track of the houses you have already counted and avoid counting some of them twice.

The only danger of systematic sampling is that if there is some kind of *systematic* variation in the cases that coincides with your sampling intervals, then the sample will not be representative. If every tenth household is in some way different – for example, if houses are arranged in blocks of ten and the houses on the end of each block are bigger than the rest – then a sample based on a sampling interval of 10 will either include *only* the bigger houses or it will include *none* of them, depending on your starting-point. It is important to look out

Box 4.3 Instructions: how to take a systematic sample of households in a village

1 Decide on a target sample size.
2 Pick a number (n) between one and ten using a random numbers generator.
3 Decide on a fixed sampling interval (m), such that you will select every mth house you come to.
4 Starting at the beginning of a street, walk along counting the houses on one side (if that is easiest) until you come to the nth house. This house is the first case in the sample.
5 Once you have collected data from that house, continue down the street counting the houses and pick every mth house.
6 When you get to the end of the street, turn around and continue the count for the other side of the street if necessary, and then do the same for every street in the village. There is no need to pick a new random number and re-start the counting for each street; just continue counting in sequence for the whole village.

for any hint of periodicity of this kind and if there seems to be a problem, change the sampling interval.

4.2.1.3 Cluster sampling

In cluster sampling the population is divided into clusters, a sample of clusters is selected, and then a sample is taken separately from each selected cluster. Clusters could be schools, villages, suburbs within a town, towns or even states, or alternatively you could divide the study area into several 'blocks' on the map using geographical coordinates. Other kinds of cluster might include tourists visiting a lodge in a national park with different tour operators, or tourists visiting different tourist lodges, or different national parks within a country and so on. The clusters should be mutually exclusive (that is, no case can belong to two or more clusters) and together should cover the whole of the population of interest (see Box 4.4).

One advantage of cluster sampling over simple random sampling is that there is no need for a sampling frame for the whole study population; you just need sampling frames for each of the selected clusters. If this is not possible, cluster sampling can also be used with non-probability sampling (see Section 4.2.2), as long as the data are interpreted with this limitation in mind. A second advantage is that clustering the selected cases cuts down on travel time, and for this reason cluster sampling is common in cross-sectional surveys of very large, dispersed populations. Sometimes more than two levels of sampling are used (a process known as *multi-stage clustering)*. For example, you could cluster by town, then suburb, then street. The more the cases are grouped together into clusters, the easier the data collection will be in terms of time and logistics. On the other hand the more the cases are dispersed, the more representative the total sample should be of the study population, and therefore as a general rule it is best to have as many clusters as you can within the limits of the logistical resources available. The number of clusters selected and the sample size within each cluster are therefore a matter of compromise between theoretical concerns and practical limitations. Box 3.5, in Chapter 3, gives an example of cluster sampling where this compromise is made successfully within an undergraduate thesis.

There is one more aspect to consider in cluster sampling. If the study population were to include one village of 3,000 households and 20 villages of 600 each, then the large village would represent a fifth of the total population (of households). Therefore in order to build a *representative* sample (in which every case has an *equal* chance of being selected), five times more households should be included from the large village than from each of the other villages. A simple way to do this would be to include the large village in the first sampling

Box 4.4 Instructions: how to take a cluster sample

1. Divide the study population into mutually exclusive clusters. Some common types of cluster are geographical areas, communities, districts or suburbs within a town.
2. Make a list of the clusters (and consider weighting them – see below).
3. Take a simple random sample of clusters.
4. For each selected cluster, construct a sampling frame.
5. Take a simple random sample separately from each selected cluster.

Box 4.5 Instructions: how to take a stratified random sample

1 Define the study population and the different 'strata' that you wish to sample.
2 Decide on the target sample size and the proportions for each stratum.
3 For each stratum in turn, construct a sampling frame and take a random sample.

frame (the list of clusters) five times. This is known as sampling with *probability proportionate to size* (PPS). A slightly less satisfactory alternative would be to include the same number of people from the large village as from each of the others, but then to weight the results during data analysis. For further details and examples of PPS and weighting see Bernard (2006: 159–66).

4.2.1.4 *Stratified random sampling*

Stratified random sampling is similar to cluster sampling, but rather than dividing the population into clusters purely for the sake of practicalities, it is divided into subgroups ('strata') of cases that are different in kind (for example, men and women; people of different ethnicities or age groups; fishermen, hunters and farmers, and so on) and a sample is taken of each subgroup separately. This means that the proportions of the different subgroups in the overall sample can be set in advance – either in order to match those in parent population, if known (*proportionate sampling*), or to boost the number of cases for subgroups that are in a minority (*disproportionate sampling*).

For example, if it were known in advance that the people to be sampled were 10 per cent indigenous and 90 per cent non-indigenous, proportionate sampling could be used to make sure that the sample also had a 10:90 balance. However in a total sample of 150, there would then be only 15 cases representing the indigenous population, which is very small for statistical analysis. An alternative would be to take a sample that included equal numbers of indigenous and non-indigenous cases, so that the two groups could be directly compared during analysis. As long as simple random sampling had been used in sampling each group, it would still be possible to make inferences from the sample to the total population through weighting the results. However, stratifying the sample in terms of one variable (such as ethnicity) means that it is unlikely to be representative with regard to other variables, and therefore proportionate sampling is more appropriate if the main purpose of the study is to characterize the population as a whole, whereas disproportionate sampling is more appropriate if the main purpose is to compare different subgroups directly.

4.2.2 Non-probability sampling

Table 4.2 summarizes the different types of non-probability sampling. *Convenience sampling* (also known as *haphazard* or *availability sampling*) involves interviewing whoever you can find, and is used mainly for pilot studies or in the exploratory phase of longer studies. Quota sampling can be used as an inferior alternative to probability sampling – either when time and resources are limited or because the latter is not practicable. The remaining methods – targeted sampling and chain referral – aim to target individuals who are most relevant

Table 4.2 Non-probability sampling strategies

Sampling strategy	*Comments*
Convenience/haphazard/availability sampling: Interview anyone that you can find who fits your broad criteria.	Used in pilot studies, market research, qualitative case studies. A special case of convenience sampling is *volunteer sampling.*
Quota sampling: Define two or more subgroups (e.g. men and women) and set the proportion you want in each category (e.g. 50:50). Interview anyone you can find in each subgroup until you have reached the target sample size.	Probably the commonest sampling strategy for surveys carried out by stopping people in the street. Frequently used in pilot studies, opinion polls and other types of market research. Can also be used in more detailed studies when full probability sampling is not possible.
Targeted/purposive/judgement sampling: Seek out individuals who are most relevant to study.	Used in studies that focus on particular subgroups or specialists.
Chain referral: Snowball sampling, respondent-driven sampling: Seek out individuals who are most relevant to the study, interview them and ask if they know of others you could interview.	Studies that focus on particular subgroups or specialists, especially when they are scarce or hard to identify.

to the study rather than to sample the population as a whole. Such individuals may be people with specialist knowledge such as healers or skilled hunters, or people who participated in a particular event in the past, or those who hold formal positions that give them a specific perspective on issues and events, such as members of a particular committee, protected areas managers or officers of different non-governmental and governmental institutions. Chain referral is a particular form of targeted sampling in which each individual included is asked for suggestions of other individuals that could also be included.

4.2.2.1 Convenience sampling

Convenience sampling (also known as *haphazard* or *availability sampling*) involves interviewing whoever you can find. You might include friends, colleagues and relatives, or you might go out into the street and interview anyone who will stop and agree to take part. Convenience sampling is often used in exploratory studies or at the start of more substantial qualitative studies, because it can give you the opportunity to try out your methods and gain an idea of the issues that may come up. As a study progresses, increasing efforts are usually made to balance the sample or to target key individuals.

Volunteer sampling is a special case of convenience sampling. It is a common technique for monitoring customer feedback – for example from tourists. It may involve leaving questionnaires at strategic points for people to pick up, advertising for volunteers, or putting a questionnaire on a website and inviting visitors to the website to fill it in. Web surveys of this kind are increasingly common. The main limitation is that there is no way to gather a representative sample; indeed it is often hard even to define the population since you do not know exactly who is visiting the website. However, web-based surveys offer a very easy and flexible means to reach highly dispersed populations.

4.2.2.2 Quota sampling

Quota sampling is a form of stratified sampling, but without meeting the all-important criteria for probability sampling. Different subgroups and the proportion of cases in each subgroup

Box 4.6 Instructions: how to take a quota sample

1 Define the study population, the categories of interest, a target sample size and the proportions to be used.
2 Using convenience or systematic sampling, collect data from cases in each category until the 'quota' for that category is reached. In quota sampling of individual people, this is often done by standing in a busy street and stopping every *n*th passer-by in each subgroup.

are defined in advance (for example, a 50:50 sample of men and women). As in stratified random sampling, the proportions can be set either to match those in the parent population (if known), or to artificially boost sample sizes for smaller subgroups, or simply to collect equal samples for each subgroup for the purposes of comparison during statistical analysis. Then convenience or systematic sampling can be used to sample cases in each subgroup until you have filled the 'quota' (see Box 4.6).

The fact that participants are not randomly selected means that quota sampling does not fulfil the basic conditions for probability sampling; not all individuals in the study population have a known chance of being included. For example, if you stop people in the street, you are only sampling people who are using that particular street at that particular time. However this limitation must be balanced against practical considerations. Quota sampling can be extremely valuable in small-scale exploratory studies (including many undergraduate theses), and sometimes, when a full probability sample is simply not possible, quota sampling is the best available option.

4.2.2.3 Targeted sampling

Targeted sampling (also known as *purposive* or *judgement sampling*) involves intentionally selecting those cases – usually people – who are most relevant to study. For example, if you wanted to study the sustainability and practices of wild honey harvesting then it would make sense to target people who harvest wild honey or are otherwise involved in the honey trade. There may be written documents that name some of them, or you may just have to go to the study site and start asking (see Box 4.7).

Box 4.7 Instructions: how to take a targeted sample

1 Define the kind of cases (usually people) you want to target, and decide on a rough target sample size.
2 Identify suitable people by whatever means are available. There may be written documents that name them or list some of them, or you may need simply to go to the study site and start asking.
3 Once you reach the target sample size, review your data and decide whether you have enough, using the principal of *saturation* (see Section 4.3.1).

Box 4.8 Instructions: chain referral – how to take a respondent-driven or snowball sample

1 Define the kind of cases (usually people) you want to target, and decide on a rough target sample size.
2 Identify as many suitable people as you can by whatever means are available.
3 When you meet with each of them, ask if they know other people who fit your criteria.

   – In *snowball sampling*, take down the names and details of anyone is suggested and approach them later. Each of these people may then give you more names.
   – In *respondent-driven sampling*, ask each respondent to let others know of your interest and encourage them to approach you (possibly with a financial reward as an incentive).

4 Once you reach the target sample size, review the data and decide whether you have enough, using the principal of *saturation* (see Section 4.3.1).

In order to compare the views of those involved in the honey trade with the views of other people – for example, to see whether honey-gatherers have different perspectives from others on conservation – *case-control sampling* can be used. This involves collecting a targeted sample of honey-gatherers in the way described above, and then building a control sample of non-honey-gatherers. With small samples, this is done by identifying one individual who is as similar as possible to each of the honey-gatherers in every respect (age, wealth, place of residence and so on) except that they do not gather honey.

4.2.2.4 Chain referral

If the individuals you wish to target are hard to find – either because there are very few of them or because they are hard to identify – then *chain referral* offers a useful variation on simple targeted sampling. You still need to find a few individuals yourself at the start of the study, but then you simply ask each of them whether they know of others that you could talk to and in this way build up a sample by following 'chains' of reference (see Box 4.8).

In *snowball sampling*, you note down the names and details of anyone recommended to you and then approach them directly. However, if the identity of the target group is sensitive – for example, if they are involved in illegal activities – then people may be unwilling to give you names. An alternative is *respondent-driven sampling,* in which each informant is asked to let others know of your interest and encourage them to approach you (possibly with a financial reward as an incentive).

4.3 Sample size

One of the most frequent questions students ask me before they leave for the field is 'how many interviews (or questionnaires) do I need?'. Unfortunately there is no quick answer to this question; in most studies it depends not only upon the aims of the study and the research design but also upon practicalities, and especially in qualitative research, also on what people say once data collection has started. The way that sample size is determined is very different in qualitative research and in quantitative research. The following sections attempt to outline the principles of setting a sample size and some general approaches that you can use, first for qualitative research (Section 4.3.1) and then for quantitative research (Section 4.3.2). Chapter 16 discusses quantitative sampling theory in greater detail.

4.3.1 Sample size in qualitative research

The aim of the sampling design in qualitative research should be to make sure that enough data is gathered to give an accurate understanding of the issues under investigation and the different perspectives that are present in the study population. In participant observation and in components of research based on informal interviews, no sample size is set in advance of data collection and indeed, sample size is not usually addressed explicitly at all in the research design (although sampling *strategies* should usually be addressed). In contrast, in research based on sets of semi-structured interviews or focus groups an initial target sample size is usually set in advance and then some initial processing and analysis may be carried out in order to determine whether more data are needed.

The key principle in assessing sample size in qualitative data sets is the principle of *saturation* (Glaser and Strauss, 1967, in Bryman, 2004: 305). Saturation is reached when you can 'make sense' of the data in terms of identifying areas of consensus or other patterns, and when collecting more data produces little important new information or understanding that is relevant to your research questions. The concept of saturation is closely linked to that of triangulation (see Section 6.6), and involves reviewing the data collected up to a certain point in order to compare what different people (or methods) tell you. If everyone has given the same information on a particular issue, then there is no point in asking more people about that issue. In contrast, if different people have given you different answers, then you need to collect more data until some kind of pattern emerges. Of course, *total* saturation is rarely possible; more data will almost always reveal *something* new. The key skill is in being able to assess what is 'important' and when you have 'enough' to address your research questions (see Box 4.9).

In participant observation, the data should be reviewed repeatedly throughout the fieldwork period in order to inform the next stage of data collection (see Section 5.3.6 and Chapter 14). In research based on sets of semi-structured interviews or focus groups a target sample size is usually set in advance, and once the target number of interviews is reached, some initial processing and analysis should be carried out in order to assess the level of saturation. If you cannot yet see any consensus or patterns in the information you have collected, you may need to set a new target and collect more data.

There are no hard and fast rules for setting the target number of semi-structured interviews or focus groups in advance, because the appropriate sample size depends partly on

Box 4.9 Key term: saturation

Saturation refers to the stage in qualitative data collection when collecting more data produces little important new information or understanding relevant to your research questions.

Box 4.10 Practical tips: setting a target number of semi-structured interviews and focus groups

- Most studies involving sets of semi-structured interviews use a target sample size of between 10 and 50 interviews.
- Small focus group studies usually use a target sample size of three to five groups (which may involve as many as 40 participants).
- In setting target sample sizes, be realistic about the amount of time needed. Transcription and initial processing of the data may take longer than the actual interviews, and in the case of focus groups, the logistics of recruiting participants and organising the sessions themselves are very time-consuming (see Section 6.2.4).
- In both cases, the ideal practice is to leave enough time after the main data collection period to make an initial assessment of the data for saturation and if necessary, carry out a few further interviews or focus groups to clarify points that are not yet clear.

what people say to you. It follows that the more you know in advance about the diversity of views you are likely to come across, the more basis you have for setting a target number of interviews. However, as a rule of thumb, most studies of this kind involve somewhere between 10 and 50 interviews, depending upon the aims of the research. If you are working with a population of less than about 60 then it may be possible to interview every individual (which means you are carrying out a census rather than selecting a sample – see Box 4.1). However this would be a lengthy project, and if the population is any larger than this it would become unmanageable – especially in terms of processing and analysing the data (see Box 4.10).

Focus groups present particular problems in terms of sampling because there is no clear sampling unit. They do not provide independent data from individuals, but neither do they usually sample a population of naturally occurring 'groups'; the groups you work with are usually brought together artificially. Therefore it is not possible to apply probability sampling to focus groups. Section 6.2.4 discusses some practical considerations related to sampling in focus group studies; in terms of sample *size*, the same principles apply as for qualitative interviews – you set a target number (taking into account the time and resources available) and then examine the data and apply the principle of saturation. Morgan and Krueger (1998, vol. 2: 77) give a typical target sample size for a simple study as from three to five groups; this may sound very few, but if each group has eight participants and lasts for two hours, that means you need to recruit between 24 and 40 participants and set aside one to two weeks for transcription and initial data processing. As with sets of semi-structured interviews, if you reach the target and can find little pattern in the data, you may need to conduct a few extra focus groups until you have reached some degree of saturation. For more on sampling issues for focus groups, see Morgan and Krueger (1998, vol. 2).

4.3.2 Sample size in quantitative research

Sample size in quantitative research is to do with how many cases you need to include in order to distinguish (through statistical tests) between chance variation and more 'meaningful' patterns in the data. The research design for any study should define what patterns you will test for, and in order to decide on a sample size you need to have a clear idea of the kinds of analysis you wish to carry out. Does the research aim to make inferences from a sample to a larger population, to compare two or more contrasting populations or to test for relationships between variables (or a combination of any of these)? The central issue in sampling theory is to do with making inferences from a sample to a larger population. This is discussed in relation to data analysis in Chapter 16; Section 4.3.2.1 gives some practical guidelines on setting sample sizes prior to data collection. Section 4.3.2.2 outlines some additional points that are relevant to comparisons between different populations (or subgroups within a population), which is a very common feature of research design.

4.3.2.1 Making inferences from a sample to a population

The principle behind making inferences from a sample to a population is that provided probability sampling is used, patterns of variation in the sample should reflect the properties of the whole population in ways that can be estimated statistically. If every case in the population has an equal probability of being included in the sample, then the sample should be *representative* of the population – it should have the 'same' characteristics as the population. However no sample is *exactly* the same as the population. The difference between the two is known as the *sampling error*, and in principal, the larger the sample, the smaller the error should be. Sampling theory provides a way to estimate what sample size you need in order to make inferences about the study population with a known margin of error. The latter is usually expressed in terms of the *confidence interval:* a range within which the true population value is likely to lie. The *confidence level* (by convention usually set at either 95 per cent or 99 per cent) indicates how likely it is that the confidence interval contains the true mean.

To take a simple example, suppose you wanted to find out what percentage of people in London are in favour of zoos. If you use a population size of 7 million (approximately the population of London) and specify a confidence level of 95 per cent and confidence interval of five (plus or minus five percentage points), then it is possible to calculate the sample size that is needed. Bernard (2006: ch. 7) presents a theoretical explanation for how this is done; for the purposes of this chapter I have used a sample size calculator available on a public-access website that does the calculation for you (Figure 4.1).

Determine sample size	
Confidence Level:	95%
Confidence Interval:	5
Population:	7 000 000
Sample size needed:	384

Figure 4.1 A simple online calculator for determining sample size.

Source: The Survey System http://www.surveysystem.com/sscalc.htm. Reproduced with permission of Creative Research Systems.

If you enter the above figures, the calculator comes up with an appropriate sample size of 384 people. That means that if you sampled 384 people in London (using probability sampling) and 54 per cent said they were in favour of zoos, you would be statistically justified in stating with a confidence level of 95 per cent that between 49 per cent and 59 per cent of people in London were in favour of zoos. This may seem to be a shockingly small sample size, but there is quite a large margin of error – a confidence interval of five means a total 'window' of ten percentage points. For a more precise estimate – say, within four percentage points (which means a confidence interval of two) – a larger sample size is needed (2,401). The required sample size also increases for a higher confidence level; for a 99 per cent confidence level with a confidence interval of five, the sample size would be 4160. Thus the greater the precision and level of confidence required, the larger the sample size must be.

The greater the variation in people's answers, the larger the confidence intervals for a given sample size (in other words, the larger the margin of error). The greatest possible level of variation in responses to a two-way (*dichotomous*) choice question is a 50:50 split, and because of this, simple sample size calculators such as the one in Figure 4.1 usually assume a 50:50 split. Therefore the calculated sample size is the largest possible value that should be required. The calculator also allows you to enter to results after collecting the data in order to calculate the actual confidence intervals.

The final variable in sample size calculators is the population size itself. Counter-intuitively, if the population size is over about 5,000 then it has no effect on the appropriate sample size, because the sample will only be a small percentage of the total population. For small populations however, the population size must be taken into account. Figure 4.2 reproduces an excel spreadsheet that is publicly available on the web and which calculates appropriate sample sizes for small populations. If the population size and desired confidence level and confidence limits are entered in the table, the appropriate sample size is calculated. The terminology is slightly different here in that the confidence level is expressed as a *z* value (the equivalents in percentage terms are given in the bottom right hand corner), and the

	Enter data in green area below			
N (population size)	1,000	Required sample size =	517	
z (confidence level)	1.96			
E (+− error)	0.03			
p	0.5			
q	0.5			
Use the following values of z for different levels of confidence:			*z*	level of confidence
			2.58	99%
			1.96	95%
			1.645	90%

Figure 4.2 Estimating the required sample size for surveys measuring proportions.
Source: http://uregina.ca/~morrisev/Sociology/Sampling%20from%20small%20populations.htm.

confidence limit (E) is expressed as a decimal rather than a percentage (0.03 instead of 3 per cent). The p and q values represent the proportion of answers in a two-way choice and, for the reasons discussed above, the default is set at a 50:50 split (represented here as 0.5:0.5).

The above sample size calculators are useful tools for simple research projects, but in most projects there are many factors that must be taken into account that sample size calculators are simply not equipped to deal with. It is therefore important that you do not just use the above calculators blindly, but understand the principles on which they work. Sampling error, confidence intervals and related terms are discussed in more detail in Chapter 16.

One final point to bear in mind is that a significant proportion of people you approach in a survey may not respond. *Non-response rates* vary according to the survey methods you use, but they can be shockingly high: in postal surveys or online questionnaires they may be well over 50 per cent. If you only distribute the minimum number of questionnaires that you need in order to meet the target sample size, then if only a few people fail to return the questionnaires, you will be under target. Non-response also introduces problems in terms of sampling strategies, because people who do not respond may be systematically different from those who do. There may be an effect of class, age or gender; people in favour of zoos may be more likely to respond to a survey about zoos than those against; and for online surveys, people who spend the most time online are probably most likely to respond. Therefore a standard approach is to set the sample size at about 30 percent higher than you actually need, monitor response rates, and do all you can to minimize non-response rates. If at all possible, leave time to chase up non-respondents with further (polite) requests for them to return the questionnaire. Whatever response rate you finally achieve, you must state it in your report and if necessary, discuss ways in which it may bias the sample.

In practice, then, researchers usually decide on a suitable sample size through a mixture of mathematical theory, convention (what sample size have comparable studies used?), gut feeling and practical considerations (how much time and money is available). A typical number of questionnaires for an undergraduate thesis is probably between 60 and 150; any less than this and there are likely to be serious limitations in terms of the statistical analyses that are possible. For a Masters thesis 100 is probably a realistic minimum, but 200 or more is preferable.

Whatever sample size is set in advance, there is a great advantage to analysing the data as you go along. Much questionnaire-based research involves testing for relationships between different variables, using the golden standard of statistical significance (see Section 16.2). The sample size that is needed in order to demonstrate statistical significance for any such relationship cannot be determined in advance, because it depends on the strength of the relationship – the stronger the relationship, the smaller the sample size that is needed. However, if you carry out some initial analysis before the end of data collection and find an apparent relationship that does not quite reach statistically significant levels, you still have the option of collecting more data to see whether statistical significance is reached.

Similarly it is important to revisit sample size considerations before carrying out the main data analysis. A typical questionnaire may include 30 or more questions using a range of formats and with different numbers of response options. However good your planning, you will probably find that the sample size places limitations on the analyses you can carry out, at least for some of the questions. One common way to deal with this for questions with multiple responses is to 'lump' the responses into a smaller number of categories during data processing; for example if a question asks people to indicate which of five age groups they belong to, then if necessary these could be combined into just three age groups for analysis.

Box 4.11 Instructions: how to decide on a sample size in quantitative research

1 Define your population and sampling units (see Section 4.1).
2 Decide what types of analysis you need to carry out. Is the main aim generalisation to a larger population, identification of relationships between variables, or comparison between subgroups?
3 Check the implications in terms of sample size with reference to the accompanying text. How big a sample do you need in order to justify statistical inferences about the total study population? If you intend to compare different subgroups, how many subgroups will there be? What is the minimum acceptable sample size in each subgroup?
4 Decide on an ideal target sample size and add at least 30 per cent for non-response.
5 Assess whether the target sample size is possible within the practical resources available (time, money and resources). If not, scale down the research design by reducing the overall scope or reducing the number of subgroups that will be compared.
6 If possible, monitor response rates and variations in responses during data collection and adjust the sample size as necessary.

Box 4.11 summarizes the different steps you should take in relation to the sample size in quantitative research.

4.3.2.2 Comparisons between two or more populations

A very common structure in research is to set up comparisons between two or more different groups – either from populations that have been sampled separately (such as two villages or two schools) or from within a single population that has been sampled as a whole (such as men and women in a sample of adult residents of London). However, the more subgroups you wish to compare, the larger the required sample size, and therefore it is important to be realistic about how many subgroups you can manage. If you want to be able to generalize from men and women in your sample to men and women in the population of London, then in theory the sample *of each group* must be large enough to allow for statistically valid generalizations to be made to the relevant subpopulation. Therefore the total sample size required for a comparison between men and women would be double what it would be in order to make generalizations to the study population as a whole. Suppose you wanted to look at the differences not only between genders (two categories) but also between age classes (using five categories). If you were to draw up a contingency table – a table showing the numbers of people in each category of gender and age – it will have a total of ten cells representing the different subgroups, and in theory therefore you would need ten times the sample size that would be required for a single group – something that would probably be impractical.

In practice, many studies that focus on a comparison between subgroups do not attempt to gain a representative sample of each subgroup but instead focus on the comparison between groups in the sample. The difference here is related to the difference between internal and external validity (see Section 3.3 and Box 3.7). Internal validity is to do with the internal logic of the comparison and the degree of confidence with which you can draw conclusions about meaningful differences between the samples, whereas external validity is to do with the extent to which the results can be generalized from the samples to the larger population. In terms of internal validity, in order to meet the minimum requirements of statistical tests that are used to compare groups, Blaikie (2000) suggests at least 50 cases per subgroup, whereas Denscombe (2007) suggests at least 30; for some types of analysis such as a Chi-square (see Section 16.6.3) even smaller sample sizes are acceptable. External validity is low and therefore generalizations from the sample to the population as a whole are not statistically valid. Nonetheless, studies of this kind make up a substantial component of the literature and are extremely useful in generating hypotheses about wider differences, which can then be tested through further research.

4.4 Conclusion

It should be clear by now that whilst there are important and complex theoretical factors to take into account in developing a sampling design, in practice it is almost always a compromise between theoretical requirements and practical considerations. A perfect sampling design is by definition unattainable, because the perfect study would encompass the total study population rather than rely on a sample. Practical limitations are particularly prominent for small-scale studies such as undergraduate theses or short-term consultancies. The important thing is not to over-interpret your data but to recognize and discuss their limitations. If your samples are so small that you cannot make inferences that are statistically valid, say so – but use descriptive statistics to point out the interesting patterns that you have found. If you are unable to rule out some kind of sampling bias, discuss it – not to do so would be misleading, and it may actually demonstrate something interesting.

Summary

1 Sampling is about choosing exactly what sources you will collect data from, and is a key aspect of research design.
2 The most important division between the different types of sampling strategy is between probability sampling and non-probability sampling.
3 Probability sampling is most important for quantitative research and allows inferences to be made about the study population based on results from a sample. Probability sampling strategies include simple random sampling, systematic sampling, cluster sampling and stratified random sampling.
4 Non-probability sampling can be used in exploratory studies and is also appropriate for many qualitative studies. Non-probability sampling strategies include convenience sampling, quota sampling, targeted sampling and chain referral.
5 In many forms of qualitative research, no sample size is set in advance of data collection and sample size may not be addressed explicitly at all in the research design (although sampling *strategies* should be addressed). However, in research based on sets of semi-structured interviews or focus groups an initial target sample size is usually set in advance.
6 Whether or not target sample sizes are set in advance of data collection, the key principle in assessing sample size in qualitative data sets is the principle of *saturation* – the stage in qualitative data collection when collecting more data produces little important new information or understanding relevant to your research questions.

7 Sample size in quantitative research concerns how many cases are needed in order to distinguish between chance variation and more 'meaningful' patterns in the data. The patterns may concern the relationship between the sample and the larger population, comparisons between two or more contrasting populations, or relationships between different variables (or any combination of these).
8 Sampling theory provides a way to estimate what sample size you need in order to make inferences about the study population with a given margin of error (usually expressed in terms of the *confidence intervals*) and with a known probability of being accurate (the *confidence level* – by convention set at either 95 per cent or 99 per cent).
9 In practice, researchers usually decide on a suitable sample size through a mixture of mathematical theory, convention (what sample size have comparable studies used?), gut feeling and practical considerations – how much time and money is available.
10 In research involving comparison between different subgroups, the more subgroups you wish to compare, the larger the required sample size, and therefore it is important to be realistic about how many subgroups you can manage. In practice, many studies that focus on a comparison between subgroups do not attempt to gain a representative sample of each subgroup but instead focus on the comparison between groups in the sample.

Further reading

De Vaus, D. (2002) *Surveys in Social Research,* 5th edn, London: Routledge. [An authoritative text on all aspects of carrying out survey research, including a very well-written chapter on sampling.]

Bernard, R. (2006) *Research Methods in Anthropology*, 4th edn, Walnut Creek, CA: Altamira Press. [The seminal book on anthropological research methods. It includes three entire chapters on sampling (Chapters 6 to 8) and probably goes further than any other book listed here in explaining both qualitative and quantitative approaches to sampling in detail.]

Wilson, I. (2005) 'Some practical sampling procedures for development research', pp. 37–51 in J. Holland and J. Campbell (eds), *Methods in Development Research: Combining Qualitative and Quantitative Approaches*, Bourton on Dunsmore: Intermediate Technology Applications/Practical Action Publishing. [Presents a useful practical perspective based on Wilson's own experience in development research, be it qualitative, quantitative or participatory.]

Section II

Methods

5 Participant observation

R.K. Puri

The final goal ... is to grasp the native's point of view, his relation to life, to realize his vision of his world.

(Bronislaw Malinowski, 1922: 25)

5.1 Introduction

Participant observation is a relatively unstructured interactive method for studying people as they go about their daily routines and activities. The researcher accompanies one or more people both to observe what they do and say, and also to participate, to varying degrees, in the activities being studied. Describing how and why people do what they do is ultimately the aim of participant observation. It can be used to collect information on how people live their lives, how they think and act, and how they describe and explain themselves and their motivations. The method is unstructured in the sense that the researcher has to follow the schedule and activities of his/her informants, rather than impose a framework that interferes with their normal routine. However, the researcher does maintain some control over which activities to participate in, how frequently they participate and the kinds of questions that are driving the study.

We all participate in social activities all the time: we live in families, have work colleagues, go shopping, attend religious, leisure and other events and organizations, and so on. We may even be attentive to the characteristics and dynamics of these social events, but we do not set off to document carefully what is happening; we do not approach these situations with particular research questions in mind. Likewise we may often observe people, while sitting at a street-side café, for example, but we usually remain outside the social spheres being observed. Attending a sports event is clearly a distinct activity from being a player, but we do participate as a 'fan' or a 'supporter', and thus can observe and participate in that social group's activities. Participant observation involves documenting what people say or do during such activities, and also how the researcher feels and what he/she experiences while participating. By accumulating these experiences the researcher can build up a picture of 'the way things are done' and develop a deeper understanding of who these people are, how they think and how they differ among themselves. The information collected can be analyzed using both qualitative and quantitative methods, depending on how systematic one is in choosing events and people to accompany and the kinds of research questions being asked.

For researchers, such as anthropologists, interested in describing the life and culture of a social group in a comprehensive manner, participant observation may be required for a year or more. It usually requires living with people, often in their homes, learning their language and generally following and observing their way of life. The anthropologist Bronislaw Malinowski, marooned in the Trobriand Islands during WWI, was one of the first scientists to describe participant observation as a field research method. Prior to Malinowski, anthropologists relied primarily on interviews with *key informants* (see Section 5.3.3) or reports written by missionaries, traders and government officials. They were seldom fluent in the local language, relying instead on translators, and never spent much time actually in the communities they were studying. Malinowski recognized that there were intangible aspects of human life that motivated people and were important in explaining their beliefs and behaviours, but that these could only be grasped and understood by experiencing life as they did so, through immersion in the routine, mundane 'imponderabilia' of everyday life. He writes:

> To study the institutions, customs, and codes or to study the behaviour and mentality without the subjective desire of feeling by which these people live, of realizing the substance of their happiness is, in my opinion, to miss the greatest reward which we can hope to obtain from the study of man.
>
> (Malinowksi, 1922: 25)

Living with people is necessarily interactive, in that researchers engage in conversation and activities with people in the course of daily life, rather than creating a separate setting where people are asked to provide information and opinions via formal interviews, surveys and questionnaires. In the latter case, the researcher runs the risk that people will bias their responses to reflect what they think they should say or what they think the researcher wants to hear. They may be suspicious of unknown outsiders, perhaps have faulty memories or even be intentionally misleading.

Malinowski also understood that even among themselves, people often do not say what they mean, mean what they say, or do what they say! (See Section 6.4) There are many reasons for this, some having to do with the very nature of culture itself. We live in families, communities and societies with cultural ideals and standards of behaviour (values, norms and roles) of which we are often completely unaware because they are ingrained in daily life and thus taken for granted. Sometimes, we find we just cannot live up to these norms even though we know they are preferred. In other cases, we are subjected to laws and regulations that are restrictive, oppressive and unfair, and so we are compelled to ignore them. In the presence of certain company then, social and legal transgressions would never be revealed, and it is thus important rather than relying on what people say to observe what people are actually doing. First-hand observation and personal experience of people living their lives can remedy these unintentional and intentional biases.

In this chapter you will learn that there are many ways to use participant observation in conservation-related research and management.

5.2 Uses of participant observation in conservation

All groups of people that participate in conservation research, management, policy making, funding, public relations and activism are potential social groups, or 'communities', that could be studied by scientists using participant observation. Let us consider a situation common to conservationists where participant observation and more informal means of learning and collecting information can be used. Box 5.1 gives a description of such a situation, and the

Box 5.1 Participant observation in conservation: two communities and a protected area

Imagine two communities: a community of closely related families dependent on cultivated and wild plants and animals for their subsistence and cash income in an area to be designated a nature reserve by the government, and a 'community' of conservationists – the regional office of an international environmental organization, staffed by local and expatriate scientists, administrators and student interns. Both communities can be said to have a 'culture' of beliefs, norms, values, roles and practices that both reflect and create a 'worldview' and a 'way of life' for its members. It is through their respective cultures that each community comes to see and understand its 'environment', organizes the activities that contribute to its survival and gives meaning to their members' lives. Given the context, it is likely that these communities will eventually come into contact as the plans for the nature reserve take shape. Through the accumulation of experiences and other sources of information, each community will undoubtedly develop views and opinions of the other. Each community may in fact wish to understand better the behaviour and intentions of the other one. A community of forest-dependent farmers might be suspicious of the intentions of the environmental organization, fearing that they are going to lose access to lands and resources, or even be moved. They may even oppose the present boundaries or future management policies of the nature reserve. Or they may want to know if and how they could become allies, for instance in creating community conserved areas for eco-tourism or some other development initiative. The environmental organization may wish to understand what resources are used and why. They may not understand how resource use varies internally. They may not understand how decisions are made in the community, and how this community is linked politically, economically and culturally to other communities and to national and international institutions. What goes on in the community may in fact be driven by these outside linkages. Third parties, such as social scientists or even government policymakers, may wish to understand the interaction of local community and conservation organization and its impact, for example, on the management of the nature reserve and local development initiatives.

Whatever the focus, participant observation would be an excellent way to uncover how each community works, their 'worldviews' and 'ways of life', and the way each sees the other.

following text describes the value of participant observation in investigating it. Box 5.2 is a journal entry describing participant observation on a hunting expedition with Penan people of Borneo.

With regard to the community of resource users, participant observation provides first-hand accounts of what people do and say in certain situations. This applies not just to some people, but potentially to all kinds of people in the community, such as women and men, the really rich and the really poor, the political elites and the politically powerless, ethnic minorities, children and the disabled. You should also gain a personal sense of what it feels like to live their lives, and thus begin to empathise with members of the community.

Through shared experiences, you would learn, as a child has to when growing up, the meanings of objects, words and activities, and the underlying motivations and rationale for

Box 5.2 Extract from account based on participant observation of hunting in Borneo with the Penan

Planting the garden and cleaning skulls were put on hold today as I joined Pak Bisa on a hunting trip to the Kedayan River. We didn't catch anything, but we did have an enjoyable walk up the river – barefoot! I've always liked walking behind Bisa because then I'm closer to his amazing feet, so broad and strong, with calluses at least a centimetre thick. He walks silently across dry leaves and thorns rarely bother him. So today I tried to walk without shoes, a good idea but maybe I should have tried a shorter walk to start with, instead of a half-day of hunting on tender feet! I suppose I've massaged my whole body as a result! It was so sensual, the mud of the trails, the rocks in the river, everything hurt at first but the textures were so rich and strong and evoked such intense feelings, you could practically taste them. But my whole body aches now.

At the mouth of the river I picked up some bright yellow lemons floating in the water. No one uses them here for cooking fish or even juice, and I've been trying to convince them that it makes an excellent cure for coughs and colds! They smelled so delicious in the rattan bag on my back, every time I turned my head I caught a sour whiff that opened my nose. I felt very light and strong suddenly, unencumbered by shoes, running barefoot up river hopping easily from rock to rock and splashing through shady cool pools. There were signs of babui (wild boar) everywhere, but we had the wind at our back and they were undoubtedly forewarned of our arrival. The morning was clear and crisp and drier than usual but signs of the flood on Sunday were still visible: grass on the banks was still flattened and the branches and leaves of trees along the riverside were still brown with the silt of the flood waters. We came across Sabung's deer (that he had killed and left) on a gravel bank, reeking and covered with flies. We noticed that pigs had been eating the carcass. Bisa described the two kinds of pigs that people hunt, those that pass through on the migrations and swim the rivers and those that stay and wait for the fruit season here. So the hunting techniques for catching pigs switch from waiting (mabang satong) to searching with guns and dogs (ngasu or nyalapang) in the hills and along the river banks. The pigs that stay behave differently, they raid the garden of the Penan in Belaka and sometimes even those in Peliran – though I've not heard of any across the river (in the gardens) and no one is hunting over there. Bisa claims that pigs that swim go up over the mountains and into the Malinau river valley. They will travel from 7 in the morning till 7 at night. Those that stay forage by the river's edge in the early morning and then climb up hill to sleep during the mid day heat (usually hiding in thorny rattan thickets). At around 4 or 5 pm they are active again, some coming down to the river's edge. Gardens are typically raided between 8 and 10 at night, and you can find pigs rooting about the river's edge at midnight.

We passed several sungan (salt springs) and sat down to wait by one up on the forested bank, approaching it from down river. We could see well-worn animal trails disappearing up the steep hill behind the rocky outcrop where the salt water dribbled out. Being near midday, we didn't expect much. Pak Bisa claimed that kijang (*Muntiacus muntjak*) would come to drink mid day, so we hid quietly, spears poised, and waited. After almost an hour we gave up and turned around to face the steam and had our lunch of rice and boiled greens with hot chillies. At one point I turned around, and there was a kijang coming down the trail. We were too far now and our spears were planted on the side of the trail. Pak Bisa turning and eyeing the deer cursed in disgust, and so the deer bolted. He smiled and said, 'How come she's late? Now we have nothing to bring home!' That comment amused me for the whole trip home. Maybe we were early? Do deer really have a set schedule? Isn't Pak Bisa denying evidence that challenges his theories about deer feeding times? Is this a case of interpreting an event in terms of a pre-existing schema, cultural model, maybe even received wisdom from the ancestors?

I returned exhausted, with lots of leech bites, and made lemonade – saving the seeds for my garden – and then collapsed for the rest of the afternoon. Luckily Pak Bit brought over some fish for me. Rice, greens (daun ubi) and fish (with lemon) was a very satisfying combination.

Source: Puri (2005)

certain behaviours. It is often the case the people are not easily able to explain their behaviours or feelings, or do so in terms that are difficult to translate and understand. A researcher is in a much better position to interpret such actions and their explanations if they have gained a good understanding of the language and its use, know the people personally and have themselves experienced life in the community. All of this is only possible through extended periods of participant observation.

If the interaction between the two communities becomes confrontational or cooperative, then clearly research in both communities would be essential, but a certain amount of neutrality would be expected in order to gain trust and maintain objectivity and credibility. Participation in daily activities, lending a hand and just getting to know people is an excellent way to overcome initial fears and suspicions and to develop the rapport that is needed both to gain access to events and conversations that may be critical to understanding this interaction, and also to be able to ask difficult questions and trust the answers given.

How important participant observation is to a conservation related study depends on your research questions. For instance, research questions might well be focused on the whereabouts and behaviour of potentially endangered animals and plants. While at first glance a social science approach may not appear relevant, you may find that communities of hunters or gatherers that are interacting with these biota have specialist knowledge of it, either because they use it or because it occurs on their lands and they know of it. In these cases, participant observation offers an excellent approach to tapping this knowledge. It might involve going out with them into the environment, joining in with or at least observing their hunting and gathering activities, and chatting informally along the way (see Box 5.2). This might be combined with more structured social science methods such as semi-structured interviews and participatory mapping, and also with methods from the natural science fields of biodiversity assessment and population ecology. Such an approach would focus specifically on the knowledge and behaviour of hunters and collectors of natural resources. In contrast, understanding general patterns of resource use in the community as a whole – why it is that people need or desire natural resource products – might require a broader scope of participant observation, encompassing all resource users in the community, not just the specialist resource collectors. Thus in some cases, participant observation can be just focused on particular activities and combined with biological or other social science methods, while in others it can be the only technique used and its scope therefore much broader, involving all aspects of daily life.

In the latter case, this means living with the community as they do, eating what they do (as far as is possible) and letting their daily rhythms determine yours. Of course, a researcher cannot be everywhere at once, so there will be a selection of activities that are prioritized for participant observation, and others that may be observed only once or twice. What is very important is of course the downtime, when people are not working, but are resting or engaged in leisure activities. These are extremely valuable moments for really getting to know people, to build friendships and to engage in informal conversations and interviews, and, of course, lots of gossip. After a while, the pretences and facades erected for enacting our various social roles fall away and both researchers and hosts begin to see each other as people, each with their own characteristic manners and eccentric traits. As you get more and more familiar with individuals, you of course can begin to make better sense of what they say and what they do, especially with regard to activities that are a focus of research. This allows you to judge the reliability of informants in a way never available to researchers using questionnaire surveys or even formal interviews. You learn who actually knows most about certain subjects; who is a good storyteller with a faultless memory, and who exaggerates or even lies (and possibly why). You understand the personal histories of individuals and the complexities of social life, all of which make for a richer and probably more accurate yet nuanced description of that social group. In a sense, you learn what is going on behind the scenes during the performance of daily life.

Another important benefit of participant observation as a research method is that unexpected things are bound to occur, and by being there you will have a front row seat to experience both the events themselves and the ways people respond. Conflicts and catastrophes, while unfortunate and personally distressing for all concerned, including a researcher, can provide the context to understand much about local knowledge, social organization and human–environment relations (see de Munck and Sobo, 1998: 43).

Conflicts over access to resources, whether swidden fields, hunting territories or wild plant and animal resources, often reveal the unwritten rules by which a social group works. The rights and regulations for ownership and use of land and resources are often invoked in public hearings or arbitration that may be held informally in residences or in more formalized courts of some sort. Being able to attend these often private meetings and being able to track the causes and consequences of the conflicts may only be possible due to key social relationships and trust built up over prolonged residence and participant observation amongst a community. Extreme climatic variability, such as drought or a flood, can provoke emergency coping mechanisms, based on traditional knowledge as well as innovations, which require changes in the use of natural resources, activation of social networks, and use of political and economic capital. In these situations, you can see the way people re-organize themselves and make use of famine foods, forest reserves or sacred wells, and fall back on old technologies and political connections.

It should be obvious that during crises protected areas and species may be placed under serious and unprecedented pressure. A good example of this occurred during the Asian Economic Crisis of 1997–8, which coupled with El Niño droughts, wild fires and political meltdown, led to the violation of protected areas by desperate residents, theft from plantations, and increased poaching and sale of forest products for supplementary cash income (Donovan, 1999). Environmental organizations are also subject to contingencies; environmental catastrophes can put real strains on finances and endanger lives. In other contexts, unexpected changes in personnel and sudden changes in financial conditions due to denied grants or windfalls from generous donors can also affect the daily running of programmes and even the whole mission of the group. Conflicts between communities and NGOs, government officials or businesses are also likely to occur in conservation related contexts, and being there, and all that entails, allows you to observe and better understand what has happened and why people have responded the way they have.

Living through conflicts, accidents and other tragedies with a group of people can be both enlightening and deeply distressing. By immersing yourself in the culture of an unfamiliar society or group, you both gain incredible personal insight, but also open yourself up to potential psychological effects. There is also the fabled problem of 'going native', where the researcher comes to identify with their hosts to such an extent that the values and norms that once guided them are submerged or forgotten in favour of the new ones. In this circumstance, the researcher has lost objectivity and in some cases the original reasons for doing research have been forgotten. This is rare though; more likely is that there will be times and occasions when one is made painfully aware of the cultural relativity of social life. This is a form of *culture shock*, which often results in increasing frustration in coping with differences and a feeling of alienation and homesickness. This usually passes as language skills improve, familiarity with people increases and shared experiences accumulate. It is common, though, that prolonged engagement with another society over many years can lead to a more permanent sense of marginality, what has been described as a state of 'betwixt and between'. Researchers and others commonly claim that they know they will never be fully accepted into their host cultures, and yet they also come to see their home society and culture with new 'eyes'; an outsider's eyes. What they once took for granted is now exposed, and the sense of familiarity and embeddedness lost. So for some researchers, participant observation may prove to be too difficult to sustain over the long term, though this may be hard to predict ahead of time; your experiences as a newcomer to a new school or town or as a tourist in a foreign country may give you an idea of how well you would do during long-term participant observation, though even these circumstances are of a smaller scale. In those cases where culture shock sets in, a break from field research is recommended, and perhaps more intermittent contact, as in participant observation during specific activities or just informal interviewing, are preferable methods for the remainder of field research.

Keeping these examples in mind, we will now turn to some of the details of actually carrying out participant observation in conservation contexts.

5.3 Doing participant observation

How do you go about preparing for this kind of research and how do you get started? In this section we describe how to go about contacting a community, getting permission to work with them and then getting accepted so that you can begin to collect relevant information. We then describe the nuts and bolts of doing participant observation, the ways that interviewing and other methods can be incorporated, taking notes and other forms of documentation, and knowing when to stop. Chapter 12 will discuss the relationship between the researcher and the host community in more depth, and should be read before leaving for the field.

5.3.1 Acceptance in your study community

If you are going to participate in just a few specific activities over a short period of time, and not actually live with the study community, then it is important to plan well in advance by seeking contacts within the group who can facilitate your participation (for further discussion of this point and of how local people may perceive you, see Chapter 12). A missionary, teacher, government official, tour guide, shop owner, ex-community member, or someone else marginal to the group may play this role. They can be thought of as your *cultural broker.*

Often gaining permission to participate may involve some preliminary activity, sometimes referred to as *social entry*, which is where you are introduced and have a casual but personal encounter with one or more members of a group. You may be invited to attend a party or ceremony, or just asked round for a drink or food and a conversation. It is best to be relaxed and respectful in these first meetings, and allow your cultural broker to do the job of explaining who you are and what it is you are planning to do; you can always explain in greater detail at a later time. Basically, these meetings are often less about the details of the project and more about 'sizing you up', and deciding whether they want to pursue a relationship with you. You have to be a quick judge of the situation to know what tone to take – serious or not. Hopefully your cultural broker can brief you on this beforehand. In some cases, the group just wants to know that you are a normal human being who they can communicate with and who listens, perhaps has a good sense of humour, someone they can get along with. In others they may want to know how serious or prepared you are for what it is you are proposing to do. Can you be one of them, in speech, dress and manners? You should never be put off if these initial meetings fall a bit flat or are even met with discouraging looks and statements. Your cultural broker may be a better judge of what has happened and whether you should pursue the relationship or not.

You may befriend someone in the group right away, who may serve as a sponsor who may advocate on your behalf or agree to be responsible for you. As with any cultural broker, there may be several reasons that such people step forth to facilitate your stay and work; they may be genuinely honest, kind and interested in what you are doing, they may be compelled to do so by others, they may also expect to be hired or paid, or gain some social or political status as a result. However, it is important not to align yourself too closely with any particular member of a community, especially at first, as this may simultaneously restrict access to others in that community (see also Chapter 12). Better to remain neutral and open to all until you are more familiar with any internal social, political or religious divisions, coalitions or parties.

5.3.2 Preparing for participant observation

Once you have passed through the social entry phase, and have some personal contacts or even a new friend, then in another meeting or encounter you can set about explaining what it is you want to do and what this will require of them. You may have to convince them that you are indeed capable (mentally, physically, emotionally) of participating in an activity such as hunting. Some activities may just be too dangerous to your health or theirs and you may not be allowed to attend or participate. Others may be inaccessible because they are restricted to members of a certain age and sex. There are often specialized domains of knowledge that may be secret to most community members (such as shamanic healing rituals or the annual budget), and thus these too may be closed to you, or you may have to wait quite a while before you are given access. So, not all activities and practices are amenable to or accessible for participant observation, and you may have to find other methods (such as interviewing) to learn about them.

You must, as in all ethical research with people, ask for and receive *free, prior and informed consent* from all those that you will be working with, not just the leaders of the community or your cultural broker or your first friend (see Box 11.3 and Section 13.4.2). Often oral agreement is all that is needed, but they may want a signed agreement that lays out the terms of the research, what will be done and what is expected of you and them. You may have to discuss payments for those allowing you to participate, as well as issues of documentation, such as audio or video recordings, translators and translations, and returning results to the community. While sometimes lengthy and tedious, these preparatory steps are considered best practice today and are designed to benefit everyone involved, and avoid conflicts between researchers and communities over intellectual property rights, bioprospecting and biopiracy, and culturally inappropriate behaviour.

One of the most important reasons for engaging in participant observation is to study how things are actually done firsthand, and repeated observations of the same type of event may begin to reveal common elements and patterns. But just as important are the differences in the nature and sequence of components of these events, the kinds of things said, and variation between the participants and even in the same individual. No two participant observations will ever be exactly the same, so do not get hung up on the notion of exact repetitions. In fact the variation among people and times is just as important as the prototypical or general defining characteristics of any particular activity (more on this later) – differences matter. Hunters may have different knowledge and idiosyncrasies that have notable consequences for prey and for household livelihoods; environmentalists also may have different levels of knowledge and experience, physical conditioning and temperament that affect their ability to conduct research or manage conservation programmes. Just as critical are the varying social and ecological contexts of hunting events and activities of the conservation organization that also generate difference. As a researcher, therefore, it is vitally important to sample as many of these events as is possible in the allotted research period, recording such variation in event and context, and attempting to make sense of both pattern and difference.

There is no set formula for determining when you've been on enough hunts or attended enough meetings (but see Section 4.3.1 for some basic principles of sampling in qualitative methods). Some events, such as daily or weekly subsistence activities or planning meetings, can be sampled frequently, more so if you have a trained research assistant. Rare events, such as an annual hunt for a religious ritual or a visit to an NGO by foreign guests, including donors and dignitaries, are of course more difficult to accumulate, so a few cases may have to be supplemented with interviews about similar events in the past.

5.3.3 Incorporating interviews

If you are allowed to participate in a community activity such as hunting or a board meeting of a conservation organization, then the next step is to find out what is required and what generally is going to happen. You can do this with an *informal interview* (see Section 6.2.1), perhaps with your cultural broker, new friends or perhaps an expert that you are introduced to. Informal interviews are much like conversations, so you can expect a bit of give and take and no preset structure. So you might be hanging out at the local bar or café and start chatting with your neighbour, and something related to your topic (hunting, accounting) may come up. You might then just encourage them to elaborate, or ask them to tell you something about your topic. This is not the time to be provocative or combative; be respectful and listen, and encourage your respondent to keep talking. Chapter 6 discusses interview techniques in detail, including the use of *probes* (Section 6.3) and ways to minimize *response biases* (Section 6.4). There is no need to force the conversation in any particular direction at this point; you are looking for general information on your planned participant observation, perhaps something about personal attitudes and expertise, the latter being sometimes difficult to detect on first encounters.

A few informal interviews with people before you engage in your first participant observation should give you an idea of what to wear and bring and whether it will be feasible to record during or after the event. Informal interviews are always useful, throughout a project, both to explore local opinions and to maintain social relationships with your hosts.

As your participant observation research proceeds you will learn about subjects that will need to be explored in greater depth, perhaps through more targeted interviews with known experts. These should also start off as unstructured open ended discussions and proceed to get more focused and more structured with each encounter. You should also cultivate friendships though repeated participant observation and casual conversations with *key informants*, those locally recognized experts in your particular field of interest.

One last point: you'll also find that many informants will come to you and casually ask such things as 'how's it going?' or 'so, what do you think of our place, so far?', and while some may be simply being polite, many others will be probing you for their own inquiries!

5.3.4 Basics of participant observation and some variations

Now that you are well prepared, it is time to make an appointment or to follow up on an invitation to participate in whatever event you are studying. We recommend that you are always a bit early for these things, so that you can watch carefully the initial stages of preparation or greeting among participants; there may be valuable information exchanged at this moment and you may be able to get a sense of the social relationships of participants that may influence the way the event unfolds. If the activity is something you are not familiar with, then following what others do is often the best strategy, though it may involve some guess work as to whom you should follow. Often a minder or tutor will be assigned to you. With imitating or learning, listen and observe carefully, noting those details that seem relevant to your research questions; you cannot record everything that comes to mind. Perhaps this will include aspects of what is said, the way it is said, non-verbal signs and gestures, what material culture is used (such as tools), or the way an event is broken up into labelled, or unlabelled, tasks. At first it will seem very difficult to choose what to write, but with time you will be able to pick out the most relevant information and you will develop your own style of note taking. In some cases you may be able to take notes in a small notebook (waterproof for outdoor work), in others you may have to wait for an appropriate moment (more on this below). You may be allowed to record conversations, take photographs or even video the event, but you will have to judge when it is an appropriate moment to do so and you must always ask permission first. You will also have to judge if it is an appropriate moment to ask questions. Watching and imitating, and then watching your companions' reactions, adjusting and trying again, is the usual manner of learning a new activity.

At this point it is important to contrast some of the variants of participant observation, such as being a *participating observer* versus an *observing participant*. Sometimes an activity such as hunting may involve a group of people with rather different roles, for example, tracker, gun-carrier, dog owner, child in training (see Box 5.2). As a complete novice, you may end up playing the role of a child in training: following along, trying to keep up, staying out of danger and trying to get in a question once in a while. In these cases, you are more likely to rely on observing. In fact in many social activities, from hunting to committee meetings, many people attend and follow along much as observers, so having another one along who also just observes is not so strange. Joining a group of farmers weeding a field may be a rather different situation, as everyone is there to work and it may be uncomfortable and disruptive to have someone just sit and watch. At the other extreme, someone who normally participates in an activity, such as an expert lead hunter, may decide also to begin observing what they and the group are doing. This is often very difficult, but with time and practice can produce very authoritative descriptions of events. If possible, recruiting and teaching community members to do their own observing of themselves can be a productive way to collect data in places and at times when you are unavailable. It can also increase the interactive and participative nature of your study, if those are important goals, and serve as a means of capacity building in projects where some form of self-monitoring is a management goal. Outside researchers, depending on the activity, can expect with time to move from a predominantly observer status toward one of increasing participation, knowledge and even responsibility.

As with all skills and methods in research, practice makes all the difference. You should not be discouraged by your first attempts at both observing and participating, for trying to do both is difficult. Getting wrapped up in participating, which can require learning new skills, often requires a degree of concentration that makes note taking or even objective observing impossible – you may simply forget that you are there to learn and observe too. At the same time it may be uncomfortable to just stand outside the social or physical activity and watch others work, especially when encouraged by your hosts to jump in. There is always the fear of making a fool of yourself and losing the respect of new found hosts. However, making mistakes is a valuable part of the process for it alerts you to rules or procedures that others are following, perhaps implicitly. Often your first attempts will be met with laughter as if you were a child, which you are in a way, but you should not miss the opportunity to use the laughter to both highlight and experiment with hidden rules. Like a child testing the bounds of acceptable behaviour, by their laughter, frowns, surprised and shaming looks, you will quickly conform to what is more acceptable behaviour. Noting down these subtle changes is quite important, especially for complex skills such as farming or hunting behaviour, and even for such things as attending a meeting of elders or board of directors.

5.3.5 Taking notes and building incrementally on what you have learnt

Note taking may be possible only intermittently during an event, and you may have to train yourself to remember details and conversations so they can be written down afterwards. There are many stories of researchers dashing off to the toilet/garden/corridor to quickly jot down notes. It may be possible to record conversations, take photographs or even video an event, but you must get permission beforehand and judge how disruptive it will be to the integrity of the activity. Note taking in 'meetings' may in fact be an observable and therefore acceptable activity, while hunting with dogs in a tropical rain forest may present real difficulties for any documenting. If you are focused on a particular activity, then very quickly you will develop a shorthand vocabulary to describe each instance or event that you participate in, so note taking should become faster and more focused.

Whether you can take notes or have to wait till after the event, at the end of the day (or in the morning if you have been participating in evening activities), you should reflect on the day's activities and review what has been written, adding additional notes and beginning the process of highlighting important bits and *coding* them (see Section 14.2.3). You may also wish to consult with participants after the event to discuss what happened or get help with translation of terms or of audio recordings. It is important to do this as soon as possible – do not wait. Events and experiences begin to pile up and unless you keep to a routine of writing and reviewing them as they come they may all begin to blend together in your mind and become indistinguishable as individual events. You should also keep a second file or notebook for questions that arise during this process, for as you become more familiar with the events, and your hosts more familiar with you, you will discover those breaks or moments when it is possible to ask questions freely.

Similarly, taking time to consult with informants after the event, debriefing, for instance, to review notes or watch videos, is a valuable aspect of the whole exercise, for in understanding how your hosts view these events you are tapping into their own ways of thinking (perception and interpretation) about the world. You will learn to move back and forth between your outsider's objective description of what is happening and their insider's subjective concepts and explanations of the same events. Over time you will learn to understand these concepts and apply them yourself, thus anticipating their own explanations. In common parlance we might call this 'getting inside their heads', but of course this is impossible, and what we aim for is a level of empathy and competence that allows us to engage, function and be accepted in that social context.

5.3.6 Knowing when to stop

Often research is conducted under time and financial constraints, so you may be forced to stop collecting data due to these factors. Having a feasible research project with a well thought-out sampling plan is essential, and will determine when the data collection can stop. However, if you are researching a particular activity then you can only participate in as many such events as occur during the fieldwork period, taking into account natural variation in the participants and contexts of these events. The bottom line in deciding when to stop is – do you have enough data to address and inform each of your research questions? Some preliminary processing of data will be necessary to judge this (see Section 14.2). You may also begin to feel that you are now so familiar with these events that they are very predictable to you, and you can describe them fairly completely. This is related to the principle of saturation, which is central to strategies for qualitative sampling and is described in Section 4.3.1.

5.4 Conclusion

There are many benefits and some costs to using participant observation in research directed at conservation issues. It is important to remember that all groups involved in a conservation issue, including researchers, are possible subjects for research using participant observation. From them we can begin to understand the details of human activities involving the environment, the personal motivations of individuals, and the social and cultural contexts that may drive land and resource use and explain variation. We also gain by engaging with people as people, by trying to see the world as they do, by setting aside our values for a moment and trying to empathize with theirs. And while there may be personal costs and scientific limitations to immersing oneself in the life of another group of people, the long-term benefits for conservation of a humane and sympathetic approach to understanding the lives of affected people must surely outweigh them.

Summary

1 Participant observation is a relatively unstructured interactive method for studying all participants in conservation related activities, as well as those affected, as they go about their daily routines and activities.

2 It can be used to collect information on how people live their lives, how they think and act, and how they describe and explain themselves and their motivations.
3 Participant observation requires contacting a community, getting permission to work with them and then getting accepted so that you can begin to collect relevant information.
4 The researcher accompanies one or more people both to observe what they do and say and also to participate, to varying degrees, in the activities being studied.
5 The method is unstructured in the sense that the researcher has to follow the schedule and activities of his/her informants, rather than impose a framework that interferes with their normal routine.
6 However, the researcher does maintain some control over which activities to participate in, how frequently they participate, and the kinds of questions that are driving the study.
7 There may be personal costs and scientific limitations to participant observation, but there are few alternatives for a deep understanding of the lives of people affected by conservation.

Further reading

Bernard, R. (2006) *Research Methods in Anthropology*, 4th edn, Walnut Creek, CA: Altamira Press. [The classic anthropological methods text, witty and full of examples. Chapter 13 is specifically dedicated to participant observation.]

Robben, A. and Sluka, J. (2007) *Ethnographic Fieldwork: An Anthropological Reader*, London: Routledge. [This is a collection of old and new essays about doing ethnographic fieldwork including one by Bronislaw Malinowski, quoted in this chapter.]

Whyte, W.F. (1943) *Street Corner Society*, Chicago, IL: University of Chicago Press. [A classic example of participant observation fieldwork among Italian immigrants and gangs in Boston's Northend. Appendix A describes the fieldwork methods.]

6 Qualitative interviews and focus groups

Get people on to a topic of interest and get out of the way.
(Bernard, 2006: 216)

6.1 Introduction

This chapter discusses qualitative interviews and focus groups. Qualitative interviews are probably the central tool in the social science toolbox, and provide data in the form of words – what people say. They differ from questionnaires in that they take the form of a two-way conversation, with discussion and follow-up questions on each point, rather than a question-and-answer session with no discussion. Focus groups are included here because, in effect, they are a particular form of qualitative interview that is carried out with a group of people rather than with an individual. Compared to questionnaires and other types of structured interview, the strength of qualitative interviews is in providing background information and context, generating ideas, and providing in-depth information on each participant's views, perspectives and motivations.

I have said that qualitative interviews consist of conversations, but they differ from normal, everyday conversations in several ways. First, the informant should do most of the talking. Your task is to encourage them to talk openly and freely yet stay more or less on subject, with as little direction or interruption as possible (although beware of remaining so aloof that they cannot relate to you on a personal level – see Section 12.5). Second, as a general rule the interviewer should remain neutral and interested throughout – something that is not always easy if you are discussing issues you feel strongly about. (There are exceptions to this rule – see Section 6.4). Third, as in all field research you should show professional courtesy at all times. This includes respecting their right not to answer (so not pressing them too hard about issues they do not want to discuss) and not telling others what they have said without their permission.

An underlying premise of qualitative interviewing is that the less you direct the conversation and the more you gain the trust of the interviewee, the more 'accurate' the information that emerges – in other words the more closely it will reflect the interviewee's knowledge, attitudes and opinions. However in most types of interview some degree of structure is necessary in order to maintain a focus on the topic of research. Interviews are categorized according to the amount of structure that is imposed by the interviewer; thus there are informal, unstructured, semi-structured and structured interviews (see Box 6.1). Informal, unstructured and semi-structured interviews are generally regarded as 'qualitative' and are

Box 6.1 Types of interview

Informal interviews: 'normal' conversations with individuals or groups of people as they go about their daily lives, from which you make notes on points relevant to the research topic. The emphasis is on listening to what people say and encouraging them to say more when they mention something that is particularly relevant to your research.

Unstructured interviews: conversations arranged in advance in order to talk about a particular subject. Since the interviewee has agreed to be interviewed, you can take a greater role in directing the conversation.

Semi-structured interviews: pre-arranged interviews based on a prepared interview guide – a list of questions or topics to be covered. The interview guide may act simply as a checklist to make sure that the key points are all discussed or it may be a list of questions that are asked in sequence.

Structured interviews: interviews using fixed wording or other stimuli that are presented in exactly the same way to all informants. Structured interviews include questionnaires (see Chapter 7) and types of interview used in cognitive domain analysis (Chapter 8).

Focus groups: formal, pre-arranged group interviews, usually with between six and eight participants. They may be based on an interview guide similar to those used in semi-structured interviews, or they may involve group exercises that are designed to generate discussion.

usually analyzed qualitatively (see Chapter 14), whereas structured interviews include questionnaires (Chapter 7) and interview formats used in cognitive domain analysis (Chapter 8), and are usually analyzed quantitatively (Chapters 15 and 16). The most appropriate type of interview to use is related both to the research design and also to practicalities of the interview situation – why people are talking to you, how much time they are likely to give you, where the interview takes place (formal office setting or evening drink), and not least, how well they already know and trust you. The divisions between categories are not always hard and fast; an informal interview may develop into an unstructured interview, an unstructured interview may be conducted in something approaching the semi-structured interview format, and a semi-structured interview may use fixed wording and include a large number of quite specific points, in which case it is close to a structured interview. Any kind of interview can be with individuals or with groups, although individual interviews are most common. The term 'focus group' refers to group interviews that are pre-arranged and formal, using a semi-structured approach.

Section 6.2 describes the different types of qualitative interview outlined in Box 6.1 in more detail. Section 6.3 then discusses some basic principles and techniques that are common to all forms of interviewing, and Section 6.4 describes different sources of inaccuracy in what people say and how to deal with them. Section 6.5 introduces different ways of recording interview data, and Section 6.6 explains how to build up a set of interviews. Finally Section 6.7 draws some conclusions. Structured interviews, including questionnaires and techniques used in cognitive domain analysis, are the subject of the next two chapters.

6.2 Types of qualitative interview

6.2.1 *Informal interviews*

Informal interviews are chance conversations (with individuals or groups of people) from which you may learn something of relevance to your research topic. If you live at the field site you will interact with people on the street, in the local shop, waiting for a bus or drinking in the bar, and in the course of normal conversation, a subject may come up spontaneously that is connected to what you are studying. The participants may or may not be aware of your interests; what defines an informal interview is that the conversation is in no sense designed or planned to address those interests. It is good practice in any kind of research – even natural science studies – to chat to people about what you are doing and keep an ear open for new information. However, whereas in the natural sciences this is not generally counted as part of the research design and is not included as 'data', in qualitative social science research it may be an important component of the research. Data from informal interviews should be recorded carefully (see Section 6.5) and can be invaluable in building up a sound understanding of a situation as a whole.

The great strength of informal interviews is in providing unguarded information with minimal structuring by the researcher. Often, people chat to you without any particular expectations about what you are looking for or what you might agree or disagree with. Informal interviews are particularly useful in providing information related to the background and context of the study, cross-checking information gathered by more structured methods (a technique known as cross-methods triangulation – see Section 6.6), and making you aware of issues you have overlooked or misunderstood. They are also valuable in providing information on sensitive subjects such as illegal activities or social conflicts, which people may be unwilling to discuss in a more formal situation. Finally, informal interviews can also be useful in identifying good candidates for more in-depth interviews later on.

6.2.2 *Unstructured interviews*

An unstructured interview is an in-depth conversation, usually arranged in advance, about a specific issue. Initial meetings with your supervisor or line manager or a potential collaborator to discuss a possible research project often take the form of unstructured interviews. If you turn up with no systematic list of questions or topics you want to cover, then you are doing an unstructured interview. You may have noted down a few specific points that you want to ask in addition to the general discussion, but for much of the interview you simply talk through your ideas and invite their comments and suggestions. There is no set format for unstructured interviews, and the length of the interview and the degree to which the interviewer directs the conversation vary enormously, depending on the time available, the purpose of the interview, and the social setting. It may consist of a half hour discussion in an office or an all-night drinking session in a bar.

Like informal interviews, unstructured interviews are particularly useful at the start of a research project in order to gather background information, test out ideas and orientate yourself. This includes meetings with site managers and collaborators prior to fieldwork, and these should be recorded carefully in the same way as other interviews, either by note taking or by making audio recordings (see Section 6.5). During fieldwork itself, unstructured interviews are often carried out with key individuals who have specialist knowledge or insights into the topic. For example, these may be particular people in the local community, or they

Box 6.2 Practical tips: unstructured interviews

- Arrange a meeting to discuss an issue of interest with someone who should have relevant information or insights.
- Start with a statement or question that makes it clear you are there to listen to whatever they want to say rather than to ask a specific set of questions. For example you could start by saying: 'I wanted to meet with you to hear what you think about … Tell me, what do you think are the most important issues related to … '
- The interview itself may last anything from half an hour to several hours and may be quite formal or may take the form of a social meeting.
- An informal setting will help to make the interview flow without very much structuring and therefore if the respondent has the time, meeting informally for a drink or meal is often very productive. In this case discussion about the subject of research may be interspersed with general conversation.
- For this reason unstructured interviews are not usually recorded in their entirety and notes may only be taken intermittently during the interview.
- Individuals with whom unstructured interviews are held regularly throughout fieldwork are known as *key informants*. It is important to consider whether they should be paid or otherwise recompensed for their time.

may be members of staff of institutions working at the same site (or even other researchers). Meeting with them informally to chat about the subject at some length may reveal important information of which you were unaware, and they may be able to offer explanations for unexpected findings that had not occurred to you. As your ideas develop, further unstructured interviews are useful to clarify points of information or discuss interpretations and implications, building your understanding from one conversation to the next. People with whom you do this repeatedly are known in anthropological research as *key informants* (see Section 5.3.3), and it is important to think about what they are getting back for their help; they may just enjoy the discussions for their own sake or be equally keen to learn your own views, or it may be appropriate to pay them or recompense them in some other way. Additionally, if their thinking makes a major contribution to your research, then you should come to an agreement with them about how their contribution should be acknowledged in your research report (for related ethical issues, see Section 13.4.2).

6.2.3 Semi-structured interviews

The defining feature of semi-structured interviews compared to other qualitative interviews is that they are based on an *interview guide* that is prepared in advance. If you go to your supervisor or line manager to talk about something and you have made a systematic list in advance of the points that you want to discuss, then in effect you are doing a semi-structured interview.

The points on the interview guide may be open-ended questions (see Section 7.3.2.1) or they may simply be general topics. The wording and order of the points is not necessarily fixed in advance, and the initial response to each question can be followed up with comments, prompts and further questions so that a conversation develops. Semi-structured interviews are more targeted than unstructured interviews but more flexible than questionnaires. They are most appropriate when you know what topics you wish to cover but do not know enough about likely responses to design a set of precise questions that would be needed for a questionnaire. They are also often used in place of unstructured interviews when time is at a premium (either your time, or the time the interviewee is willing to give) – for example, in interviewing a busy executive.

A semi-structured interview may last for anything between about 20 minutes and two or even three hours and it is courteous to say in advance how long you expect it to take (although how long it *actually* takes depends very much on how talkative and interested the respondent is). One way to gain a rough idea of the timing the first time you carry out a semi-structured interview on a new topic is to break down the proposed interview length according to the number of topics or questions you wish to cover, and think about how long you want to spend on each question. As a general rule, simple factual questions take less time than questions that ask about experiences, opinions or suggestions. In a one hour interview, allowing for introductions and closing conversation, you have a maximum of about eight minutes for each of six topics or four minutes for each of 12 topics. You may also need to allow five minutes to ask for some basic, factual information about the participant.

During the interview itself, introduce the first item on the guide, listen to the response, probe for more detail or ask follow-up questions as appropriate, and then when you have exhausted the topic, move on to another item. Each item may be covered in turn, or alternatively the interview guide may be used simply as a memory aid – a checklist you look at from time to time to see if there is anything that has not been covered yet – in which case there is no need to raise items that have already been covered sufficiently. When all the points on the guide have been covered, ask if the informant would like to add anything, thank them for their cooperation, and close the interview.

Semi-structured interviews are used for a range of purposes, from elicitation of information from a specific person with specialist or privileged knowledge to exploring the views of a sample of people from a particular population. The former is often the case when interviewing officials, staff of non-governmental organizations or the leader of a community;

Box 6.3 Practical tips: semi-structured interviews

- The interview guide may consist of five or six general topics or a much longer list of more specific points or questions.
- Questions should be open-ended and should be worded so as to invite detail.
- Unlike questionnaires, initial responses to each question can be followed up with further questions, prompts or comments if further information is desired.
- Where semi-structured interviews are held with specific individuals with specialist or privileged knowledge, a separate guide must be prepared for each respondent.
- Where they are held with a sample of people from a given population, the guide may either be completely standardized (fixed wording and order) – in which case it should be piloted – or may evolve from one interview to the next based on the information that has emerged.

Box 6.4 Semi-structured interview guide on villagers' perspectives on tourism, Zanzibar

Information about the informant: Date of interview:

- Name
- Age
- Gender

Questions:

1 Tell me about what you do and how you make a living.
2 What do you think brings tourists to this area? (What do they like to do here?)
3 What kinds of relationship does the village have with the hotels?
4 Can tourism help the village? (If so, how?)
5 What can the village offer to tourists?
6 On the whole, would you say tourism is good or bad for the village? (and why?)
7 If you could write some guidelines for tourists in the area, what would you include?

(Source: adapted from Slade, 2007)

their institutional role means that they should have privileged knowledge and a unique perspective. In this case, a different interview guide is often needed for each informant to reflect their area of knowledge. Where a sample of people from a particular population is interviewed, the same interview guide can be used. In some studies the points on the guide are standardized to make comparison across the sample easier; that is, standard wording is used and the points are asked in a fixed order. In order to do this the guide must be piloted before the main data collection phase (that is, it should be tested on a small number of respondents and edited as necessary until it works as intended). In other, more inductive studies, the guide may evolve from one interview to the next as different issues arise over the course of the data collection.

Boxes 6.4 and 6.5 give contrasting examples of interview guides. The first is for an interview that was carried out with local villagers in Zanzibar to explore their perspectives on tourism (adapted from Slade, 2007). Note that in addition to the main questions for discussion, there are also some initial short factual questions about the respondent. The second is an extract from a detailed interview with one specific person – the co-director of a UK-based organization called Common Ground, which has played a major role in encouraging the creation of community orchards in England (Johnson, 2008). Here, each point is much more specific and the whole interview guide consisted of 15 questions, many of which asked for factual information about the organization's activities. A pre-condition of carrying out the interview was that the researcher had read several of Common Ground's publications on orchards so that she would not need to spend time going over basic information. In both examples, the points in brackets are possible follow-up questions – only to be asked if they are not addressed spontaneously in the conversation.

Box 6.5 Extract from semi-structured interview guide with the co-director of Common Ground (http://www.commonground.org.uk)

- How did Common Ground decide to work for the conservation and creation of orchards?
- Were there any community orchards before Common Ground came into existence?
- How often is Common Ground directly involved with helping people to make decisions about their community orchards?
- Does Common Ground have any sort of network set up for orchards to contact one another? (If yes, how does it work; if no, why not and do they foresee setting one up?)
- Is land tenure a consideration in an orchard being designated a community orchard? (Is an orchard that is privately owned but open to community members for events considered a community orchard?)
- Is it more common to see parish councils or organisations or trusts managing the orchards?
- What do you see as the primary values that orchards have for people in England?

(Source: Johnson, 2008)

6.2.4 Focus groups

Focus groups are pre-arranged group interviews that usually follow an interview guide similar to those used in individual semi-structured interviews. A typical focus group involves between six and eight people, lasts for from one to three hours, and covers from four to twelve questions or topics for discussion. Focus groups are widely used in policy and marketing research and have been viewed by some academics with a degree of scepticism, but as long as they are used appropriately they are a valuable tool in the academic toolbox. Their great strength – and weakness – is that they are based upon discussion among a group of participants rather than independent statements by each individual. A successful group discussion brings out contrasting views, encourages reflection and often makes people state the reasoning behind the views they express. The researcher may need to ask only the broadest exploratory questions in order to stimulate discussion. On the negative side, focus groups present particular problems in terms of sampling because there is no clear sampling unit. They do not provide independent data from individuals, but neither do they sample a population of naturally occurring 'groups'; the groups you work with are usually brought together artificially (see Section 4.3.1 for further discussion of sampling in focus group research). Therefore focus groups are excellent in generating ideas and opinions and in revealing the reasoning behind those opinions, but poor at providing generalizations to a wider population.

The key issues in designing a focus group are how many participants it should have, how long it should last and how many questions you should aim to cover. These three issues are closely related: the more participants and the more questions, the longer the session will take. There is an almost universal fear amongst novice interviewers that they will run out of things to ask, and for this reason, a common error in designing focus groups is to be too

Box 6.6 Practical tips: focus groups

Preparation

- Recruit 'similar' people to take part in each focus group.
- There should be no more than eight people per focus group.
- If you have arranged the session well in advance, contact the participants a day or two in advance of the event to make sure they are coming.
- Keep the design simple. If you are unsure how much material you can cover, keep a few relevant but inessential topics in reserve.

Facilitation

- Lay down some ground rules before introducing the first topic – for example that everyone will have a chance to speak and that people should try not to talk over the top of one another.
- State that there are no right or wrong answers and you are genuinely interested in hearing the participants' views.
- Use an 'icebreaker exercise' to help put people at ease with one another.
- Introduce each topic and then keep your comments to a minimum.
- Stay neutral and smooth over any tensions. If two participants start contradicting one another do not take sides, but try to find a way to recognize the validity of each point of view and either encourage further discussion or move the discussion on to another point.
- Summarize the discussion at the end of each point before moving on. The summary should only be a few sentences and should recognize any differences in opinion that were not resolved.
- Do not try to take notes at the same time as facilitating the discussion. Either employ a note taker or record the session (or both).
- If recording the session, ask each participant to state their name at the start of each interjection.

ambitious in terms of the amount of material to be covered. However, everything takes a lot longer when working with a group of people than with individuals. In a two hour focus group with eight questions, then allowing for introductions and closing the session, there are about 12 minutes per question. That may sound quite a lot, but if you have six participants, then on average they will be able to talk for significantly less than two minutes each per question (allowing for your introduction to each question, your interjections during the discussion and a final summary). If the questions aim to probe attitudes in depth then this is not very much. If you are unsure how much material you can cover in the available time, a useful strategy is to keep a few relevant but inessential questions in reserve so that you can use them if necessary.

In addition, group size affects the dynamics of the discussion. If there are too few people it may not be possible to generate an interactive discussion, whereas if there are more than about eight people it is likely that a few people will dominate the discussion and the dynamics of the group will be difficult to manage.

In terms of logistics, focus groups take much more preparation than individual interviews. First, you need to put considerable effort into recruiting suitable participants. Second,

Box 6.7 Structuring a focus group session

1 Introduction/ice-breaker exercise.
2 Introduce a topic or question and 'facilitate' discussion.
3 When appropriate, close that topic and introduce the next one.
4 When all topics have been covered, thank the participants and close the session.

you need to find a venue that is suitable and arrange facilities, which usually include light refreshments. Third, you need to do everything you can to make sure that all the participants actually turn up. If they have been recruited some time in advance, it is wise to contact them again one or two days before the session to remind them and make sure they are still planning to come.

In recruiting participants the aim should be to bring together a group of people in each focus group that will interact easily. As a general rule, people talk most freely in front of others who they perceive as 'similar' to themselves – especially people they would expect to have similar views on the matter under debate. Therefore focus group research usually aims to recruit similar people within each group and put people who may be expected to have very different views in separate groups. In a study of local attitudes towards wildlife tourism, it makes sense to run separate groups for those who work in tourism and those who do not, or perhaps for women and men (since they often have different roles in tourism), or for young people and elders. Dividing the groups like this also allows for comparison between different groups of people during analysis and is therefore closely related to the research design.

The actual process of running a focus group is similar to that for semi-structured interviews but with an added set of issues connected to group dynamics. As the facilitator your job is not only to keep the discussion flowing and on subject, but also to put people at ease with one another, make sure that everyone can have their say and that no one dominates the conversation, and defuse any tensions or conflicts that develop. It is useful to start by laying down a few guidelines – people should try not to interrupt each other or talk over the top of one another – and then start with an exercise that allows them to begin to get to know each other and relax in what is probably an unfamiliar situation (this kind of exercise is known as an *ice-breaker* exercise). You could simply go around the group and ask each person in turn to say their name and a little bit about their background or about their own experiences of the subject of research. What they say can act as a starting point for the subsequent discussion. However, it is essential to make it clear how much detail you want in the initial statements or people may start off on extended monologues. One way to do this is to ask each person to talk for a specific time – say, one minute – which gives you the 'right' to stop them politely if they keep talking for too long. After the ice-breaker it may be possible simply to invite comments on what the different members of the group have said, or else you may need to introduce the first question or topic on your list. You then facilitate the discussion and introduce additional topics in turn much as you would do in a semi-structured interview. When all the points have been covered you close the interview (see box 6.7). If you need information about each participant for your records, you can get them to fill in a written form before starting the group session or after it has ended.

6.3 Interviewing techniques

We now get down to the detail of what you should be doing during an interview. At the start of any pre-arranged interview, you should greet the interviewee(s), thank them for coming, and state the aim of the interview. If the interview will be recorded, you must ask for their consent. If necessary, you should also assure them that any information that they give will remain anonymous. However it is important to match the formality with which you open the interview to the degree of structuring you wish to impose: the more formal you are, the more the respondent will expect you to take the lead in the conversation. The same goes for the setting: as a general rule the more formal the setting, the more they will expect you to structure the interview. If you arrange a one hour meeting at the office of someone you do not know then they will expect you to be well prepared and efficient, and therefore you should have a well-prepared interview guide and, if relevant, should already have familiarized yourself with some basic information about the institution they work for or the field site you want to discuss. On the other hand if you arrange to meet someone in a bar and tell them you want to discuss a particular topic with them, they are more likely to take an active role in directing the conversation. They may also expect to spend some time relaxing socially, and may even bring along a friend or colleague. Obviously in this situation it would be inappropriate to be too formal, and you should be ready to be flexible. In fact in any interview, engaging with the respondent as a person will make things run much more smoothly and is a major factor contributing to success (see also Section 12.5). Even in a formal interview you should make eye contact, treat them with courtesy, and be prepared to chat a little when you meet them or at the end of the interview, unless they make it clear that this is unwelcome.

In all qualitative interviews, the first substantive question should be broad and uncontroversial and you should narrow down the subject area as the interview proceeds (Box 7.17 and Section 7.5.1 present some other practical tips about the order of the questions). If you start with a string of questions that invite short answers, it will be more difficult to get the respondent to talk at more length later on. One good way to start is to invite the informant to tell a story about themselves or about something that happened in the past; if you do this it makes it quite clear that you do not just want short answers, and can set the tone for the rest of the interview. The aim at this stage is not only to elicit information but also to put the interviewee at their ease and start the conversation flowing; then as the interview progresses and the respondent becomes more at ease you are likely to get a better response to more difficult, thought-provoking or sensitive issues.

However, whatever you ask first, people's responses to the first question are likely to vary in length from a few words to a long, rambling statement full of irrelevant information, depending on how you word the question, how at ease the interviewee feels, whether they have a strong view on what you have asked them and whether the subject matter is sensitive, and simply whether or not they are a natural 'talker'. Perhaps the most important skill in interviewing is to encourage people to open up and give you the level of detail you want while keeping them on subject. A response of a few words puts the burden back on you to elicit more detail, whereas a long rambling statement may eventually require a (courteous) interruption. The rest of this section describes some techniques for dealing with either of these cases.

First, think carefully about how you word the first question. Short, specific questions are likely to elicit short, specific answers. Longer questions give respondents time to think, remember, and consider their answer, and have been shown to improve reporting of past events (Cannell and Marquis, 1972, cited by Fowler, 1995: 23). If you want a detailed answer, use longer questions that make this clear. Compare 'Tell me about how you became involved' to 'how did you get involved?' The former is more likely to give an in-depth response; it invites the participant to tell you a story rather than just give a quick answer to the specific question. As a general rule, avoid simple 'how', 'what', 'why' questions if you want in-depth answers. If appropriate, preface your question with some background information or comments. Alternatively, ask several questions about the same thing, worded in slightly different ways (but not to the extent that the interviewee gets annoyed). It gives respondents time to think about the issue or try to remember details and suggests that you are particularly interested.

Second, make use of silence. If you leave a few seconds' gap, the respondent is likely to say something more without any prompting from you. Resist the feeling that you must have something to say the moment they stop speaking; if you find this difficult, try counting slowly up to five before you say anything. However the silence should not be so long that it becomes awkward. There are cultural differences in what is acceptable, so if you are working in a culture that is not your own, try to observe how people interact in everyday conversation so that you can fit in with the cultural norms. Box 6.8 lists some additional neutral probes that can be used to elicit extra information without leading the conversation in a particular direction (see Bernard, 2006, for more detail). It is surprising how well simple probes of this kind can work.

In addition to neutral probes, you may need to ask specific follow-up questions to what the respondent has said. If they use terms that you are unfamiliar with or if there are points you did not quite understand, ask for clarification. If they mentioned something that you would like to know more about, ask specifically about that point. Ask extra questions on events or items in a category that are likely to be overlooked. For example, if asking about 'holidays', ask specifically about weekends away to make sure that these are recalled as well as longer holidays. Finally, if trying to find out when an event happened, ask about preparations for the event, follow-up and consequences. Alternatively, especially for events in the distant past, ask about family circumstances or other aspects of daily life that might be easier to pin down to a year *(Did you already have children? Were they at school? Where were you living then? Which job were you doing then? Was it before or after your father died?)* Keep using probes and follow-up questions as necessary until you have enough information or there seems to be little more to say. Then move on to the next question or topic on your interview guide that has not yet been discussed.

Box 6.8 Some useful neutral probes

[silence]
Tell me about…?
That's interesting – Tell me more about that
Mmmm…
In what way?
I see – so your brother got you involved. Can you tell me more about how that happened?… [Repeat back last few words (mirroring) or summarize key points, and ask for more detail.]

The interview usually becomes easier as it goes on, but if this does not happen and the answers continue to be short and awkward, do something to try to put the informant at ease. If necessary, break off and chat off the subject for a couple of minutes. Move on to a different topic in case what you have been asking about is sensitive; you may be able to come back to it more successfully later on. If nothing works it is not necessarily your fault; there may be some hidden reason why the person doesn't want to cooperate, or alternatively they may genuinely have very little to say on the subject you're interested in. They may even be someone who never says more than a few words at a time to anyone about anything, in which case they will never make a good informant!

At the other extreme, what should you do if you ask an opening question and the interviewee is still talking 20 minutes later? The answer depends on how much time you have, whether what they are saying is relevant to the study, and whether you will have a chance to interview them again if time runs out before all the planned points have been covered. It also depends on how long you can concentrate for and whether you can remain patient – stop the interview at once if you are getting irritated. If you have all evening ahead of you and will be able to talk to the same person at another time if necessary, then there's an argument for just sitting back and listening, even if much of the material is irrelevant. You will get to know each other and they will probably be more open with you in subsequent meetings. However, if time is limited you may have to interrupt them – politely – and bring them back to the subject with a more specific question. A useful way to do this without causing offence is to say something like 'that's very interesting, and I'd like to come back to it another time. But first, can I ask you ... '.

You should finish the interview when you have enough information, when the conversation begins to flag, or when either you or the interviewee is losing concentration. You can always make an appointment for a second interview if necessary. Announce that you would like to stop, ask the participant if there is anything they would like to add or to ask, thank them for their time, and close the session.

Box 6.9 Practical tips: interviewing techniques

- Match the way you arrange the interview, the setting and the way you open the interview with the level of formality and structure that you want.
- Start with something broad, uncontroversial and interesting that invites a lengthy answer; for example, invite them to tell a story of something that has happened.
- Word questions in a way that invites detailed answers. Avoid direct, simple questions beginning with 'what', 'how', 'why'.
- Do not assume knowledge (unless interviewing a specialist). Avoid technical terms and very formal language.
- Make use of silence and other neutral probes.
- Ask about anything that is unclear, and ask for further details of anything they say that is particularly interesting.
- If you need to stop them, do so politely.
- Stop the interview if either you or the informant becomes tired, irritated or bored.
- After the 'formal' end of the interview, invite informal conversation as you prepare to leave. 'Off the record' comments at this stage can be very revealing.

One final hint: some of the most interesting statements are often made *after* the 'end' of the interview when the recording device has been switched off and the notebook has been put away. It is amazing how many people will say something at this point along the lines of 'off the record, actually I think ... ' – especially if they were being interviewed in their professional role (for example, as a government official or NGO employee). Note down what they have said the minute they have left. You will not be able to quote them directly without their permission, but you may be able to work the substance of their comment into your report in some other way, or explore it in further interviews with other people (without mentioning the source).

Interviewing is something of an art. If your first interviews go badly, review what has worked and what has not worked and try to work out why, but do not assume it is all down to you. Even the most experienced interviewers sometimes fail to make an interview 'flow'. The truth is that some people will talk for hours at a time with minimal prompting whereas others are almost impossible to move beyond answers that take the form of single words and short phrases. Similarly some people will keep to the point, organize their thoughts as they speak, and explain any terms they think will be unfamiliar to you, whereas others will ramble on, assume you know as much as they do, miss out vital pieces of information and contradict themselves. However, do not be daunted by this – many people find that interviewing comes naturally, and it is possible to carry out a perfectly respectable study within the confines of an undergraduate project.

6.4 Sources of inaccuracy and strategies to deal with them

A common concern among biologists when confronted with interview data is 'but how can we believe what they say?' They are quite justified: according to Killworth and Bernard (1976, in Bernard, 1994: 234) between one-quarter and one-half of what people say about their behaviour is inaccurate. There are many different reasons why this may be the case. However, there are various tools for reducing the level of inaccuracy – discussed in this section – and for cross-checking between different sources in order to assess what inaccuracies may remain (see Section 6.6).

Errors can be non-directional or directional (biases). Box 6.10 gives some of the common reasons for non-directional errors. Often people make a genuine mistake: they simply misremember or do not know the answer. Most people are uncomfortable at repeatedly answering 'I don't know' and therefore if you ask a lot of questions that put them in this position, they may start guessing or even make things up. Try not to ask strings of questions that ask point blank for specific points of knowledge, and if appropriate, make it clear at the start that 'I don't know' is an acceptable answer. If someone obviously cannot remember something, try to jog their memory by repeating back what they have said so far. However, there comes

Box 6.10 Some common causes of non-directional errors in what people say

- Memory error and lack of knowledge.
- Generalising between events.
- Reports of what they suppose happened rather than what they actually saw.
- Distortion of memory to conform to their own prejudices.

a point where you must accept that people really do not know. If you probe too hard you will only elicit false answers.

Another common tendency is for people to generalize between different events or issues, or report what 'people' say rather than anything based on their own experience. I am amazed how every year when our students hand in their course evaluation forms, a significant minority rate or comment on a session they did not attend. When questioned about this, they usually say either that they assume it was similar to other sessions that they did attend, or that they are basing their answers on what another student has told them. Obviously the information they provide on these bases is not a reliable source of feedback.

In addition to non-directional errors, responses may be directionally biased. Bias can be affected by the question wording or interview structure; by the interviewee's preconceptions about what is the 'right' or 'good' answer; or by the wish to give a particular impression to whoever is listening – which includes you, the interviewer! In a minority of cases, participants may consciously try to hide the truth or give misleading information. Box 6.11 lists some of the main types of response bias together with strategies for how to minimize them.

Box 6.11 Types of response bias

Yes effect: Most people will say yes to a neutral question.
Strategy: Avoid yes/no questions or ask the same thing more than once, changing the wording to reverse the likely bias.

Self-esteem effect/deference effect: People tend to give answers (consciously or unconsciously) that reflect well on themselves, either to maintain their own self-esteem or to appear in a good light to others.
Strategies:

- Emphasize that you are not looking for 'right' answers but are genuinely interested in what the individual thinks.
- Get to know respondents and put them at ease.
- Remain neutral: avoid showing disapproval.
- For sensitive issues, if necessary try asking what 'people' do or think rather than what the participant does or thinks.
- If necessary use research assistants who are less likely to elicit a response bias – for example people who are similar to those being interviewed (locals of the same gender, ethnicity and so on) or people who are not perceived to be associated with conservation organizations (if asking about attitudes to conservation).

Order effect: People's answers later on in the interview will be influenced by what has been discussed up to that point.
Strategy: Start broad and narrow down. This is known as the 'interview funnel', and will be discussed further in Chapter 7 as it is particularly important in designing questionnaires.

Audience effect: People may answer differently according to who is listening.
Strategy: If you suspect that the 'audience' is having an effect, try to catch the interviewee alone on another occasion.

There is one more technique that you can try if you are finding it impossible to break through a wall of politically 'correct' answers' and that is to break the rule of keeping your questions neutral. *Directed probes* are probes that suggest certain answers. If everyone tells you how wonderful conservation is and you suspect that they are just saying what they think you want to hear, one way to try to find out their true views is to make some outrageous statements in the other direction – 'so I hear your lives have been really messed up by conservation activities around here'. If they really are positive about conservation they will contradict you passionately and give you examples to prove their point; if this is not the case their objections may be more muted, or they may now agree with you. You should use this technique with extreme care and only after you have exhausted more neutral approaches – after all, you are leading your witness. A short answer expressing agreement or disagreement should not be relied upon, but once you have expressed a strong view against what people have been saying to you, they may open up and discuss it more freely.

6.5 Recording data from interviews: note taking and audio recordings

In informal interviews, the participant is often unaware they are being interviewed and it is regarded as unethical to record them secretly (see Chapter 13). Therefore it is common practice to wait until you have parted company and then – as soon as you can – write down everything you can remember in a pocket notebook that you should keep with you at all times. In all other kinds of interview, either note taking or audio recording (or both) during the interview itself is the norm. Recording is the only way to make a complete, accurate record of the interview, but it does have drawbacks. It may make the interviewee much more guarded and self-conscious, and each recorded interview takes considerable time to process. Note taking is less intrusive but less accurate; it is very likely that there will be some errors and there will always be omissions. The following two sections discuss note-taking and recording in more detail.

6.5.1 Note taking and initial processing of notes

The first time you try to take notes it may seem like an impossible task. Every time you look down at your notes you must break eye contact with the informant, and it is hard to listen to what they are saying while writing notes on their previous statement. However, with a little practice it quickly becomes easier – your memory improves and you get in the habit of picking out key points as you listen. You should also develop your own 'shorthand' – standard initials and abbreviations for commonly occurring terms. If necessary it is acceptable to ask the interviewee to pause occasionally so that you can catch up with your notes – but not for so long that the thread of conversation is lost.

Your notes will normally consist of summaries or paraphrasing of most of the conversation, with occasional verbatim (word for word) quotes of particularly pertinent phrases, and notes of specific facts and figures. You should also write down your own follow-up questions or comments where these direct the conversation. Mark your notes so that you can recognize these different types of information. I enclose direct quotes in quotation marks, use the first person (but no quotation marks) for close paraphrasing of specific statements, and the third person for my own summaries. I also use square brackets to mark my own questions and comments. The aim of note taking is to record data, so do not process the information any more than necessary as you write it down; use the same vocabulary as the informant

Box 6.12 Practical tips: note taking

- Before you introduce the first topic, write down the date, the name and/or role of the person interviewed, and any other general details that are relevant.
- While you are making notes, make sure you continue to engage with the informant – make regular eye contact and keep the conversation flowing.
- Do not try to write everything down verbatim (word for word). Most of the time it is sufficient to summarize or paraphrase.
- Develop your own 'shorthand' – standard abbreviations – to save writing time.
- Do write down particularly pertinent phrases verbatim that may be useful as quotes. Put them in quotation marks so that it is clear when you have done this.
- Also note down your own follow-up questions and comments if they lead the conversation in a particular direction. Make sure they are easily identifiable – for example by putting them in square brackets.
- Also write down facts and figures in full.
- Feel free to ask the informant to wait for a moment if you need to catch up with your notes.
- Check through your notes as soon as possible after the interview and 'clean them up' so that they will make sense if you look at them a month or more later. If necessary, write out a 'clean' version, adding any extra detail that comes to mind as you do so.

(including colloquialisms or slang) and do not try to reorganize the information into a more logical sequence at this stage. Processing comes later on.

The extent to which you take notes depends both on your own experience and confidence in doing so and also on the level of formality of the interview. Taking notes inevitably makes the situation more formal, and the more intensively you do so, the more this is the case. Therefore the need to record the conversation accurately must be balanced against the effect note-taking will have on the 'feel' of the interview. Notes are not usually made in informal interviews until after the interview has ended; it could be extremely off-putting to the interviewee if you suddenly start writing things down in what, to them, is a casual conversation. In unstructured interviews, the level of note taking varies enormously, depending both on the social setting and on how much new information the interviewee gives you. I often begin an unstructured interview without taking out my notebook, but if the interviewee begins reeling off facts and figures or starts describing something in detail then I will ask 'Do you mind if I just note that down?' and get the notebook out. No one has ever objected. Thereafter we can continue to talk informally, and whenever anything comes up that I want to note down, I can do so without interfering with the flow of the conversation. In semi-structured interviews, note taking needs to be more continuous because the conversation should be more concise and directed. Indeed, if you begin by taking notes intensively then the interviewee may become concerned if you suddenly stop, and assume that you are not interested in what they are saying. In this situation, note taking can actually encourage them to say more rather than interfering with their flow.

As soon as possible after the interview you should read through your notes carefully. It is common for inexperienced interviewers to find that they have missed out key words or links in the conversation, or simply that they cannot read their own handwriting. If you check for this quickly you should be able to correct the notes and fill in most of the gaps from memory, whereas if you leave it for a couple of days the whole interview may be useless. I follow a rule that I must check all interview notes on the same day as the interview. Unless they are already full and clearly presented I write out a 'clean' version that should be understandable weeks or months later without relying on memory. That means neat handwriting (or typing into a computer), minimal abbreviations and so on. It also often results in much more detail than the original notes because as you go over them you remember points that you did not have time to note down fully during the interview itself.

6.5.2 Audio recording and initial processing of audio recordings

Recording can be used either alone or together with notes in order to reach a balance between accuracy and efficiency (see Box 6.13). The most appropriate balance depends both on practicalities and on the level of analysis you intend to carry out. If you rely on a recording alone, bear in mind that transcribing a one hour interview in full (that is, writing it out word for word) may take three to five hours. Another option is only to transcribe selected sections that you identify either by listening right through the tapes or, if you take at least brief notes during the interview, by referring to the notes to look up what subjects were covered in what order in each interview. Perhaps the most efficient approach, if the informant is comfortable with it, is to take full notes but also to make a recording so that you can use it to check what you have written and transcribe quotes in full.

In focus groups, the data should include not only what was said but also who said it. If you are recording, ask people to say their name at the start of each statement so that this information is not lost (this is not as intrusive as it sounds once people get used to it). If you

Box 6.13 Options for processing of audio recordings

Recording alone

Full transcription (writing out the interview word for word): Allow three to five hours per hour of interview.

Indexing, summarizing or transcription of different sections: Index the subjects that are covered in the recording and for each relevant section either write a summary or if it is really key to what you are researching, transcribe it. Also transcribe phrases that may be useful as verbatim quotes.

With accompanying outline notes: Use the notes as an index so that you can find different topics easily in the recording. Summarize or transcribe different parts of the interview as needed.

To complement extensive notes: Use the recording to check any sections where your notes are unclear, add more detail as necessary, and transcribe phrases that may be useful as verbatim quotes.

As a backup to extensive notes: If your notes include everything you need, you may not use the recording at all, but it is useful to keep it as a backup in case you need to check some points later on.

rely on note taking you should employ an assistant to make the notes rather than trying to do it yourself while facilitating the discussion.

Finally, a warning: make absolutely sure you are familiar with the recording equipment well before you use it for your first interview. Almost every researcher has stories of interview data that was lost because the recording failed – for example, because the microphone was not switched on, the batteries were low, there was too much background noise, or they accidentally recorded over the top of it. If you do rely on a recording then check the recording has worked as soon as possible after the end of the interview. That way, if not, you should still be able to remember enough detail to write some useable notes.

6.6 Building up sets of interviews: triangulation and research design

It often happens at the start of field research that you talk to someone who seems extremely knowledgeable and gives you a wealth of useful information. You feel you have learnt a lot and already answered some of your research questions. Then you interview the next person and they tell you something totally different, and you are back where you started. This can be very disheartening the first time it happens, but it is a normal part of the research process. Just as in the natural sciences you cannot take the first measurement as absolute, so in the social sciences you cannot take the data from the first interview as correct or accurate. You must check the information by collecting more data in other ways and from other sources. This process is known as triangulation.

In geography, triangulation is the process of taking bearings on a distant object from two or more different locations. It is then possible to calculate the location of the object with a high degree of accuracy, using the geographical coordinates of the locations, the compass bearings and the point at which the lines of sight cross one another. Similarly in the social sciences, triangulation involves 'taking bearings' on a certain item of information from different perspectives. If you ask someone the same question on two occasions, do you get the same answer? Do different people say the same thing? Does the information from informal conversations agree with what people say during formal interviews or focus groups? Box 6.14 lists the different types of triangulation. The underlying principle is that the greater the number of different sources that give the same answer, you more trust you can put in the findings.

One very effective form of within-subject triangulation is to conduct a second interview with the same person and check your notes back with them – either through further questions

Box 6.14 Types of triangulation

1 *Within-subject triangulation:* check with the same person more than once.
2 *Between-subject triangulation:* ask several different people about the same thing.
3 *Cross-method triangulation:* use different methods.
4 *Cross-researcher triangulation:* where interviewing is carried out by several research assistants, check for consistency by getting two or more of them to take notes on a single interview or by sitting in on interviews or holding review sessions.

and probing or by reading out a summary constructed from your notes on the first interview and asking for comments and corrections. I did this in a study of mobility among rural communities in northern Amazonian Peru. During a second round of interviews, when I read back people's accounts of their lives – where they had lived, what work they had done – very often they would interrupt to correct a date or the name of a person they had mentioned, or to add further details ('but there's something missing there – before I did that I did this ... '). Thus the second interview served both to catch inaccuracies (both in what they had said and in what I had noted down) and also to provide more detail that had not been mentioned before. The fact that there was a gap of weeks or in some cases even months between the two interviews gave people time to think about what they had said and bring up further memories. It is also possible to use this method electronically; for example, after interviewing officials from non-governmental organizations about fair trade issues I sent each of them a summary by email of what I understood them to have said and asked for comments. Again, this elicited not only corrections but further factual details and more considered opinions.

Triangulation is at the heart of building up and analyzing a set of interviews and is closely connected to the concept of *saturation* in qualitative sampling, which is used to determine when you have gathered enough data (see Section 4.3.1 and Box 4.9). In much inductive research triangulation must be done throughout data collection. This process of checking for consistencies and inconsistencies between different sources allows you to follow up on what you find from one day to the next. In more deductive research where the aim is to carry out a set of standardized semi-structured interviews, triangulation is less important during data collection itself, but even then, it is still useful in identifying points that you need to clarify or pay special attention to in future interviews. The role of triangulation during data processing and analysis is discussed further in Chapter 14.

Triangulation is most relevant for factual information. It would be ridiculous to cross-check people's attitudes and opinions according to consensus and interpret inconsistencies in terms of inaccurate data; different people have different attitudes and this is often of interest in itself. Even differences in factual accounts may be interesting in revealing different perspectives and alliances. During the study in Peru described above, I asked people to tell me the history of their community and its relationship to the nearby protected area. Most people were cooperative but inevitably there were differences between their accounts. Through various forms of triangulation – further interviewing, group discussions and, where possible, checking against archival records – it was possible to resolve the major factual discrepancies, but some points of detail continued to be disputed. For example, accounts differed subtly with respect to the roles of local people versus external NGOs in initiating the creation of the reserve, what proportion of local people supported it, and *why* they supported or opposed its creation. Far from being problematic, these differences were useful in building an in-depth understanding of the multiple perspectives of local people on the reserve's creation and continued existence, which revealed different subgroups within the community reflecting different social and historical factors. The lesson is that inconsistencies are not always a bad thing; they may actually be a useful source of information. What is important is that you explore inconsistencies thoroughly and try to understand what is causing them.

A final situation in which triangulation is particularly important is when you are dealing with sensitive issues, such as illegal activities. Suppose you want to find out why there has been an increase in the number of gorillas killed by local people at a particular site. The first person you interview may tell you that it is not local people who are doing the killing, but unscrupulous traders who are funded by international backers. However, as you ask more people you may get some very different stories – some intentionally deceptive, others based on rumour, and others from people who know what is going on and really want to tell you. In this case you need to explore the issue further until you understand why there are discrepancies. Be tactful in how you do this – it would be highly inappropriate simply to say 'but so-and-so told me the opposite of what you have just said'. Instead you probe, observe and triangulate further until you are confident that you know what your data represent. If the story about unscrupulous traders is just that – a story, designed to put you off the right track – it is very unlikely that you will hear it consistently from several people. If, in contrast, a lot of people tell you the same thing, it is *more* likely to be true. This reliance on consensus is by no means failsafe because unless the people you interview are strangers to one another they are not entirely independent in what they say – they talk to each other! A high level of consensus could reflect a widely repeated story, rumour, or concern to say the politically correct thing, but still be wrong. Ultimately, the only way to know the truth for sure is to observe people killing gorillas, but this is not usually possible. Interviewing people and listening to what they say to each other in casual conversation is the next best option (although see Section 13.4.1 for ethical issues in investigating illegal activities). The more time you spend in the field, the more likely you are to find out what is really going on, and studies of illegal activities are therefore not generally suitable for short-term research projects.

6.7 Conclusion

If you feel overwhelmed by the apparent complexity of interviewing, do not be. Qualitative interviews are simply a formalized version of everyday conversation, and once they take the plunge, most people quickly develop their interviewing skills. Taking notes while maintaining the conversation is certainly challenging, but most people soon develop their own technique and if you have trouble in doing so, there is the option of making an audio recording instead. Triangulation requires a similar skill to essay writing; it involves comparing and synthesizing different sources of information. Focus groups are only really suitable for postgraduate or higher-level research as they take considerable time, resources and skill to do well. However, a small number of qualitative interviews can form a significant and very rewarding component of an undergraduate thesis.

Summary

1 Different kinds of interview are classified according to the degree of structure imposed by the interviewer; thus there are informal, unstructured, semi-structured and structured interviews.
2 Informal, unstructured and semi-structured interviews are generally regarded as 'qualitative' and are usually analyzed qualitatively. Structured interviews include questionnaires and interview formats used in cognitive domain analysis, and are usually analyzed quantitatively.
3 Informal interviews are casual conversations; unstructured interviews are conversations arranged in advance to talk freely about a particular subject, and semi-structured interviews are more formal conversations arranged in advance that follow a pre-defined interview guide.
4 Focus groups are pre-arranged formal group interviews that use a semi-structured approach.
5 In carrying out interviews, it is important to match the level of formality with which you arrange and begin them to the degree of structure that you wish to impose.

6 Always engage personally with the interviewee – the more you do so, the more they are likely to speak freely.
7 Begin with a broad question and narrow down the topic as the interview proceeds.
8 Use prompts and follow-up questions to elicit further information.
9 Look out for potential response biases and try to word your questions so as to minimize them.
10 Finish the interview when you have enough information, when the conversation begins to flag, or when either you or the interviewee is losing concentration.
11 In all forms of interview other than informal interviews, either note taking or audio recording (or both) during the interview itself is the norm.
12 Notes should consist of summaries or paraphrasing, with occasional verbatim quotes and lists of facts and figures as appropriate. Notes should be checked and 'cleaned' as soon as possible after each interview.
13 Recording can be used either alone or together with notes in order to reach a balance between accuracy and efficiency.
14 Information from different interviews should be reviewed regularly using the principal of triangulation. Triangulation can be within-subject, between-subject, cross-method or cross-researcher.
15 Focus groups are only really suitable for postgraduate or higher-level research as they take considerable time, resources and skill to do well. However a small number of qualitative interviews can form a significant component of an undergraduate thesis.

Further reading

Bernard, R. (2006) *Research Methods in Anthropology*, 4th edn, Walnut Creek, CA: Altamira Press. [The seminal textbook on anthropological research methods. Chapters 9 and 10 give a detailed and very readable account of qualitative interviewing techniques, full of examples.]

Bryman, A. (2004) *Social Research Methods*, 2nd edn, Oxford: Oxford University Press. [Chapter 16 gives a comprehensive introduction to focus group research that goes into more detail than is possible here.]

Harper, D. (2002) 'Talking about pictures: a case for photo elicitation', *Visual Studies* 17(1): 13–26. [A stimulating article introducing the use of photographs to elicit information in qualitative interviews.]

Morgan, D. and Krueger, R. (1998) *The Focus Group Kit*, Thousand Oaks, CA: SAGE Publications. [Written in the style of an instruction manual, this 'kit' consists of six slim volumes, each of which deals with a different aspect of running focus groups. It is certainly not concise, but is full of useful tips on practical as well as theoretical issues.]

Spradley, J. (1979) *The Ethnographic Interview*, New York, London: Rinehart & Winston. [Gives an in-depth account of in-depth interview techniques from an anthropological perspective.]

7 Questionnaires

7.1 Introduction

Questionnaires consist of a series of specific, usually short questions that are either asked verbally by an interviewer, or answered by the respondent on their own (*self-administered*). The principle behind questionnaire design is that each question should elicit information on a specific, quantifiable *variable* that you have defined in advance (see Box 7.2). Each question is asked in the same way of each respondent. The interviewer, if there is one, may clarify points when the respondent has not understood properly, but other than this there are no prompts and no follow-up questions. This is the definitive difference between a questionnaire and a qualitative interview: a questionnaire is not a conversation, but a question-and-answer session. In fact it is not only the questions that are standardized but also most of the answers: the majority of questions are *closed*, which means that the respondent has to choose from a set list of possible answers rather than answering in their own words. The questions are also asked of different respondents in a fixed order and as far as possible, under the same conditions. The advantages of this level of specificity and standardization are that (1) you can gather highly targeted data; (2) you can compare the responses of different people directly; and (3) you can easily carry out statistical analyses to look at patterns of variation in the results.

Questionnaires are probably the most widely used social science method in conservation (for a review of their use, see White et al., 2005). They are the most structured of the range of social science methods and therefore give the researcher the highest level of control over the form and content of the data collected. This, and the fact that they can easily be used to generate quantitative data, means that they fit easily into the deductive, quantitative approach to research that is dominant in the natural sciences. This is both their greatest strength and their greatest weakness. It means that provided that you know exactly what you want to ask and what range of answers are likely to be given, you can collect very precisely targeted data. The fact that questions are standardized across respondents simplifies the process of summarizing across populations and increases the validity of comparisons between different groups of people. In addition, the fact that most answers are picked from pre-defined lists of options makes it easy to quantify the results for statistical analysis. Questionnaires are also quick to administer to a large number of people – especially people who are spread over a large geographical area – and when self-administered, they allow complete anonymity, which is particularly valuable when dealing with sensitive issues. In many ways, then, questionnaires have major advantages over other social science research methods. However, designing a valid questionnaire is not simple; it requires careful thought and attention to detail. The reason why it is important to standardize the wording of the questions (and potential responses in closed questions) is because the exact wording will have an effect on the way people respond. The wording must be unambiguous, socially and culturally appropriate, precise enough to let the respondent know *exactly* what kind of information you are asking for, and as neutral as possible. Therefore questionnaires are not appropriate unless the research objectives and questions are very specific (so that you can identify the precise items of information on which you wish to collect data), and unless you know enough about the background and the social and cultural context to make sure that the individual questions are appropriate. A last potential problem with questionnaires is that in many parts of the world, people associate them with government surveillance or some form of outside interference and may therefore give very guarded or outrightly misleading answers. Box 7.1 summarizes some of the advantages and disadvantages of using questionnaires.

This chapter describes the process of designing and administering a questionnaire. Section 7.2 outlines how you should define exactly what factors or topics should be included in the questionnaire in order to address your research objectives. Section 7.3 discusses how to draft the actual questions to include on the questionnaire and describes the common question formats. Section 7.4 explains how to code the answers to different types of question; this involves allocating standardized numerical values to each possible answer. Section 7.5 is concerned with how to assemble the questions into a questionnaire, and Sections 7.6 and 7.7 discuss how to review, pilot and administer the questionnaire (collect the data). Finally, Section 7.8 draws some conclusions.

First a note on formatting in this chapter: in boxes giving examples of questions, *italics* are used for all text that would actually appear in the questionnaire, while [regular text in square brackets] is used for my annotations and comments in the context of this book.

Box 7.1 Some strengths and weaknesses of questionnaires

Strengths

- Precise targeting of data collection to match research focus.
- Fixed wording allows comparability between respondents.
- Fixed answers allow easy quantification for statistical analysis.
- Good at characterising populations and testing for relationships between variables.
- Quick to administer.
- Self-administered questionnaires allow for anonymity.

Weaknesses

- Results are sensitive to precise question wording.
- Must be well contextualized or may be off target.
- No flexibility – cannot make changes once data collection has started.
- Poor at providing in-depth description and understanding.
- In self-administered questionnaires, there is no way to clarify misunderstandings or correct errors by the respondent.
- Response rates are often low.
- May be inappropriate in some cultures.

7.2 Defining what you need to ask about

7.2.1 Defining the subject matter

Rather than plunging right away into drafting questions for the questionnaire, it is important first to make a careful list of precisely what information you wish to collect. Otherwise you are likely to end up with a questionnaire that is broadly 'about' the research topic but is not well targeted to address your aims and objectives.

Questionnaires are often discussed in terms of four different types of information:

1. Information on respondent *attributes* – basic characteristics of the respondent.
2. Information on knowledge – including both abstract knowledge and memories or experiences.
3. Information on behaviour.
4. Information on subjective states such as attitudes.

Respondent attributes usually include basic socio-demographic factors such as age, gender, education or income level. These may be of interest in their own right, or they may be used in analysis as *independent variables* (see Section 7.2.2 and Box 7.2) to test for significant differences in knowledge, behaviour and attitudes among different types of respondent.

Box 7.2 Key terms: operationalization, variables and coding

Operationalisation: The process of developing a definition of a concept in terms of one or more variables, each with a known set of possible numerical values.

Variable: 'A concept which can have various values, and which is defined in such a way that one can tell by means of observations which value it has in a particular occurrence.' (Stinchcombe, 1968: 28–9, cited in Blaikie, 2000: 134)

Examples of variables:

- age in years;
- gender;
- safari lodges you have visited;
- attitude towards ecotourism on a scale of 1 to 5.

Constant: A concept that has only one possible value (for example, the freezing point of water under standard atmospheric conditions).

Independent or *explanatory* variable*:* A variable that you suspect may have an effect on the dependent variable – the potential cause.

Dependent variable*:* A variable that you suspect may be affected by the independent variable.

For example, for the null hypothesis 'there is no difference in attitudes between men and women', the *independent variable* would be gender and the *dependent variable* would be a measure of attitude.

Coding: The process of allocating numerical values to each possible response to a given questionnaire question when the responses take the form of words.

They may also include more specific attributes that are particularly relevant to your study (such as membership of a conservation organization, vegetarianism, or whether or not the person has been on a safari holiday). Sometimes one or more questions of this kind are used to screen respondents – to see whether they are suitable candidates to answer the entire questionnaire. For example, if the questionnaire is about people's experiences of safari holidays, then it will be inappropriate for anyone who indicates that they have never been on a safari holiday.

Other kinds of information are more specific to each particular research project, and may concern the respondent's factual knowledge, behaviour, and/or subjective states. Factual knowledge includes both abstract knowledge and knowledge based on experience, such as memories of events. Information on people's behaviour is also factual, and concerns what people do, how they do it, with whom and how often. Information about subjective states can include their beliefs, emotions, opinions, attitudes and moral values, and so on. It is different in kind from other types of information in that there is no objective 'right' or 'wrong' about the answers, although a particular respondent may give an answer that reflects their own view more or less accurately. The information of each of these kinds to be included in the questionnaire must be identified by continuing with the process described in Chapter 2 of narrowing down the objectives or questions in the research design.

7.2.2 Developing a list of variables

Each item must be defined specifically enough so that it can be broken down into one or more variables on which data will be collected. This process is known as *operationalization* and produces an *operational definition*, which is far more specific than a *conceptual definition* of the type described in Section 2.3.3. In a separate but related step, one or more questionnaire questions are then designed to gather data on each particular variable. Thus the questionnaire questions are *not* the same as the research questions, but must be much more specific.

In the case of closed questions, each possible response to each question must be assigned a numerical value – a process known as *coding* (see Section 7.4). This process is fundamentally similar to the coding used in qualitative analysis (see Chapter 14) but it is far more specific and the codes *always* take the form of numbers. In some cases, it is straightforward. For example 'age' is usually expressed as a number (age in years) and therefore no further narrowing down is needed. In cases where the answers take the form of words rather than numbers, each possible answer can be assigned a number for analysis purposes. For example, 'gender' is usually treated as a variable with two values, male 'M' and female 'F', which can be represented as M = 1 and F = 2.

Coding is closely linked to the exact format of the questions, and as you develop the precise questions and the codes, you may find it necessary to go back and make fine adjustments to the operational definitions or even to some of the objectives and research questions you developed earlier in the research design process. Therefore, while it is best to develop a list of information you wish to collect before you get embroiled in the detail of the question wording, it is usually necessary to go back and forth between these three steps – defining what information to include, developing the questions and coding.

7.2.2.1 Complex concepts: developing composite indices

Some concepts or items of information cannot be expressed as a single measurable variable, and therefore in order to operationalize them it may be necessary to break them down further into several different subfactors. Suppose you are interested in people's use of natural resources. One of your research questions might be 'How reliant are people in community x on their local natural resources?' This is a question about quantity – how much – but a moment's reflection should tell you that it is not appropriate as a questionnaire question. If you asked people directly 'How much do you rely on your local natural resources?' you would get answers in many different forms and reflecting many different assumptions on their part about what you are asking. Their answers might reflect how they interpret the concepts of 'use' and 'local natural resources': they might only think about direct use, or aesthetic values, or ecosystem aspects such as water, shade or fresh air. Similarly they might base their answer on specific resources such as fish or meat and not think about medicines or building materials. In addition, they might make different assumptions about what you mean by 'how much' – for example, their answers might reflect how *often* they use natural resources or how many *types* of natural resource they use or how much *time* they spend harvesting natural resources or what proportion of their daily needs is met from local natural resources versus those that are bought in (probably the nearest interpretation to the implicit sense of the original question). If you do not define your terms and specify exactly what you want to know, then when you come to analyze the data you will not have a clue what they mean – and different respondents may mean different things.

One way to deal with this is to break the concept of resource reliance down into a set of far more specific variables, each of which acts as an *indicator* of resource use. You could do this by making a list of different types of use or different types of resources, and then develop questions on each of these. However it would be impossible to make a *complete* list, and therefore you must either develop a list that you think is adequate to reflect the original concept sufficiently, or you must narrow down the original concept. During analysis and writing up, the results can either be reported separately for each variable or they can be combined across variables for each respondent in order to calculate an *index* of local resource use – a number representing each respondent's overall use of natural resources.

Indices of this kind (sometimes referred to as *multiple indicator measures*) are very common in questionnaire research and there are many standardized sets of indicators that have been developed over many decades, especially in relation to attitudes. For example, the New Environmental Paradigm scale is designed to produce a standardized index of 'pro-environmental orientation' – an aspect of attitudes towards the environment (Dunlap et al., 2000). However, just because a particular index has been in use for decades does not mean that it is well designed or that it is appropriate for your particular research project. All an index of this kind can tell us is how each respondent answered the specific questions in the questionnaire, which, especially in a different social setting, may not be the most relevant questions to ask and may even be nonsensical. Obviously this is particularly so when

Box 7.3 Key terms: indicators and indices

Indicators: specific variables, each of which is designed to measure some aspect of a complex concept.

Index (pl. indices) or multiple-indicator measure: a numerical value calculated by combining the data across several indicators for each respondent. The index is used as a proxy for the underlying concept.

Box 7.4 Instructions: how to develop a list of variables

For each research question or objective:

1 Identify the factors or **concepts** you need to collect data on.
2 For each factor or concept, identify one or more **variables** – measurable factors on which you could collect data.
3 Make a list of possible **values** of each variable – the possible response categories. The values may be either words or numbers.
4 If the values are words, assign a number to each value (**coding**).
5 Use the list of variables and their operational definitions as you draft the questions and develop a code list. Check back against the operational definitions repeatedly and make fine adjustments as necessary.

working across different cultures; sets of indicators rarely translate well (either linguistically or in broader cultural terms). Therefore do not use standardized scales of this kind just because they are there, but think carefully about whether the individual questions really tell you something that is of relevance to your own research questions. Box 7.4 summarizes the initial steps involved in developing a list of variables.

7.3 Drafting the questions

7.3.1 Importance of wording

The next step is to design the questions that will be included in the questionnaire. Question design is almost certainly the biggest source of error in questionnaire surveys (Sudman and Bradburn, 1974, in Fowler, 1995: 150), and yet most faults can easily be avoided through applying some of the basic principles described here (see Fowler [1995] for a far more detailed account of question design).

The questions on a questionnaire must be framed so that respondents can understand them easily, know exactly how they should answer them, and can do so in a way that provides meaningful information. Other than defining the topics and items of information to be included, according to Fowler (1995) there are four challenges to writing a good question:

1 The question should be something that all respondents can answer – something they know about.
2 The question must be understood consistently, and as intended by the researcher, by all respondents.
3 The question should be framed in such a way that respondents can answer it in the terms required.
4 The question should be about something the respondents are willing to answer accurately.

Making sure that the question is something that all respondents know something about is partly a matter of sampling. There is no point in asking non-specialists highly technical questions about conservation policy, for example. However it is also partly a matter of thinking through the range of possible answers and making sure that they are all represented in the list of options the respondents can choose from. In many cases, therefore, a 'don't know' option is included. This will be discussed further in the following sections.

The example about natural resource use in Section 7.2.2 is a good illustration of the significance of point 2 above. In order to avoid confusion or misunderstanding, both the question and the list of possible answers should be reviewed in order to see whether there is any possible ambiguity in interpretation. Terms that may be unfamiliar to some of the respondents should either be replaced with more commonly known terms, or defined in the text introducing the question, or if the definition is complex, broken down into more specific aspects that are addressed in separate questions. This applies to many terms that we as conservationists take for granted; for example non-specialists may be very vague about what is meant by a protected area or a national park, even though these terms have precise (though not entirely consistent) definitions in international policy and national legislation of different countries.

Many points about question wording apply to qualitative interviews as well as questionnaires, and are discussed in Section 6.3 (interviewing techniques) and Section 6.4 (sources of inaccuracy and strategies to deal with them). Both of these sections should be read before you try to draft the questionnaire questions. However, question wording is especially critical in questionnaires because of the need to standardize between respondents and because of the lack of discussion following each answer.

In addition to the points in the sections mentioned above, it is particularly important to avoid double-barrelled questions (questions that ask about two things at once). Box 7.5 gives some examples.

One last issue of question wording that is relevant to many conservation-related studies is to do with the frequency of particular behaviours connected to natural resource use – hunting, fishing, collecting wild products and so on. Box 7.6 gives some guidelines on the wording for these kinds of question, based partly on Fowler (1995: 22–5, 37–9).

Finally, questions should be about something that the respondents are willing to answer accurately. One obvious step that can help with this is to avoid wording that suggests that a certain answer is more socially acceptable than others. An initial assurance of confidentiality, or better still, the use of self-administered questionnaires will also help. However, there

Box 7.5 Examples: avoiding double-barrelled questions

'Did you go on safari last year, and if so, with which company?' ________________

Better:

1. *Did you go on safari last year [Y / N]*
2. *If so, with which company?*____________________

Zoos are important because of their value in awareness-raising. [tick-box to show agreement, or agree/disagree scale]

Better:

1. *Do you think that zoos are important? [Y / N]*
2. *If so, why is this?* [Either open question or a closed checklist of possible reasons, including an 'other' option].

Box 7.6 Practical tips: asking about the frequency of a particular behaviour

Asking what people 'usually' or 'frequently' do or how many times they do something 'on average' does not work well. 'Usually' is vague and subjective, and so is 'average' in common usage. Moreover the frequency of a particular behaviour may be highly variable, in which case 'average' is not a very informative measure – and may not be something they can calculate with any degree of accuracy. It is better to ask about behaviour in the immediate past:

1 *Have you been fishing during the past three days?*
2 *If so, how many fishing trips have you made? _____ trips*

The shorter the time-span you ask about, the more accurately the respondents will be able to recall the information needed to give an accurate answer. Therefore a short time-span is better if you wish to calculate the overall number of fishing trips for a sample of people. However, the number of fishing trips taken over three days is not a reliable indicator of an individual's overall fishing behaviour; a slightly longer time-span is better if you wish to find out which individuals are the most frequent fishers. Answers will be less accurate but will give a better measure of that particular respondent's frequency of fishing (within a particular season). However, if the time period is too long people will stop thinking about recent events and revert to their idea of what is 'average'.

1 *Have you been fishing during the past fourteen days?*
2 *If so, how many fishing trips have you made? _____ trips*

NB: 'fourteen days' is better than 'the past two weeks', because the latter is ambiguous. It could mean the past fourteen days, or the two calendar weeks (from Monday to Monday including or excluding the current one).

is a limit to how far questionnaires can produce meaningful results when asking about highly sensitive information such as illegal or socially unacceptable activities; qualitative methods (particularly participant observation) may be more appropriate. On this and other sources of inaccuracy, see Section 6.4.

7.3.2 Question formats

This section describes the most common formats for questionnaire questions (for a more detailed account, see de Vaus [2002]). The main division in types of questions is between *open-ended questions* (where the respondent can answer in their own words) and *closed questions* (where they choose between fixed items that are presented to them). The most common forms of closed questions are closed checklists, ranking tasks and rating scales (see Box 7.7). Closed questions provide data that are precisely targeted and easily quantifiable whereas open-ended questions invite greater detail and can also catch the unexpected.

Box 7.7 Key terms: common question formats

- *Open-ended question:* the respondent answers in their own words.
- *Closed question:* the respondent chooses from one of several predefined answers.
    - *Closed checklist:* a question with a list of possible answers (Yes/No; longer lists).
    - *Ranking:* the respondent is asked to *rank* a list of items in order – for example according to preference or importance.
    - *Rating scale:* the respondent is asked to rate an item on a numerical scale between two alternatives.
        - *Semantic differential scale:* the alternatives are two opposing terms.
        - *Likert scale:* the alternatives are 'agree' or 'disagree'.
        - *Horizontal scale:* the alternatives are two opposing statements or viewpoints.

Most questionnaires are composed mainly of closed questions, but it is useful to include some open-ended questions as well to provide complementary information (see below).

7.3.2.1 Open-ended questions

The simplest form of open-ended question asks for a number (see question 2 in Box 7.8 for an example). Other than this, Fowler (1995: 177–8) gives five ways in which open-ended questions can be used:

- When the range of answers is too great to cover in a closed list.
- When descriptive answers are required that cannot be expressed in a few words.
- In determining people's knowledge.
- In order to find out the reason for a particular viewpoint or behaviour.
- To learn about a complex situation.

In all these cases, open-ended questions can be used to complement closed questions. Box 7.8 gives several examples of the use of open-ended questions in this way (questions 3, 5, 7 and 8).

7.3.2.2 Closed checklists

A closed checklist consists of a list of items from which the respondent must choose either one or as many as apply. The items may be things, activities, views, categories and so on. The simplest form of closed checklist is a yes/no question; longer lists can be taken from previous published studies, developed by brainstorming, or developed by using an open-ended question during initial piloting and building up a list based on the responses. If some respondents may not know the answer, a 'don't know' option should be included; this is most likely to be the case for questions about factual knowledge or issues of which they may

Box 7.8 Examples of the use of open-ended and closed questions

[Example 1]

1 *Have you ever taken part in a wildlife safari?* **(PLEASE CIRCLE YOUR ANSWER):**
i. Y ii. N **(IF NO, GO TO QUESTION 4)**
[closed question]

2 *If so, how many times? . . . . . .*
[open-ended question asking for a number]

3 *Please name the company that arranged your last trip:*

- -

[open-ended question asking for a specific name / instance; it would not be possible to cover all the possibilities in a closed list]

[Example 2]

4 *Tick **ONE** statement below that best expresses your view:*

i ☐ *All international trade in ivory should be banned*
ii ☐ *International trade in ivory should be permitted only for stockpiled ivory*
iii ☐ *International trade in ivory should be permitted only from countries where elephant populations are healthy*
iv ☐ *International trade in ivory should be completely unrestricted*
v ☐ *I don't know enough about the ivory trade to give an opinion*

[closed question]

5 *Please explain the reason for your answer:*

- -

[open-ended question asking for the respondent's reasoning]

[Example 3]

6 *Have you ever experienced an incident in which visitors to the site have disturbed nesting birds?* **(PLEASE CIRCLE YOUR ANSWER)**
i. Y ii. N **(IF NO, GO TO QUESTION 8)**

[closed question]

7 *Please describe the incident:*

- -

[open-ended question asking for description]

[Example 4]

[at end of questionnaire:]
8 *Please use the space below to add anything else you would like to say.*

- -

have no personal experience (for example, '*Does the safari company you travelled with have a policy about environmental performance*?') Questions 1 and 6 in Box 7.8 are both Yes/No questions about the respondent's own experience, and therefore you can assume that they know the answers. Question 4 is an *attitude statement choice* (with a list of statements representing different attitudes). If you do not include a 'don't know' option, you will push the respondent to tick a box when they simply do not know the answer or do not have an opinion, which will produce an invalid response. Attitude statement choices should be used only with extreme care, as it is very difficult to design a set of statements that are unambiguous and straightforward and that will include something most people can relate to.

Box 7.9 gives two examples of closed checklists that might be used to collect visitor feedback at a tourist lodge in the Amazon rain forest. Note that in the first of these examples the respondent is instructed to tick only one option, whereas in the second example they are asked to tick all the options that apply. Both forms are valid, but it is essential that you tell the respondent which of these they should do; otherwise some respondents will do each, and the results may be unusable. The instructions should be very prominent on the layout as otherwise people will miss them. Note also that neither list includes a 'don't know' option for the reasons explained above, but both lists include an 'other' option to catch answers that were not foreseen.

Box 7.9 Examples of closed checklists: visitor feedback at a tourist lodge in the Amazon

[Example 1]

What was the main reason for your visit? **(PLEASE TICK ONE ONLY)**

i ☐ *To experience the Amazon rainforest*
ii ☐ *Birdwatching*
iii ☐ *Botanising*
iv ☐ *To see monkeys in the wild*
v ☐ *To visit an Amazonian tribe*
vi ☐ *To learn about traditional medicines*
vii ☐ *Other (please specify): ....*

[Example 2]

Which of the following activities did you participate in during your visit?
(PLEASE TICK ALL THAT APPLY):

i. ☐ *Hiking*
ii. ☐ *Canoeing*
iii. ☐ *Fishing*
iv. ☐ *Visiting the local people*
v. ☐ *Tour of a medicinal plant garden*
vi. ☐ *Use of canopy walkway*
vii. ☐ *Observation from hides*
viii. ☐ *Other (please specify): ....*

Box 7.10 Practical tips: closed checklists

- Closed checklists can be developed from previous published studies, from answers given to open-ended questions during piloting, or by brainstorming.
- Unless it is self-evident, indicate whether the respondent should tick only one option or as many that apply.
- Unless the question is about personal experiences, consider including a 'don't know' option.
- Unless the list is logically all-encompassing, include an 'other' option.

Again, a follow-up open-ended question is useful in checking why people have ticked a certain option and how they interpret it.

7.3.2.3 Ranking

Ranking also involves closed checklists, but rather than just checking those that apply the respondent ranks them in order – usually according to preference or perceived importance or relevance. Apart from their use in questionnaires, ranking exercises are an important component in other types of structured interview (Chapter 8) and can generate a lot of discussion in focus groups or workshops.

Ranking is best done in two stages – people are first asked to tick those that apply, then to rank only those that they have ticked. Box 7.11 shows how this approach, known as combined checklist and ranking, could be used for one the examples from Box 7.9. Note that the direction of ranking is indicated (1 = most important), and also that there is an 'other' category that can also be ranked.

Box 7.11 Example of combined checklist and ranking

1 *What were the reasons for your visit?*
(PLEASE TICK THE BOXES FOR ALL THAT APPLY)

i ☐ *To experience the Amazon rainforest* --
ii ☐ *Birdwatching* --
iii ☐ *Botanising* --
iv ☐ *To see monkeys in the wild* --
v ☐ *To visit an Amazonian tribe* --
vi ☐ *To learn about traditional medicines* --
vii ☐ *Other (please specify): ....* --

2 *Using the lines on the right hand side of the page, please rank the items you have ticked in the above list in order of importance (STARTING WITH* ***1= MOST IMPORTANT****).*

Ranking exercises can be extremely valuable in determining people's priorities as long as they are used carefully. In phone interviews the list of items must be extremely short because respondents will not remember a long list read out one by one. In self-administered questionnaires errors are particularly common in ranking tasks: respondents may rank all the items rather than only those items that apply, or they may just tick those that apply and fail to rank them, or they may invert the direction of ranking (using 1 for the lowest rank rather than the highest). They may also make different assumptions about whether two items can have the same rank. Therefore the instructions need to be both precise and prominent.

7.3.2.4 Rating scales

While ranking exercises involve comparing different items in a list and putting them in order, a rating scale involves rating a single item on a numerical scale. The scale stretches between two opposite alternatives and usually has either five or seven points; obviously, seven gives you a finer tool for measurement. Using an odd number of points ensures that there is a neutral midpoint.

Box 7.12 gives examples of three types of rating scale:

- *Semantic differential scale:* the alternatives are two terms with opposite meanings (semantic = related to meaning).
- *Likert scale:* the respondent has to indicate to what extent they agree or disagree with a given statement.
- *Horizontal scale:* the alternatives are two opposing statements or viewpoints.

Rating scales are often grouped together in a block. Blocks of rating scales are extremely quick to answer, especially in written form. Box 7.13 shows a battery of five-point Likert scales from a national survey of attitudes, knowledge and behaviour in relation to the environment in the UK. The different items in a block like this are frequently designed to be used as indicators and combined into an index (see Section 7.2.2) – in this case, to do with attitudes towards personal action on environmental issues. Note that there is a neutral mid-point on the scale and also a sixth option of 'can't choose' – roughly equivalent to 'don't know'. Also note that each and every point on the scale is 'labelled' – a descriptive phrase is printed above each point. Including text labels of this kind helps to standardize the respondent's interpretation of the scales, as long is the labels are absolutely clear and unambiguous.

In self-administered forms, as with ranking tasks, there is a danger that some respondents will invert the scale, and therefore the headings must be prominent and clearly worded. Fowler (1995: 65) criticizes the terms 'agree strongly' and 'disagree strongly' on the basis that 'strongly' implies an emotional response rather than a rational one; this wording is common, but perhaps a better alternative is to use 'completely agree', 'mostly agree' and so on. A second danger is that if the block of scales is very long, the respondents may get impatient and begin to tick boxes without taking the time to think about their response properly. Finally, people tend to avoid using the extremes of the scale, especially in face-to-face interviews. Using a seven-point scale rather than a five-point scale makes this less problematic because there are more 'middle' points, but it becomes harder to find clearly distinct 'labels' for each point on the scale.

Box 7.12 Examples of rating scales

[Example 1: semantic differential scale]

How would you describe the immediate environment where you grew up?
(PLEASE CIRCLE ONE NUMBER ON EACH LINE)

1	*Rural*	*1*	*2*	*3*	*4*	*5*	*Urban*
2	*Natural*	*1*	*2*	*3*	*4*	*5*	*Built up*
3	*Attractive*	*1*	*2*	*3*	*4*	*5*	*Unattractive*
4	*Rich in wildlife*	*1*	*2*	*3*	*4*	*5*	*Poor in wildlife*
5	*Economically prosperous*	*1*	*2*	*3*	*4*	*5*	*Economically poor*

[Example 2: Likert scale]

How much do you agree or disagree with each of the following statements?
(PLEASE CIRCLE ONE NUMBER ON EACH LINE)

		Completely Agree					***Completely Disagree***	
1	*Zoos are essential for raising awareness of conservation issues*	*1*	*2*	*3*	*4*	*5*	*6*	*7*
2	*Zoos involve unacceptable cruelty to animals*	*1*	*2*	*3*	*4*	*5*	*6*	*7*
3	*The only justification for zoos is captive breeding*	*1*	*2*	*3*	*4*	*5*	*6*	*7*

[Example 3: horizontal scale]

Who do you think should be responsible for regulating fish catches on community lands – government, local people, or a mixture of both?
(PLEASE CIRCLE THE NUMBER THAT BEST REPRESENTS YOUR VIEW)

Government should have sole responsibility				***Local people should have sole responsibility***	***Don't know***
1	*2*	*3*	*4*	*5*	*6*

7.4 Building a coding list

This section describes the process of building a *coding list*, which consists of a list of the variables associated with each closed question on your questionnaire, together with the possible numerical values for each one (see Box 7.14). Where necessary the list should also include short notes to remind you exactly how you have defined the terms.

Coding is extremely important because in quantitative analysis it is the numerical values that are analyzed rather than the original responses in words. For this purpose, an electronic

Box 7.13 Example of a battery of Likert Scales

8. How much do you agree or disagree with each of these statements?
PLEASE TICK ONE BOX ON EACH LINE

	Agree strongly	**Agree**	**Neither agree nor disagree**	**Disagree**	**Disagree strongly**	**Can't choose**
a. It is just too difficult for someone like me to do much about the environment	☐	☐	☐	☐	☐	☐
b. I do what is right for the environment, even when it costs more money or takes more time	☐	☐	☐	☐	☐	☐
c. There are more important things to do in life than protect the environment	☐	☐	☐	☐	☐	☐
d. There is no point in doing what I can for the environment unless others do the same	☐	☐	☐	☐	☐	☐
e. Many of the claims about environmental threats are exaggerated	☐	☐	☐	☐	☐	☐
	(1)	(2)	(3)	(4)	(5)	(6)

(Source: National Centre for Social Research 2000)

spreadsheet is set up in a programme such as Excel, with one column for each variable and one row for each respondent. During data entry, a code is entered in the correct cell for each variable in turn from each completed questionnaire; thus as you enter data from a single questionnaire you fill in successive cells in a single row.

Coding for the different question formats is described in Box 7.15. For closed lists where the respondent can only select one item, coding is straightforward – each item in the list is assigned a different number. For rating scales, the rating itself is used as the code. If someone circles the number 3 on a rating scale, their answer is coded as '3'. The types of question where coding gets more complex are closed lists where the respondent can select more than one option and questions that ask for the most relevant items in a list to be ranked. The problem in both of these cases is that the answer contains more than one piece of information, and therefore you cannot allocate a single, unique number to each possible answer for entry into the corresponding cell on the spreadsheet.

Box 7.15 gives instructions for the most common ways of coding responses to questions of these kinds. In the case of closed lists where multiple responses are allowed, usually a separate variable is created for each item in the list and given two values – ticked and

Box 7.14 Key term: coding lists

A coding list includes:

- A list of variables, each with a name (e.g. income).
- Where not obvious what measurement you need to take, a specific quantifiable definition for each variable (e.g. annual gross income in £000s).
- Where not obvious, a list of values (possible responses) (e.g. 'To experience the Amazon rain forest' – an item in a closed list of possible responses to the question 'What were the reasons for your visit'?).
- Where the 'values' are words, a numerical code for each one (e.g. 'To experience the Amazon rain forest' = 1).

Box 7.15 Instructions: coding

Closed checklist where only one item can be selected: the response is dealt with as a single variable, and each item in the list is allocated a code.

Closed checklist where more than one item can be selected:
In the *multiple dichotomy method*, each item in the list is treated as a separate variable with two possible response values – ticked (1) and not ticked (2).

Ranking: There are two common ways to assign codes:

1. Each *item* in the list is treated as a separate variable and the rank allocated to it by the respondent is used as the code.
   Or
2. Each *ranking position* is treated as a separate variable and each item in the list is allocated a code. Thus the value given to the variable 'rank 1' for any respondent is the numerical code related to the item that the respondent has ranked the highest.

Rating scales: The rating itself is used as the code. Thus, code values usually range from 1 to 5 or 1 to 7.

Open-ended questions where the answers are given in words: Either define a list of codes in advance or develop a list of codes after data collection, as for qualitative interview data.

All question formats: Additional codes are assigned for 'don't know'/'other'/not applicable options if these are included. Codes are also assigned for missing values.

not ticked. In the case of ranking, there are two commonly used forms of coding – one in which a separate variable is created for each item, and the other in which a separate variable is created for each ranking position. The method you select should reflect what exactly you want to know, and has implications for the kinds of statistical analyses that will be appropriate. In both these cases, the response to a single question will take up more than one column in the spreadsheet, and therefore the number of columns in the spreadsheet for any questionnaire reflects the number of *variables,* not the number of *questions.*

If there is an 'other' option on a closed list (as in Boxes 7.9 and 7.11), then it is given a code value in exactly the same way as the other items. If there is a 'don't know' option in any type of question, it may be given a value that is not sequential with the rest so that in analysis it is clear that this value is not part of the range of choices. For example the code for 'can't choose' (more or less equivalent to 'don't know') in Box 7.13 is 8 to make it clear that it is not connected to the five-point scale itself.

The final kind of question is the open-ended question. Open-ended questions do not *have* to be coded numerically and analyzed quantitatively; the responses may be left as text and analyzed qualitatively instead (see Chapter 14). However with large sample sizes it makes sense to use quantitative analysis. The responses can be coded numerically simply by developing descriptive codes in the form of words or abbreviations, and then allocating a number to correspond to each textual code. Just as with qualitative interview data, the list of codes can be constructed either in advance as part of the research design process or after data collection has ended, by looking through people's answers and identifying recurring or particularly relevant topics from what they have said.

What should you do if someone skips a question or messes up the form and answers in the wrong format? If you just leave the corresponding cell for that value blank in the spreadsheet, the programme will assume that there was no respondent and therefore calculate the sample size incorrectly. Instead you should allocate a numerical value in the code list for each variable for '*missing values*'. Obviously you must use a number that could never represent a valid response; by convention, the numbers 9, 99 or 999 are used. In an analysis programme such as SPSS you can specify the codes allocated to missing values and the programme will calculate how many missing cases there are (which may be important in assessing response rates and sampling validity) but exclude them from further analysis. If appropriate, different numbers may be assigned for different *types* of missing value, such as cases where the question was not appropriate and the respondent was instructed to skip ahead versus cases where the respondent should have answered but did not do so. For further detail of all aspects of coding and data processing prior to analysis, see de Vaus (2002: chs 9 and 10).

7.5 Assembling the questionnaire

Once you have developed a set of questions there are several factors to consider as you put them together into a questionnaire. What order should you put them in? What instructions should you include in order to make sure that people understand exactly what they need to do? How should it be set out on the page? How long can it be? This section gives guidance on these issues.

7.5.1 Putting the questions in order

Questionnaires are usually divided into different sections on different topics, reflecting the different areas of information that were defined by the process described in Section 7.2.1. One section – usually at the beginning or the end – should include all the questions asking about the respondent's *attributes* (see Section 7.2.1). There is no logical order to the questions in this section; they simply need to fit neatly and clearly onto the page. Box 7.16 gives some examples of factors that might be included and indicates the most common question

Box 7.16 Information on respondent attributes

Demographic/socioeconomic factors:

For example:

Nationality	[closed list]
Sex	[closed list: M/F]
Marital status	[closed list]
Number of children	[open question asking for a number]
Age	[closed list of categories (age ranges) or open question asking for a number]
Annual gross income	[can be an open question asking for a number but more usually a closed list of income categories]
Occupation	[usually open-ended question]

May include additional relevant factors, such as:

Programme of study (students)	[closed list or open question]
Membership of conservation organizations	[either Y/N question or closed list of different conservation organizations]
Vegetarianism	[Y/N or closed list with different types of vegetarianism]
Participation in a safari holiday	[Y/N question – see example 1 in Box 7.8]

formats for each one. Questions about age and income can ask for a response either as an exact number or by ticking one of a list of categories; an exact number gives you more precise information, but in many cultures these are sensitive questions, and therefore people may be more willing to answer a question that uses categories.

Box 7.17 offers some further guidelines related to the order in the final questionnaire. A general principle in ordering both the other sections and the questions within each of them is to apply what is known as the *interview funnel* – start with broad questions and narrow down to specifics as you go along. The reason for this is that if you started with specifics, you would prime the respondent to think of the things you have already asked about. For example, if you ask '*do you think whales should be a priority species in conservation?*' followed by '*which species do you think should be prioritized in conservation?*' it is likely that many people would include whales in their answer who would not otherwise have done so, simply because they are uppermost in their minds at that moment. You should also start with topics that are easy, unthreatening and interesting, so that the respondent is put at ease and encouraged to continue. Save difficult or sensitive topics or questions until later on, when the respondent should be more engaged and relaxed.

Within each section, the questions should be arranged in order so that each question leads on naturally to the next one. Box 7.8 gave simple examples of two-question 'strings' made

Box 7.17 Instructions: how to put the questions in an order

- Sort the questions in to groups (sections), including one section on the respondent's attributes.

  Ordering the sections:

  - Put the section on attributes at the beginning or the end.
  - Start with a topic that is easy, unthreatening and interesting.
  - Leave sensitive or complex topics until later on.
  - Apply the interview funnel – start broad and narrow down.

  Within each section:

  - Sort the questions into subgroups if necessary.
  - Apply the 'interview funnel' – start broad and narrow down.
  - For each subgroup think in terms of a line of enquiry – a 'string' of questions where each question leads naturally to the next one.

up of a closed question followed by an open question; questionnaire sections usually include longer strings of questions, each on a slightly different aspect of the same topic. To some extent the order may have been clear as the different items of information were defined in the first place (see Section 7.2.1) but even if so, it needs to be reviewed and thought through carefully at this stage.

7.5.2 Layout and instructions

The questionnaire should begin with a very brief statement of the general topic and purpose of the questionnaire, an assurance that answers will remain confidential, and any general instructions. There may be similar, very brief statements at the start of different sections ('The following section is about...'); respondents are more likely to be cooperative if they understand why each section is relevant to your study. Lastly, at the end of the questionnaire there should be a statement thanking the respondent for their cooperation.

The most important rules for the layout are that the questionnaire should be easy to read – do not use tiny font and do not overcrowd the page – and that it is immediately obvious what needs to be done at each stage. There should be instructions connected to each question or set of questions stating *exactly* what needs to be done: to tick one box or as many boxes as apply, circle the number corresponding to the most appropriate response, write in their own words, and so on. Do not ever assume that it is obvious; people will get it wrong. Instructions should be in large font and capitals (see the examples given earlier in this chapter), or people will miss them. Closed lists should be laid out down the page, not across it, and must not go across a page break. Tick boxes or lines should be provided for each response; otherwise you may have trouble telling which response a large tick is meant to refer to. All questions and sub-questions should be numbered separately in order to assist with data entry, when each variable will correspond to one column on the spreadsheet. If the questionnaire will be self-administered, then only print on one side of the paper – it is amazing how many people will overlook any questions on the reverse.

Including the response codes in the layout of the questionnaire makes data processing much easier and reduces errors during data entry. In multiple checklists, a common option is to number the items in the list rather than just use bullet points; then the numbers can be used as the code values. If the items are also to be ranked then particularly careful attention must be paid to the layout, because otherwise the item numbers can influence the ranking numbers allocated by the respondent.

It is also important that the item numbers are not written in the same format as the question numbers or it is easy to get confused between them during data entry. In the examples of closed checklists in Boxes 7.8 and 7.9, I have used conventional numbers (1, 2, 3) for question numbers and (*i, ii, iii*) for coding values. I have continued this convention in the rating scales in Box 7.12, where the numbers down the left-hand side of the page are question numbers rather than coding values. I prefer this to the format in Box 7.13, where a single number is given to the whole set of scales and letters (a, b, c) are allocated to each scale. Each scale is in fact a completely separate question and the responses must be entered separately for each question during data entry, even if they will be combined into an index prior to analysis. However an alternative, rather than trying to include codes on the printed questionnaire, is to leave space on the page to write the codes in for each response after data collection (for example, by leaving a column on the right hand side for this purpose). This should be done before data entry begins; then there will be far fewer copying errors than if you attempt to do both things at once. Online questionnaires have an enormous advantage in this respect because there is no need to enter the data manually. As well as saving time, this means that there is one less source of error.

A last aspect of the layout that needs careful scrutiny is to make sure that there are clear instructions indicating any places where the respondent should skip some questions. It is rare that every respondent should answer every question, because often, there are follow-on questions that are only relevant for those who give a certain response. In Box 7.8, if the respondent answers 'no' to the first question, then the next two questions are irrelevant. Therefore there is an instruction at the end of question 1 stating '*if no, go to question 4*'. Most questionnaires include quite a lot of instructions of this kind, telling the respondent where to go next according to what answer they have just given. Using arrows in addition to

Box 7.18 Practical tips on questionnaire layout

- Do not use tiny font and do not overcrowd the page.
- Include instructions for each question stating *exactly* what needs to be done, in large, bold font.
- Closed lists should be laid out down the page, not across it, and must not go across a page break.
- Tick boxes or lines should be provided for each item in a closed list.
- ALL questions and sub-questions should be numbered.
- Include the response codes if you can, or else leave room to write them in prior to data entry.
- Only print on one side of the paper.
- Include instructions telling the respondent when they should skip questions.

the text instructions makes it even clearer what they should do. You must check the questions again and again to make sure that whatever their responses, the respondent will never be faced with a question that is inappropriate, or they are likely to lose interest and, if the questionnaire is self-administered, stop filling it in. Online questionnaires have a major advantage here too: they can be set up so that the respondent cannot skip questions that they should answer and so that they will automatically miss out questions that are not relevant, according to the answers they have given so far.

The examples given in this chapter are all presented in a format that is suitable for self-administered questionnaires. For face-to-face interviews, precise instructions to the interviewer should be written out as well. Fowler (1995) discusses this in detail and gives many examples of how a slight change in wording can influence the respondent's answer. The instructions, which are often put in capitals, should give the *exact* wording that the interviewer should use, both for the initial introduction and any instructions along the way, and also in asking each question.

7.6 Reviewing the questionnaire

Once you have drafted the full questionnaire, check through everything one last time for weaknesses. Boxes 7.19 and 7.20 present two checklists – one for the individual questions and the other for the questionnaire as a whole. Check the points in Box 7.19 for each question. Then check the points in Box 7.20 for the questionnaire as a whole. This may seem a long-winded process but it will save you much time in the next stages, and will result in higher quality data. If you do not find anything that needs changing, then you deserve a prize.

Once you have checked the questionnaire as thoroughly as you can, it is also valuable to ask a small number of other people to provide comments – either on the whole thing or on particular questions and aspects that you are doubtful about. This can be useful initially even with colleagues, fellow students, friends and family; anyone who comes fresh to the questions may notice ambiguities or other problems that you have overlooked. Fowler (1995)

Box 7.19 Checklist for reviewing the questions

- Have you included precise instructions (e.g. circle a number, check one or several items in a list, answer in your own words)?
- Is it possible that respondents will not know the answer? (If so, include a 'don't know' option).
- Does the range of answers cover all the possibilities? (If not, consider adding an 'other/not applicable' option).
- Is the question (or some part of it) ambiguous?
- Is the question (or any part of it, or any of the possible responses listed) double-barrelled – does it ask two different things at the same time? If so, re-word or split it.
- Is the question 'leading'? Could the wording elicit an unintentional response bias?
- Does the question use technical terms unnecessarily or assume knowledge that respondents may not have?
- Have you written down a list of codes for the possible answers?

Box 7.20 Checklist for reviewing the questionnaire as a whole

Related to subject matter:

- Last check: is there anything you have missed out?
- Is there a section on attributes of the respondent?
- Would it be useful to add any open-ended questions to follow up specific closed lists?
- If the questionnaire is very long:
    - Are there any questions that ask the same thing?
    - Are there any questions or sections (and the related variables) that are not essential to your research objectives?

Related to order:

- Is there an introduction at the beginning, thanks at the end, and instructions as necessary throughout?
- Does the questionnaire start with a topic that is easy, unthreatening and interesting? Are sensitive or complex topics placed near the end?
- Does the order 'make sense' and flow smoothly? Do the sections move from broad to narrow? What about within each section?
- Could the answer to one question be influenced by the previous questions? If so, consider re-wording or changing the order.
- Could the answer to one question make the next question irrelevant? Is there anywhere where the respondent (or interviewer) might be unsure where to go next? If so, add 'signposting'.

Related to layout and presentation:

- At a glance, does it look clear and straightforward, or complicated and daunting?
- Is the layout easy to follow and the font easy to read?
- Are closed lists laid out down the page rather than across it?
- Is there enough space on the page for answers to open-ended questions?
- Have you included the numbers you will use as codes on the questionnaire layout, or else left space in the right-hand margin to write the codes in later?
- Try going through it and imagine answering the questions. How long does it take you?

describes more formal procedures involving individual in-depth interviews (preferably with colleagues who know something about questionnaire design) or focus group discussions. In individual interviews, people may be asked to answer selected questions in turn and then to describe their thought processes as they did so – how they interpreted the question and any uncertainties they had about the meaning of the question or the way in which they should answer. It is not practical to ask focus group members to actually answer the questions and then tell you their thought processes, but specific questions can still be presented for discussion, feedback and suggestions.

Many of the above points are concerned with *measurement validity*. In Chapter 3, validity was defined as concerned with 'whether the evidence which the research offers can bear the weight of the interpretation that is put on it' (Sapsford and Jupp, 1996: 1, in Bell, 2005: 117–18) and discussed three types of validity to do with the overall research design (*internal*, *external* and *ecological* validity). The process of reviewing and piloting the questionnaire is concerned with the validity at the level of data collection: do the specific data collection tools measure what you want them to measure?

There are many different types of measurement validity (see Box 7.21) and unfortunately, different researchers use different terms for them. The definitions used here follow Bryman (2004: 72–4), but other researchers use *construct* or *content* validity in place of measurement validity, or use each of these terms to mean something different again.

Face validity can be assessed during your initial review of the questionnaire. After piloting, it may be useful to check for *concurrent validity* of certain questions – did certain people score as you would predict from other information you know about them? (Obviously this is only a reasonable question to ask where there are very concrete grounds for your prediction). *Convergent validity* is usually assessed after the main data collection as part of the analysis of the results. In the case of composite indices you may also need to test for *internal reliability*, preferably after extensive piloting or else after the data collection. Internal validity is said to be high when the different indicators vary in roughly similar patterns across respondents. If they do not, then it suggests that they are not measuring aspects of the same thing, and therefore you are not justified in joining them together into a single index. In this case you can either delete the questions that show very different patterns in responses, or else analyze them separately as variables in their own right. Internal reliability can be assessed informally by eye, or more rigorously by means of Cronbach's alpha (for further details see Bryman, 2004: 71–2).

Box 7.21 Key terms: measurement validity

Measurement validity: whether specific variables (or indices) measure what you want them to measure. Includes (but not limited to) the following:

Face validity: based on subjective evaluation of the match between your original concept and the variables (or questions) that you have ended up with.

Concurrent validity: based on a comparison of the variation in responses to variation in other factors that should be related. These may include respondent attributes. For example, do people who have a degree in wildlife conservation score more highly on an index of 'knowledge about conservation'? If not, there is probably something wrong with the index.

Convergent validity: compares the results from your draft questions against results from other methods. For example, does information from questionnaire questions on people's behaviour coincide with observational data on their actual behaviour?

Internal reliability: the degree to which different variables being used as indicators that you wish to combine into a single index vary in similar patterns across respondents.

7.7 Field piloting and administering the questionnaire

Some people regard the above measures as part of the piloting process – asking someone to look through and comment on the questions, or trying out specific questions on a small number of individuals, or asking for feedback in individual interviews or focus groups. However, there must also be a formal round of field piloting, which involves administering the questionnaire to a small group of people similar to those you will include in the final study. The importance of field piloting cannot be overemphasized; you will almost always find that there are questions that people fail to understand or interpret in different ways, places in the questionnaire where they are not sure where to go next, and questions that turn out simply not to elicit useful information. If your field site is distant from your home base you must leave time for proper piloting in the fieldwork schedule.

The questionnaire should be administered exactly as you will do in the final study so that you can see whether it all goes smoothly, how long it takes to complete, and whether the respondents seem to get bored or lose concentration before the end. The respondents should not be aware that it is still in draft, and unless it turns out to be very problematic, they should not be encouraged to comment as they fill it in.

Questionnaires can be administered either interactively (in person or by phone) or by the respondent alone (known as 'self-administered'), either on paper or electronically. Each method of administration has advantages and disadvantages, which are summarized in Table 7.1.

Table 7.1 Advantages and disadvantages of different methods of questionnaire administration

	Advantages	*Disadvantages*
Self-administered on paper	Allows for anonymity Takes little time to distribute	Errors in completion cannot be corrected Clarification cannot be given You cannot be sure exactly who has completed the questionnaire Poor response rates Not feasible if there is a low rate of literacy
Electronic	Allows for anonymity Takes relatively little time to set up Good at reaching dispersed populations Can be set up to control incorrect response formats, question 'skips' and the order in which questions are answered Manual data entry is unnecessary	Clarification cannot be given You cannot be sure exactly who has completed the questionnaire Not feasible if there is a low rate of literacy or computer use
Face-to-face	The interviewer can guide the respondent, explain terms, and check that they answer in the appropriate format Open-ended questions work best by this method Better response rates than by other methods The only option if literacy and phone ownership are low	People may be less willing to answer sensitive questions honestly than with other methods More time-consuming than other methods

Table 7.1 (*Cont'd*)

	Advantages	*Disadvantages*
By phone	The interviewer can guide respondent, explain terms, and check that they answer in the appropriate format Open-ended questions work better than in self-administered methods Better response rates than for paper questionnaires Good at reaching dispersed populations	People may be less willing to answer sensitive questions honestly than with self-administered methods Ranking tasks and long closed checklists can be problematic when the informant cannot see the whole list at once

It is useful to observe the pilot interviews even if they are self-administered, because you may pick up useful tips from the respondent's reaction to each question. If several respondents pause and look puzzled, or in a face-to-face interview ask you to repeat or explain a question, it suggests that the question needs to be re-drafted. Similarly if several respondents give a verbal answer to one question in a way that makes the following questions redundant, you need to look again at the way the questions fit together. Finally, if people seem willing to spend time talking further with you once they have completed the questionnaire, tell them that you are still working to improve it and ask for their comments and suggestions.

Based on the results of the piloting you may decide to change the wording of some questions; add items to closed lists to reflect responses that come up repeatedly in the 'other' category; split a single question into two or more questions; delete questions that do not work very well or that do not add anything to what is covered elsewhere, or even add totally new questions to cover aspects you had not thought of. You may also need to change the

Box 7.22 Practical tips: piloting

Things to look out for:

- misunderstandings;
- answers in forms you did not expect;
- puzzlement or hesitation in answering;
- need to ask for clarification;
- not sure where to go next;
- questions that people feel are irrelevant or don't allow them to give their 'real' view;
- questions that ask for information that several people have already given verbally earlier in the questionnaire;
- recurring answer in 'other' category (add it to closed list);
- length – do they get bored or impatient?

The final stage of piloting should be on people from population you will be studying.

order and layout, or even grit your teeth and cut out whole sections if the questionnaire turns out to be far too long. The amended questionnaire should then be piloted on a few more people until no more changes are needed.

The appropriate total length of the questionnaire depends upon how it will be used. If you will be stopping people in the street, then you cannot expect them to spend more than about three minutes to complete it. If you are mailing it out or leaving it in public places for people to pick up then it can be longer (up to 10 or 15 minutes is often assumed to be appropriate). If it is targeted at a population who have a special interest in the topic, and especially if you contact people in advance and arrange a face-to-face interview, it can be longer still. However there is little hard data on the effects of questionnaire length on response rates or quality of the data. The best way to check whether it is an acceptable length is to pilot it on a few people exactly as it will be used and check whether they fill it out readily or seem to get bored or impatient part way through.

In administering questionnaires face-to-face or by phone, the most important rule – as for qualitative interviews – is to be courteous and professional at all times and avoid showing disapproval or frustration. Accept that not everyone you approach will agree to be interviewed, and do not push them so hard that they get annoyed or upset. Paper copies of self-administered questionnaires may be distributed to a pre-selected sample of people (usually with a covering letter), or they may be left at strategic points where people can pick them up – in shops, hotel rooms, airports, at the entrance to a nature reserve or zoo, and so on. You can either arrange to collect them from the individual or venue at a later date (say when you will be back to do so), or include a self-addressed envelope so they can be mailed back to you.

Sampling strategies and sample sizes for the actual data collection are discussed in Chapter 4. Non-response rates are problematic in terms of sampling because those who fail to respond may be atypical of the sample as a whole (see Section 4.3.2.1). Therefore time should be allowed to follow up on non-respondents (courteously) in the days or weeks following your initial approach. A follow-up letter, email, phone call or knock on the door after a week or so later may be enough to encourage many of those who have failed to respond to do so.

Once the completed questionnaires begin to be gathered, they should be checked periodically so that you know how many correctly completed questionnaires you have. Ideally, the data should be entered onto a spreadsheet periodically during data collection so that you do not have to spend weeks doing this when you return from the field. If you use a programme such as Excel, the data can then be exported directly to a statistical package such as SPSS for analysis. Data analysis usually includes some descriptive analysis (for example, to describe the socio-demographic profile of the respondent sample) and then selected inferential analyses to address your original research questions and test for statistical significance of other patterns in the data that appear to be present. The analysis techniques themselves are described in Chapters 15 and 16, using examples from the complete questionnaires in Appendices 1 and 2.

7.8 Conclusion

According to Fowler (1995: 154), 'there is a long history of researchers designing questions in a haphazard way that do not meet adequate requirements', and unfortunately poor design or inappropriate use of questionnaires continues to be common. Most people have had the experience of starting to fill in a questionnaire, only to come to questions that are inappropriate or ambiguous or that try to force an opinion when you do not really have one. In order to produce a sound questionnaire a lot of time and thought must be put into its design. Even then, it is rarely possible to design a 'perfect' questionnaire simply by editing it at your desk, and therefore time must be allowed for adequate piloting and redrafting. However, when designed carefully, and under the right circumstances, questionnaires are precise and powerful tools for collecting an enormous amount of carefully focused information from a large number of people.

Summary

1 Questionnaires consist of a series of specific, usually short questions that are asked in exactly the same way of all respondents.
2 Before designing a questionnaire, an *operational definition* must be developed for each concept or type of information, such that it can be expressed in terms of one or more quantifiable *variables*.
3 Next, the questions themselves should be designed. They must be framed so that respondents can understand them easily, know exactly how they should answer them, and can do so in a way that provides meaningful information.
4 Questions can be *open-ended* or *closed*. The most common forms of closed questions are *closed checklists* and *ranking* or *rating* tasks.
5 As the questions are designed, a *code list* should be developed for each variable, consisting of numerical values that are assigned to each possible answer.
6 Questions should be assembled into a questionnaire in sections on different subtopics, starting broad and narrowing down and making sure the questions in each section flow naturally from one to the next.
7 The questionnaire layout should be easy to read and make it immediately obvious what needs to be done at each stage. Instructions should be in large font and bold.
8 The draft questionnaire should be reviewed carefully both by yourself and by other people, and should then be thoroughly field-piloted.
9 It can be administered interactively (by phone or face-to-face), or self-administered by the respondents alone (on paper or electronically).

Further reading

Czaja, R. and Blair, J. (2005) *Designing Surveys: A Guide to Decisions and Procedures*, 2nd edn, London: SAGE Publications. [See particularly Chapter 9, on reducing sources of error in data collection, and Chapter 10, on ethical issues and on how to report the methodology.]

De Vaus, D. (2002) *Surveys in Social Research*, 5th edn, London: Routledge. [An authoritative text on all aspects of carrying out survey research. There are substantial chapters on developing indicators, ethics and data collection, sampling, constructing and administering questionnaires, coding and data processing; and six chapters on data analysis.]

Fowler, F. (1995) *Improving Survey Questions: Design and Evaluation*, Applied Social Research Methods Series Volume 38, London: SAGE Publications. [A gem of a book, giving a detailed, realistic and very accessible account of issues of question wording, evaluating and piloting, with many practical tips and examples – especially for face-to-face administered questionnaires. The appendices are excellent reference sources in their own right, detailing many specific formats for wording and the use of graphics in questionnaires.]

Nichols, P. (1991) *Social Survey Methods: A Field Guide for Development Workers*, Oxford: Oxfam. [Gives a straightforward, practical account of how to carry out applied surveys in a development setting.]

8 Documenting local environmental knowledge and change

R.K. Puri

The small part of ignorance that we arrange and classify we give the name of knowledge.
(Ambrose Bierce, 2007: 209)

8.1 Introduction

This chapter describes a set of methods that take place within the context of a structured interview. Some of the methods are better conducted with individuals while others can also be used in focus groups and workshops. The first set of methods, collectively known as *cultural domain analysis*, is aimed at uncovering local knowledge, its structure and variations (see Section 8.2). The second set of methods is used for documenting changes in communities over time, from seasonal cycles to family trees to historical events (Section 8.3). Conservationists need to be aware of both kinds of information, because local environmental knowledge underpins in part the choices people make about what and where to hunt, fish and gather plant resources. Part of this knowledge includes an understanding of natural cycles in the environment, such as weather, phenology and migration, as well as an understanding of the history of land use, conflicts over resources, and the particular circumstances and events that have led to the current situation. Besides this *oral history*, people may also have some knowledge of the future, in the sense that they may see trends in their environment and have an idea of where things are going and what they expect to happen. Understanding of and expectations for the future can have powerful effects on current behaviour, so the second set of methods can also be used to investigate possible futures of communities and environments.

8.2 Cultural domain analysis

Cultural domain analysis (CDA) is used to understand how people in a society think about and define their world. Since all cultures use some system of categories to order experience, the researcher tries to determine what categories are important to people, how they are arranged and what values are attached to them. These important categories are known as *domains*, each of which is composed of a group of elements or *items* organized according to rules or criteria that are culturally determined and may be culturally specific. In other words, domains are the stuff of culture. Identifying and understanding them leads to a better understanding of cultural organization (Weller and Romney, 1988).

For example, the domains 'edible foods' or 'medicinal plants' or 'hunting rules' or 'kin' can vary significantly from culture to culture and items can be included or excluded

Table 8.1 Steps in a cultural domain analysis

Step	*Structured interview techniques*
Identify its components	Freelisting, identification
Discover their arrangement	Triads, pilesorts
Identify rules for arrangement	Paired comparisons, rating, ranking
Explore the associated values	Weighted ranking
Explore variability in a group	Consensus analysis
Explain variability	Factor analysis

for a variety of reasons. What would you consider 'inedible' and why? What are other domains of interest to conservationists? Domains and what they include are learned as you grow up in a society and are necessary for perceiving, interpreting and communicating your experiences with the world and people. Domains are the starting point for studying people's perceptions of the natural world. Thus they are highly relevant to social scientists and conservationists.

Many of these domains are important aspects of what we might call 'local knowledge' (LK), 'indigenous knowledge' (IK), 'traditional ecological knowledge' (TEK) or 'indigenous technical knowledge' (ITK). Identifying domains and their structure can be a starting point for understanding knowledge systems at a much deeper level, and comparing the knowledge of different people in a social group. For example, how does the domain 'forest products' differ between men and women?

CDA can be broken down into a series of steps (see Table 8.1) and particular methods can be used to elicit the information related to each step. In this chapter I will introduce some of these methods, describing how to use them and analyze the data by hand. If you are interested in more sophisticated quantitative analyses then you can use the software program ANTHROPAC (Borgatti, 1996), which is specifically designed for CDA, and in addition to the techniques discussed here also includes *consensus analysis*, which is a powerful way of studying variation among informants.

Throughout the sections on CDA, the domain of *fruits* will be used to exemplify the methods at each step. If you would like to follow along, before reading further please list all the fruits you know on a separate sheet of paper.

8.2.1 Freelisting

Freelisting is used in the course of an interview with an individual to elicit the *items* of a domain. It is used to check whether the domain is locally *salient* or meaningful, which items are included and which of the individual items is most salient. Freelists can also be analyzed to compare informants' different perceptions of what items are important, determine a consensus of what items belong to the domain and, critically, identify potential experts or knowledgeable amateurs. If you have written down the fruits you know, then you have just completed a freelist.

Freelists are often conducted early on in a study, but not until you have got to know someone and they are comfortable being questioned by you. At certain times of the year certain domains are more relevant than others. For instance, 'fruits' may be on people's minds during the fruiting season and their answers may be conditioned by recent events such as harvesting a particular fruit. It may be critical to repeat the exercise at different times of the year to see if the freelist changes. You will also need to pick a representative sample of respondents from the community you are working with. You may wish to include men and women, old and young, and people with specialist occupations.

Prior to collecting freelists, you first need to determine exactly what the domain is called in the language you are working in and whether it is a recognized and important domain for the people you are working with. Collecting wordlists in order to study the language is a good starting point. Casual conversations with a translator may also be useful. Once you have your domain you can formulate the freelist question and begin to collect freelists (see Box 8.1).

If you are interested in the *boundaries* of a domain (that is, all the items included in the domain), then you should continue to interview people until you get no new items listed. Often that requires somewhere between 10 and 20 interviews for a typical domain among a fairly homogenous community. But if you are testing hypotheses concerning differences in knowledge among subgroups, such as men and women, then you need samples sizes that reflect those populations (see Section 4.3.2).

Freelists can give us some clues as to the importance or *salience* of certain items in the domain. Salience reflects meaningfulness – their centrality or typicality as exemplars of the domain. For most northern temperate dwellers (and city folk everywhere), apple, pear, orange and even the tropical banana are considered to be the most salient items in the domain of *fruits*. This does not imply that they are eaten most often or are the most valuable economic products, though they might be. The most salient items are the ones that capture the essence of what the domain means for a group of people, and are often the first items one

Box 8.1 Instructions: how to do a freelist exercise

Freelists are conducted with individuals, either orally or by asking respondents to complete the freelist themselves in writing.

1 Ask the same question of all respondents: *'Please tell me (or write down) as many X (domain name) as you can think of'* or *'Please list all the X (domain name) you know'*.
   a If they come to a stop, encourage them to remember more items by reading over the list they have produced so far.
   b You may wish to limit the list to 10, 15 or 25 items depending on your goals.
   c Please note that a freelist asks people for what they *know*, not what they *use* or *do*! Those are different questions, often asked later.
2 Write down or record answers in the order they are given and in the language being used, exactly as they are spoken (this is where the translator may help). Audiotape or digital recordings allow you to check your transcriptions. Do not 'correct' the list yourself – you must do that with informants (see main text).
3 Follow up the freelist with an interview to discuss in greater detail the items they have mentioned, collecting synonyms, descriptions, uses and the significance of each item to the informant.

thinks of when mentioning the domain – and therefore the first items that are likely to be produced during freelisting. This suggests a means for analyzing freelists; making use of the order in which items are mentioned, as described below.

Freelists can be analyzed by hand, but they need to be arranged in a particular way, which involves the use of a *matrix* (a table of rows and columns) (see Box 8.2 and Table 8.2). In fact all CDA methods rely on matrices to order and analyze data.

Analyses of freelists can focus on either the items (to tell you about the nature of the domain) or the informants (to compare different people's knowledge of the domain),

Box 8.2 Instructions: how to analyze a freelist

1. Create a matrix of items (rows) by informants (columns). You can do this on a spreadsheet very easily. As you enter the data for each informant, add any extra items (as successive rows) that are not yet listed. Use codes rather than names for the individual informants (see Table 8.2).
2. For each informant, fill in the rank number of the item as it occurs in their list (1 for the first item mentioned, 2 for the second, etc.).
3. Determine the frequency of occurrence for each item: the number of times it occurs across all respondents divided by the total number of respondents.
4. Determine the average rank for each item: add up all the ranks within a row and divide by the number of informants that listed that item (note: NOT the total number of informants).
5. Plot average rank (y axis) against frequency (x axis) (see Figure 8.1). Where are the most salient items? Are there clusters?

Table 8.2 Freelist matrix

		F21	*F22*	*F27*	*F33*	*M21*	*M22*	*M23*	*M33*	*Frequency*	*Average rank*
1	Orange	1	0	0	0	0	3	6	0	0.38	3.33
2	Apple	2	1	1	1	1	1	1	3	1.00	1.38
3	Banana	3	2	2	0	4	2	2	8	0.88	3.29
4	Pineapple	4	9	0	5	0	0	5	0	0.50	5.75
5	Grape	5	0	8	8	2	4	0	0	0.63	5.4
6	Mango	6	0	0	9	0	8	4	1	0.63	5.6
7	Coconut	7	6	0	0	0	7	0	5	0.50	6.25
8	Papaya	8	0	0	0	0	9	0	9	0.38	8.66
9	Guava	9	10	0	10	0	0	0	4	0.50	8.25
10	Avocado	10	0	0	0	0	0	0	0	0.13	10
11	Strawberry	0	3	10	0	5	5	0	0	0.50	5.75
12	Blueberry	0	4	0	0	7	0	0	0	0.25	5.5
13	Raspberry	0	5	0	0	6	6	0	0	0.38	5.66
14	Melon	0	7	0	0	0	0	9	0	0.25	8
15	Pear	0	8	3	2	0	0	0	0	0.38	4.33

Source: University of Kent Workshop in Practical Methods in Conservation Social Science.

Note: Informant codes show sex and age. Informants were asked to list fruits known, but were stopped after ten fruits. The table is truncated – it shows only the 15 most frequently mentioned fruits.

or preferably both. If you are interested in the items, you can simply compile all the items mentioned, and see which ones were mentioned most frequently and which ones occurred highest up in the list. The assumption is that the more salient an item is in a domain, the more likely it is to be mentioned sooner and the more people will mention it.

Items mentioned only once may be questionable members of the domain or they may be older synonyms, names from different dialects or languages, or just mistakes. Freelists should always be checked carefully with knowledgeable informants. Never delete the raw data, as idiosyncratic answers may in fact prove to be interesting and important. For example a long list of items mentioned by just one person may point to an expert with memory of past terms now lost to the majority of people in the community. Some domains are more commonly known, such as 'fruits', while others may only be fully known by specialists, such as 'medicinal plants'. Similarly, never edit a list for spelling differences (grape – grapefruit) or synonyms (pawpaw – papaya) until you are sure that you know what people are referring to. Sometimes plurals are used for fruits in everyday language (for example, grapes) while the name of the source plant may in fact be singular (grape). You may have to use *identification exercises* to finally discern the identity of items on a freelist (see Section 8.2.3).

You can compare informants by adding up how many names each person mentioned and which items they mentioned first. More advanced analyses use *similarity* tests to determine the level of agreement (or consensus) among a group of informants (Borgatti, 1996).

Figure 8.1 shows a plot of average rank (y axis) against frequency (x axis) for a simple freelist of fruits. The plot shows more salient items in the lower right hand corner; 'apple' and 'banana' were mentioned more often (higher frequency) and listed higher up in the freelist (lower average rank). At the other end, in the top left corner, 'avocado' was mentioned only once and was last in the list, suggesting it is not a very meaningful item in the domain; indeed it may not be considered a 'fruit' at all by many people. Perhaps with more

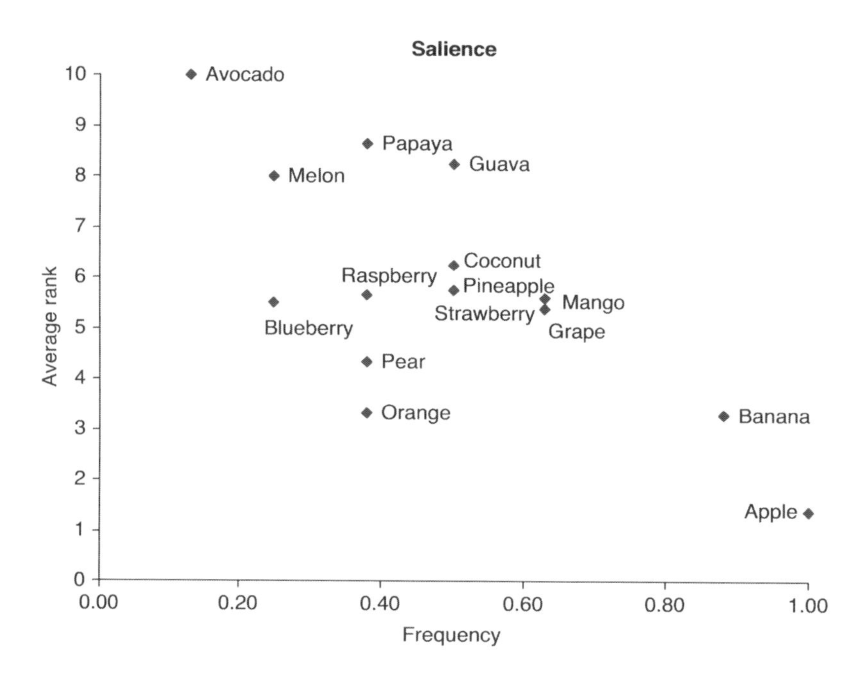

Figure 8.1 Salience of fruits.

informants, 'pear' and 'orange' would move to the right, as the frequency of mention increased. Note also the clusters of similar items – the 'berries' appear close to each other, as do some of the 'tropical' fruits, such as papaya, guava and melon. This usually results because thinking of one item triggers memory of other 'similar' items, so they appear close to each other in the individual freelists. These clusters can be thought of as subgroups, or categories, within the domain. The *pile sort* exercise is used to explore the structure of the domain more directly (see Section 8.2.3); here we have just a hint that there may be complexity in the domain.

Figure 8.1 shows a measure of salience in terms of frequency and average rank in simple graphical terms. A more sophisticated measure is Smith's S (Smith, 1993; see Borgatti, 1996). This can be calculated by hand as described in the following text, but for longer lists ANTHROPAC is much faster.

To calculate Smith's S, first take the freelist matrix and for each informant, calculate the salience of each fruit listed. This is done by inverting the rank (for example, 1 becomes 10, 2 becomes 9, and so on) and dividing by the total number of items in that particular informant's list. Table 8.3 shows this calculation for the first freelister in Table 8.2.

In Table 8.4, the salience index or S value is then calculated as the sum of all individual saliencies divided by the total number of informants (not just the informants that mentioned the fruit, as was the case in the average rank in Table 8.2).

Having a quantitative value for salience of each item in a domain makes comparison within and between communities possible. You could test the hypothesis that men and women think of the domain of fruits differently by comparing Smith's S for just the male and just the female informants in Table 8.4. In this dataset, 'orange' and 'mango' are slightly more salient for the men, while 'pineapple' is more salient among the women.

Note that including all the zeros for fruits not mentioned means that the salience value is much lower than it would otherwise be, and the differences between the fruits appear much greater than is seen in the plot. 'Orange' seemed to be a very salient fruit in the plot because of its low average rank, but in fact the distance between 'orange' and 'apple' is shown to be roughly the same as that between 'apple' and 'pineapple', 'strawberry', 'coconut' or 'pear'. However, as mentioned above, the graph helps you to see that despite a similar salience index, 'orange' is more closely clustered with 'pear' than with the others.

Table 8.3 Calculating Smith's salience index for freelist respondent 1

	Rank	*Inverted rank/ total listed*	*Salience*
Orange	1	10/10	1
Apple	2	9/10	0.9
Banana	3	8/10	0.8
Pineapple	4	7/10	0.7
Grape	5	6/10	0.6
Mango	6	5/10	0.5
Coconut	7	4/10	0.4
Papaya	8	3/10	0.3
Guava	9	2/10	0.2
Avocado	10	1/10	0.1

Table 8.4 Calculating Smith's S for all respondents

	F21	*F22*	*F27*	*F33*	*M21*	*M22*	*M23*	*M33*	*Sum*	*Smith's S (sum/8)*
Orange	1	0	0	0	0	0.8	0.5	0	2.3	0.29
Apple	0.9	1	1	1	1	1	1	0.8	7.7	0.96
Banana	0.8	0.9	0.9	0	0.7	0.9	0.9	0.3	5.4	0.68
Pineapple	0.7	0.2	0	0.6	0	0	0.6	0	2.1	0.26
Grape	0.6	0	0.3	0.3	0.9	0.7	0	0	2.8	0.35
Mango	0.5	0	0	0.2	0	0.3	0.7	1	2.7	0.34
Coconut	0.4	0.5	0	0	0	0.4	0	0.6	1.9	0.24
Papaya	0.3	0	0	0	0	0.2	0	0.2	0.7	0.09
Guava	0.2	0.1	0	0.1	0	0	0	0.7	1.1	0.14
Avocado	0.1	0	0	0	0	0	0	0	0.1	0.01
Strawberry	0	0.8	0.1	0	0.6	0.6	0	0	2.1	0.26
Blueberry	0	0.7	0	0	0.4	0	0	0	1.1	0.14
Raspberry	0	0.6	0	0	0.5	0.5	0	0	1.6	0.20
Melon	0	0.4	0	0	0	0	0.2	0	0.6	0.08
Pear	0	0.3	0.8	0.9	0	0	0	0	2	0.25

Note: To calculate Smith's S for each fruit, the sum of saliences for each individual informant is divided by number of respondents ($n = 8$).

8.2.2 *Identification exercises*

All of the above analyses can be done with just the lists, without knowing what the names of individual items refer to. In order to move along in the analysis of the domain of fruits, we now need to identify what exactly these things are. For those of us that speak English and live in the UK, perhaps all of the fruits mentioned above are well known and we could identify them in a market. But if you are working in areas where you are not familiar with either the language or the fauna or flora, the terms in a freelist may be unknown.

Identification exercises (ID) involve a structured interview where respondents are shown different items of a cultural domain and asked to identify them with local names and, in many cases, answer a series of follow-up questions about them. The basic objective is to determine the 'correct' answers for local names of items and then score respondents on how close they come to providing these answers. In some studies, you may choose simply to assess an answer as true or false. One can infer an informant's knowledge of a domain from the results. Scores can be used to test hypotheses about variation and change in knowledge – for example by testing for correlations with age, sex, occupation, education and other factors. However, it is often the case that there is more than one correct answer, and unless you are sure what the correct answers are you can also not be sure that the consensus answer is indeed the correct one. Working with recognized experts in a community beforehand is a good way to establish an answer key for an ID test. Alternatively, you may choose to compare answers given by different people directly, assessing the *pattern of agreement* in answers rather than devising a *score*. Hypotheses about the factors responsible for these patterns of agreement can then be developed and tested using other data collection and analytical tools. One can also examine which items are not well known to which informants by looking at the pattern of 'mistakes', in terms of both informant and item characteristics.

Using props as an elicitation device is an excellent way to get people talking more generally about the relationship between people and the domain in question. An identification

Box 8.3 Instructions: how to do an identification exercise

1 Locate or collect items to be identified.
2 Lay items out in a random order on tables; give each item a reference number.
3 Ask a sample of respondents, one by one, to identify the items and write the name(s) of each item on a sheet of paper together with the reference number of that item. Alternatively, accompany respondents and write down, or record, their answers for them. You must specify a language, or ask for all names in all languages known.
4 If you are accompanying them, then for each item you may also ask what criteria were used to make the identification.
5 Ask additional follow on questions about uses, value, origins, etc.
6 Score answers as correct or not according to an answer key, or analyze *patterns of agreement* across respondents.

exercise can also identify experts who might be interviewed at length at a later date or hired as field research assistants for other research projects. On the other hand, the technique requires a concentrated effort on the part of respondents, not all items of a large domain can be used, and some people may choose not to participate in what they may perceive as a test of their competence.

Box 8.3 describes the basic steps of an ID exercise. However, there are many variations on this basic structure. If you are working with plants and plant related domains, you may use photographs, collected plant voucher specimens (including fruit, seeds or flowers), objects made from plant materials, or show people plants in situ – for example in a home-garden (Boster, 1986; Vogl et al., 2004), a permanent plot (Bernstein et al., 1997) or along a specified plant trail (Puri and Vogl, 2005; Stross, 1969). Those working with animals may use live specimens (at a zoo), dead specimens (in a museum), photographs, videos or even cultural objects and artefacts produced from animal parts (Boster, 1985, 1986). You could collect soil specimens around the community for an ID task on soil categories. Berlin and Kay (1969) used colour chips from the Munsell Colour Chart to identify colour terms in several languages.

Before you can begin the interviews, you will need to determine what items are to be included in the test and who the respondents will be. Researchers are free to vary the number of items and the number of informants interviewed. The number of items used usually depends on how many questions are to be asked and, ultimately, on how long informants have to participate and how long you can hold their attention. Experience suggests one hour uninterrupted is probably an ideal target, but with breaks you may be able to extend the interview. The number of people you interview may be limited by their availability, and also by the amount of time the researcher has available. You must also plan for a period of testing and fine-tuning of the interview protocol to fit local languages, customs and other conditions before you begin data collection in earnest.

In the case of fruits, you could set up an identification task by visiting a market and buying all the things that people say are fruits or that you think are fruits. You could then set up a few tables in a common room, label the specimens with numbers and ask people to walk around identifying them orally or by filling out an answer sheet (see Figure 8.2).

Figure 8.2 Fruits identification exercise in South Africa.

Alternatively, you could walk around a community and find trees in fruit, label them and then create a *tree trail* so people can identify them in situ; again you can lead informants around and interview them orally or let them go on their own. In both cases the aim is to establish a link between a local name and a biological individual; if you find high agreement about the name of a specimen, then you can set about collecting a botanical voucher specimen and getting it identified by professional taxonomists (see Martin, 2004).

Table 8.5 is a score sheet for an ID exercise carried out among African students (A–F) to identify 18 fruits and vegetables (1–18) found in a supermarket in South Africa. From their answers, can you determine the 'correct' answer for each specimen?

By simply eyeballing this small dataset you can probably come up with answers based on majority responses (for example, 'avocado' for no. 1), though you might have difficulty with numbers 4, 7, 8, 15 and 18, and be dead wrong on number 17, which was indeed a prune, despite four out of six informants labelling it a date. This is an understandable mistake since African students are more likely to have seen dates than prunes. You'll also notice that one student has answered in French. While translation to English is possible, and most of the answers would be in agreement with others, it is important to know that some of your respondents are using a different language and therefore may have different conceptions of that domain. Still, you might not use that respondent for further analyses but seek additional ones – also to get more accurate identifications for the other undetermined specimens. It may also be the case that the large variation in names is due to a problem with the specimen itself, which may not be a typical example and therefore may be somewhat ambiguous in appearance.

There are other interesting data to be gleaned here. In some cases, we are given additional names, such as 'avocado pear', which suggests that this fruit is a type of 'pear'. Other names suggest varieties, such as 'red apple' or 'pink apple', or 'baby marrow', 'baby pumpkin' and 'baby tomato'. All of these are somewhat ambiguous from an English speaker's

Table 8.5 Identification exercise data sheet

Respondent Fruit	*A*	*B*	*C*	*D*	*E*	*F*
1	Avocado	Avocado	Avocado pear	Avocat	Avocado	Avocado
2	Kiwi	Chestnut	Fig		Kiwi	Sapota
3	Dried apple			Anacarde	Dried apple	
4	Lemon	Lime	Lemon	Citron	Lime	Limon
5	Banana	Ripe banana	Banana	Banana	Banana	Banana
6		Pumpkin		Zucchini	Pumpkin (baby)	
7	Cucumber	Baby marrow	Courgette	Courgette	Squash	Cucumber
8		Cojeti				
9	Apple	Red apple	Apple	Apple	Pink apple	Apple
10	Pineapple	Pineapple	Pineapple	Ananas	Pineapple	Pineapple
11	Greenpepper	Sweet pepper	Green pepper	Poivron	Capsicum	Green pepper
12	Orange	Orange	Orange	Orange	Orange	Orange
13	Pear	Pear	Pear	Poire	Pear	Pear
14	Tomato	Tomato	Tomato	Tomate	Baby tomato	Tomato
15	Squash	Yellow baby Marrow				
16	Dried pear		Fig		Pear	
17	Prune	Date	Date	Prune	Date	Date
18	Dried peach	Zucchini	Fig		Dried mango	

point of view: are they proper varietal names or just descriptive adjectives? 'Baby' could refer to the small size or an under-developed fruit. Similarly, 'dried apple' is the name of a product, made from 'apple', sold in packets where the apples have been sliced and dried. Perhaps in this case, the researchers need to specify that the identification wanted is of the fruit and not of the product or how it looks. All of these examples reflect problems in local classification that have to be recognized in order to comprehend a local domain and how it can be constructed. The next method, pile sorts, is particularly suited to this task.

8.2.3 Pile sorts

Pile sorts are used to discover how respondents think of the internal structure of a domain by showing how items are related to each other, and how they are classified in subgroups. Of particular interest are the criteria used to classify items and the underlying logic governing the arrangement of subcategories.

One can see local classification systems at work in the way store owners arrange and display goods, such as the famous spice and herb sellers in the souks of North Africa or in Chinese pharmacies. Understanding the way these items – or any foods – are organized is important for understanding dietary choices, such as which foods can or cannot be cooked or eaten together. Another example is that classification systems help immigrants assess the probable uses of plants and animals they come across in their new surroundings. Conservationists may also be interested in the composition and internal structures of domains such as hunted prey (see Figure 8.3), firewood or fodder sources, income opportunities, or environmental threats and risk.

Box 8.4 describes the basic steps in conducting a pile sort exercise. Often it is done as a group exercise, and stimulates lively conversation as respondents try to come to a consensus

Figure 8.3 Pile sort of Bornean mammals.

about which criteria to use in creating the piles. Observing and listening to these discussions is an excellent way to learn more about the domain and your respondents.

Box 8.5 outlines the different ways in which pile sorts can be analyzed. Analyzing a pile sort qualitatively can involve simply discussing the results with the respondents, focusing on the criteria they used and any difficulties they had in placing items in piles. Asking how

Box 8.4 Instructions: how to do a pile sort exercise

In interviews with one or more respondents:

Put names of items on to scraps of paper, or paste photos or drawings onto stiff paper. You may use actual items if small enough. Create a number code for each item and label the paper, card or item.

Ask respondents to make piles of *similar* items.

1. Be careful how you translate 'similar' and always pilot your question with several respondents. If they respond by asking 'What do you mean by "similar"?' Tell them that you are interested in what *they* think that means.
2. You can have a pile of one, but you can't have each item in a pile by itself or they would very likely not be part of the same domain.

For each pile made, ask if there is a name for that pile, and why the items are placed in that pile.

Record the number code of all items in each pile. Write down any names or sorting criteria mentioned and if you can, photograph or draw the sort as a whole so that you record how the piles were arranged.

Variations:
You may give slightly different instructions in order to produce different types of pile sort, as follows:

1 In *'unconstrained'* pile sorts, you give no instructions as to what criteria to use.
2 In *'constrained'* pile sorts, you ask informants to sort by some specified criterion such as use, morphology, habitat, and so on.
3 In *successive* pile sorts, respondents are asked to break down the piles from the initial sort into further sorts. Repeat until all items are left as single piles. Alternatively, ask respondents first to sort all items into piles and then ask them to group them into larger piles, until you end up with one pile.

items in one pile *contrast* with a neighbouring pile helps to clarify the characteristics used for inclusion. Sometimes multiple criteria may be used to sort all items into piles. If this is confusing to you or your respondents, then conducting a *constrained* pile sort (see Box 8.4) for each criterion separately may be useful.

Noting the names of all subcategories (and some that may not be named) and any synonyms or disagreements in names is also important. Photographs or drawings of different pile sorts made by individuals or groups can also be compared side by side to get some idea of the different ways the subcategories are arranged to create an overall structure of the domain.

Box 8.5 Instructions: how to analyze pile sort data

Qualitatively

1 Among respondents or groups, compare the number of subcategories, the items included, and their names and criteria, by 'eyeballing' the data:

  a How many basic categories are there? How many larger, or more inclusive, categories are there? Are all categories named? Do any items not fit in any categories? Why?
  b Did the criteria used to classify items change?
  c How are the piles arranged spatially, does this have any significance?

Quantitatively

2 Create a *proximity matrix* with the list of items along the top and down the side.
3 For each cell in the matrix, add a tally mark every time that the two items the cell refers to occur in the same pile. You can do this either for each respondent (an *individual proximity matrix*) or across all the respondents (an *aggregate proximity matrix*).
4 Divide the final tallies by the total number of pile sorts completed, to get a percentage of agreement.

Box 8.6 Quantitative analyses of pile sorts

Data

Pile sort 1: AB C DE [based on origins: temperate, Mediterranean, tropical]
Pile sort 2: AB CDE [based on origins: temperate, tropical]
Pile sort 3: AB C DE [based on origins: temperate, Mediterranean, tropical]
Pile sort 4: AB CDE [based on origins: temperate, tropical]
Pile sort 5: AB C DE [based on origins: temperate, Mediterranean, tropical]

Aggregate proximity matrix

	A (apple)	B (pear)	C (orange)	D (banana)	E (mango)
A(apple)	–				
B (pear)	5 (100%)	–			
C (orange)	0	0	–		
D (banana)	0	0	2 (40%)	–	
E (mango)	0	0	2 (40%)	5 (100%)	–

Note: Proximity matrix shows number of times each fruit is in a pile with every other fruit. Matrix top half would mirror bottom half.

Cluster diagram: **Per cent of Agreement**

```
A---B   C     D---E        100%
  |     |       |
  |     |_______|           40%
  |         |
  |_________|                0
```

Venn diagram:

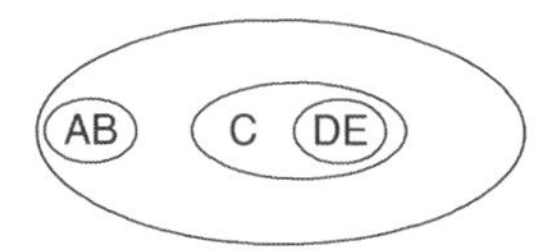

For instance, while a pile sort of mammals among Bornean hunters produced subcategories of similar looking animals, with some categories equivalent to 'bats', 'monkeys', 'deer', 'mice' and 'otters', the piles were then arranged spatially by habitat, with flying and tree dwelling categories at the top, land dwelling categories in the middle and water dwelling and underground categories at the bottom (see Figure 8.3).

The basic steps in quantitative analysis are also outlined in Box 8.5, and an example showing the format of the results is given in Box 8.6. If each respondent does one sort for all items, then each item can occur either no times or just once with any other item, so in the individual proximity matrices, the values in each cell can only be one or zero. Aggregate proximity matrices compile all these individual matrices to produce a group answer, which when compared to each individual proximity matrix allows you to test for consensus and variation among respondents.

Matrices can be analyzed qualitatively and quantitatively using specialist software (for example, ANTHROPAC, UCINET, SPSS). Box 8.6 contains the results of five pile sorts done by informants on five fruits, shown as an aggregate proximity matrix, and two ways of displaying the information it contains: a cluster diagram (or tree diagram) and a Venn diagram. A simple cluster or Venn diagram of items (for each respondent or the group as a whole) is useful in showing the internal structure of domain. In the example, informants made either two or three piles, all based on the criterion 'origins'. Three informants distinguished oranges as Mediterranean, while two lumped them together with banana and mango as tropical fruits. In the cluster and Venn diagrams we can see that the pairs (AB and DE) are physically close together because they were deemed similar 100 per cent of the time, while C is shown to be more distant. All the items are included in the domain of 'fruits', hence the link of all items at 0 per cent agreement in the cluster diagram and the large oval encircling all items in the Venn diagram. This is a simple example to demonstrate what can be done. For further details, see software instructions and works by Borgatti (1996).

The diagrams in Box 8.6 merely show patterns of agreement among informants, where similarity (agreement) is expressed as closeness and difference as distance. This is a simple case where the underlying criteria are understood as 'origins' and where differences among informants are analyzable. However when the number of items and the number of informants is increased, patterns may emerge that are not easily explained, even when informants claim they have used set criteria. Sometimes respondents will indicate that two or more items go together, but not be able to articulate why; they may claim to use their instinct or a gut feeling in forming the piles. This is more likely when dealing with more abstract domains than 'plants' or 'animals', such as 'kin', 'emotions', 'illnesses' or 'rights'. Thus, there may be hidden dimensions that underlay the ordering of complex domains, which when revealed may indicate deeply held cultural logics, beliefs and values. You can generate hypotheses about what these dimensions may be by examining the reasons given for the pile sorts as well as eyeballing the results and looking for patterns.

In the next sections then, you'll learn about *paired comparison* and *weighted ranking* exercises, which can be used to test how well noted criteria and hidden dimensions actually fit the pile sort data.

8.2.4 Paired comparisons

In a paired comparison exercise (PC), a respondent, or group of respondents, is asked to compare two items at a time from the same domain according to some predetermined criteria or attributes (also known as dimensions), such as importance, size, value, traditional character, healthiness and so on. The data generated allow you to place the items along the dimension in question – for example from least to most 'important'. In general, paired comparisons are used following pile sorts, but they may be used outside of cultural domain analysis in evaluating conservation or development options with individuals or groups or in valuation studies of plant and animal resources (more on this later).

Box 8.7 gives simple instructions for conducting a paired comparison, and Box 8.8 shows an example of the resulting data and their analysis.

In the pile sort example in Box 8.6, based on the criteria given by informants you might hypothesize that the fruits displayed in the Venn diagram are arranged from left to right along a dimension of increasing 'exoticness' (from the perspective of a northern European). Box 8.8 shows an example of data collected using a paired comparison exercise to test this idea. If you take the final scale from the paired comparison and superimpose it on top of the Venn diagram from the pile sort (Box 8.6), it seems to fit very well; the order is the same and the distances seem also proportionate. However 'orange' and 'banana' are closer on the exotic scale, while the pile sort puts 'mango' and 'banana' together more frequently. This would indicate that exoticness alone is not sufficient to explain the Venn diagram. Perhaps 'distance to grower' might be a better fit dimension. Using a type of multiple regression called Property Fitting (PROFIT), ANTHROPAC can analyze the fit and show how statistically significant the relationship is.

Identifying the criteria people use in organizing a domain such as fruit trees might be important for understanding why people plant species in certain proportions in their orchards. Some may grow fruit that is highly valued as a food source, or has a high market price, or is drought resistant or produces good fodder. Perhaps people plant a variety of trees so that overall, they maximize all these values and reduce risks. An understanding of these types of value would be very important when attempting interventions in local agricultural practices. This will be discussed in greater detail in the next section, which is on weighted ranking. Paired comparisons can also be used in other kinds of applied research – for instance to determine a value scale for natural resources such as wild plants or animals.

Box 8.7 Instructions: how to do a paired comparison

With an individual or group:

1 Select a set of items and on a sheet of paper, make a list of all possible pairs.
   a Duplicates (AB and BA) are often included to check reliability.
   b If you have a very large domain you may eliminate the reverse order pairs, or do the exercise several times with different subsets of the domain.
2 For each pair, ask the informant 'which is more [important, or whatever the dimension you are interested in] – A or B?' or 'Is A more [important] than B?'
3 You can ask either orally, or in a written questionnaire where you list all the pairs in a random order, and then ask respondents to circle the one of each pair that is more (or less) [important]
4 Compile answers for each respondent in an individual proximity matrix, tallying the number of times A is said to be more [important] than B, and vice versa.
5 Add up the responses from all informants in an aggregate proximity matrix to compute an aggregate scale that places all items along a scale from less to more [important].

Box 8.8 An example of a paired comparison

Five respondents were each asked 20 questions (four for each of five fruits) about whether one fruit was more 'exotic' than another. The answers were either yes (1) or no (0) and were recorded in the rows of a matrix, aggregated answers are shown below. Can you tell whether 'apple' is regarded as more exotic than 'pear'?

Question asked: Is x (row) more exotic than y (col)?

	A	B	C	D	E	Total
A Apple	X	0	0	0	0	0
B Pear	5	X	0	0	0	5
C Orange	5	5	X	2	0	12
D Banana	5	5	3	X	0	13
E Mango	5	5	5	5	X	20

Note: In this aggregate matrix top and bottom halves are not mirrors of each other.

All five agreed that apple is *not* more exotic, so the square A (row) x B (column) has zero in it. In fact the whole of row A contains zeros, meaning that no one said that apple was more exotic than any of the four other fruits. It has a total of 0 and is therefore the least exotic. On the other hand 'mango' (E) was judged by all five informants to be more exotic than all the other fruits, with a final score of 20 it is at the other end of the scale from 'apple'. 'Pear' (B) received a total score of 5, because all informants thought it more exotic than 'apple' but less exotic than the other three fruits. 'Orange' (C) and 'banana' (D) are very close on the scale, because two informants judged 'orange' to be more exotic than 'banana' while three thought the opposite. The final results can be presented as a scale of exoticness, as shown below.

Final scale: **'Exoticness'**

Less exotic 0____5____10____15____20 More exotic
Fruit: A B CD E

8.2.5 *Weighted ranking*

In a weighted ranking exercise, respondents score (or vote for) each of a set of items on one or more attributes using counters (for example, maize kernels, beads or stones). In a typical ranking exercise the respondents would list items from most to least (for example, 'most important' to 'least important'), but in a *weighted* ranking exercise respondents do more than this – they have a set number of counters and distribute them across the items, giving more counters to items they rank more highly. In short, ranking provides order of preference but no indication of relative magnitude; weighted ranking gives information on both of these factors. If the criterion were 'importance', as in the importance of fruits in the diet of a group of people, a weighted ranking allows researchers to say *how much* more important (in relative terms) is one food item over another. Box 8.9 gives instructions for carrying out a weighted ranking exercise.

As with paired comparisons, the end result is a scale for each attribute showing how similar or different items are. Box 8.10 describes exactly how to analyze a weighted ranking.

Box 8.9 Instructions: how to do a weighted ranking

With an individual or group:

1 Write the names, draw, or paste photos of the items to be ranked on separate sheets of paper or index cards.
2 Decide what attributes or characteristics they will be ranked by. For instance, fruit trees could be ranked according to their importance for uses such as food, fuelwood, fodder, marketability, etc. Freelisting may be used to determine these attributes: you would simply ask *'Please list all the uses of fruit trees'*.
3 Create a data collection sheet in the form of a matrix, with the items to be ranked in the rows and the attributes in the columns. The instructions to respondents might look like this:

Please distribute 100 counters between the fruits in proportion to their rank for each of the attributes listed. For example, if the attribute were 'weight', you'd score fruits according to how heavy they are, with very light items receiving a few counters and heavy ones receiving relatively more. Remember that 30 votes is three times as heavy as 10. You may give an item zero votes. Thank You.

	Taste	*Fuelwood*	*Marketability*	*Storability*
Apple	______	______	______	______
Pear	______	______	______	______
Orange	______	______	______	______
Banana	______	______	______	______
Mango	______	______	______	______

4 Ask respondents to shuffle and arrange the items in rank order according to the first attribute. Respondents then score items using the counters. Repeat for each attribute. (Note: 100 counters make for easy calculation of the results by hand. Some groups will prefer to work together to decide how many counters should be placed on an item, while others may choose to distribute the counters evenly among themselves and then vote individually.)
5 Listening to the discussion of each group is useful to understand if and how consensus is achieved, as well as to gain a better idea of variation in local opinion.
6 Record the number of counters on each item on the data collection sheet in the column of the attribute being assessed. Repeat for each attribute. (See also IIRR, 1996; Sheil and Liswanti, 2006; Sheil et al., 2003).

Box 8.10 Instructions: how to analyze a weighted ranking

1 Examine the total scores for each item on each attribute. You can also add across the attributes (columns) to get a total score for each item, thus:

	Taste	Fuel	Sale	Storage	Total
Apple	16	50	25	40	131
Pear	14	15	15	28	72
Orange	22	10	24	10	66
Banana	10	0	26	17	53
Mango	38	25	10	5	78
	100	100	100	100	

2 Draw a linear scale for each attribute and plot the items along the scale. As with the scale created using the paired comparison test (Box 8.8), the scale can also be used to analyse patterns in pile sort data.

For example, the scale for *taste* looks like this:

0__________10__________20__________30__________40

Ba Pe Ap Or Ma

3 The information on each attribute or alternatively for all attributes together can also be displayed graphically using a pie chart:

Sum of values for all attributes

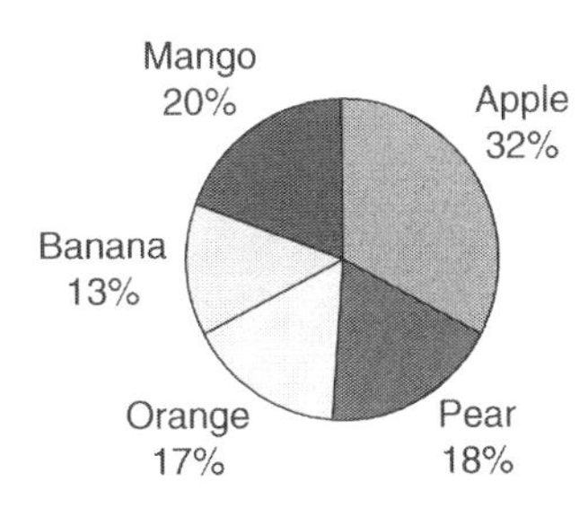

4 Results of repeated weighted ranking exercises with different individuals, groups and communities can then be compared using analytical statistics.

To demonstrate the use of this method in conservation, Box 8.11 shows the results of research in Indonesia to understand the importance of land types for hunting among two ethnic groups in a rainforested valley in the lowlands of East Kalimantan, Indonesia. Weighted ranking exercises were conducted on a variety of plant use categories with 26 focus groups made up of people of different age categories and sexes among two ethnic

Box 8.11 An example of a weighted ranking: the importance of land types for hunting places in Indonesia

Note: values are mean weighted ranking across focus groups.

Land/forest type	*Punan*	*Merap*
Village ground	0.19	0.00
Abandoned village	7.06	4.67
Horticulture	9.06	4.17
River	16.69	11.67
Swamps	4.94	10.33
Swidden	6.63	8.75
New fallow	4.19	6.33
Old fallow	12.19	18.58
Forest	39.06	35.50
Total	100.00	100.00

Punan hunting places

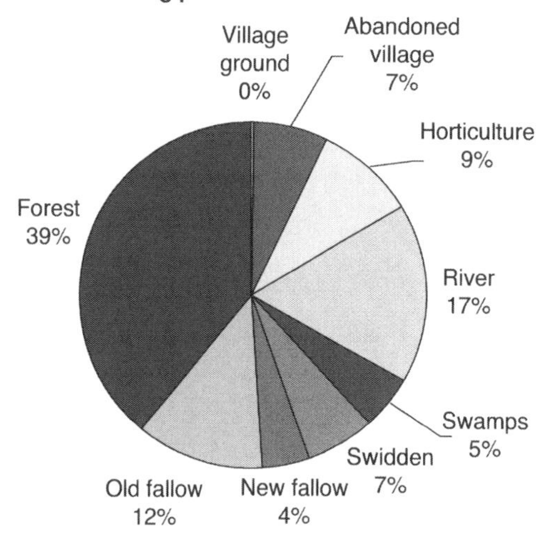

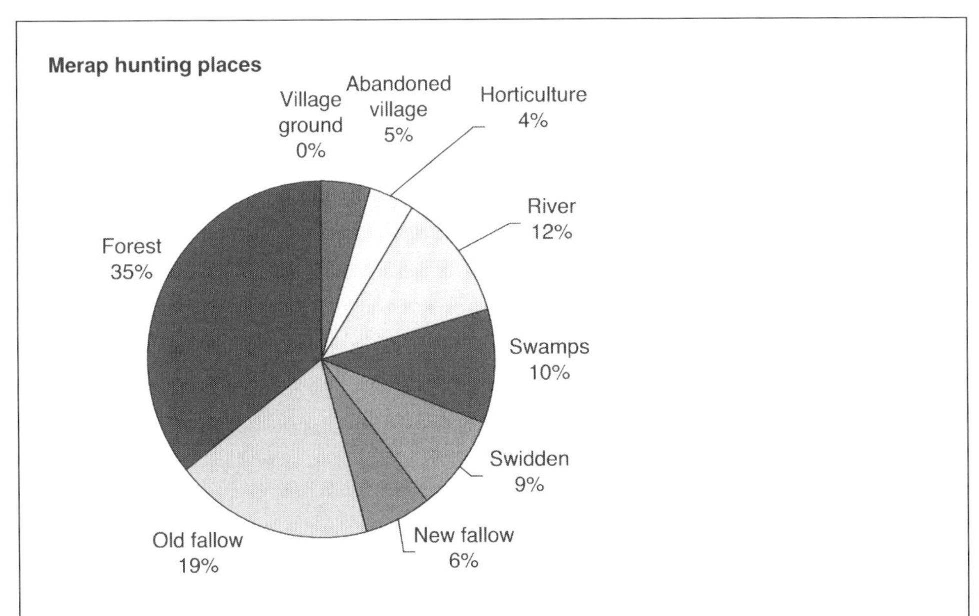

Source: MLA Project, CIFOR (see http://www.cifor.cgiar.org/mla). See also Sheil et al. (2003, 2006).

groups (the Punan and the Merap) living in seven villages. The results are compiled to show Punan foragers' versus Merap farmers' mean scoring of land types in terms of their 'importance' for hunting locations, which are often near fruiting trees. These percentages can be displayed as a pie chart, which shows similar results for the two groups, though it appears that Punan view forest, rivers, gardens and abandoned villages as most important for hunting, whereas the Merap farmers have scored fallow forest, swamps and cultivated areas more highly. This result is understandable given that Punan communities live upriver in less populated areas and are less engaged in agriculture, while Merap farmers live downriver in larger villages with larger settlements and more commercial activities nearby, such as logging, mining and plantations.

Though not pictured here, you could also break these results down by other groups of people – men and women or young and old members of each community – to compare how they value the different land types for hunting (see Sheil et al., 2006).

8.2.6 Concluding remarks about CDA

Freelisting, identification exercises, pile sorts, paired comparison and weighted ranking are methods conducted in structured interviews that can provide both qualitative and quantitative data on how people see and organize domains of importance to their livelihoods, cultural practices and their relationships with their environment. Conservationists can use these methods alone or in combination to compare individuals and groups' understandings and values associated with landscapes, natural resources and their uses. For those who want to pursue a more comprehensive understanding of local people's, or indeed conservationists', knowledge of particular domains, these methods are a simple yet richly productive way to begin that exploration.

8.3 Seasonal calendars, trends and timelines

This last section describes ways to investigate local knowledge of processes of change occurring in and around communities. Examples of environmental change might include seasonal differences in weather, flowering and fruiting, animal migrations and, perhaps, the timing of mating and fish spawning. Local people may know of all of these events and may also schedule their own economic activities in response to them. If you're interested in human–environment relationships, then understanding the annual agricultural cycle of farmers or mobility patterns of hunters, fishers and pastoralists is essential, for patterns of resource use vary in time as well as space. Since we rarely have time to observe a location or community for a whole year we often have to rely on local people's knowledge and understanding (see Selener et al., 1999).

A *seasonal calendar* can be constructed to display these patterns and to investigate relationships between the various annual cycles (see Box 8.12).

Box 8.12 Instructions: how to make a seasonal calendar

With individuals or in groups:

1 On a large piece of paper, construct a table of 13 columns and at least five rows, but leave space for more rows. Some people will find it easier to think of time as a circle; in this case construct a round calendar of concentric circles instead of rows.
2 In the cells in the first column, list the themes and categories you wish people to describe, such as weather (rain, temperature, winds), flowering and fruiting of plants, animal migrations, agricultural activities, hunting activities, gathering activities, economic activities (e.g. markets), religious and ritual activities.
3 In the cells across the top row of the next twelve columns, put the name of the months in the local language (or whatever divisions are suitable – some societies rely on a lunar calendar, others may not even use months).

   a Ask several key informants when the year begins; it may be at the beginning of the rains, or for many farmers the harvest festival signals the end of one year and the beginning of another.

4 Starting at the beginning of the year, for each theme or category, ask respondents to list all the general characteristics of that month, including all events that usually occur then.

   b Be careful to distinguish between unique events of the past years and generalizations about what 'normally' happens.
   c Check relationships of events by moving between different categories occurring at the same time; for instance, ask *What farming activities occur just after the rains start?*

5 Work with several individuals or groups (especially men and women) to capture variation in activities and experience.
6 Display results publicly, get comments and feedback from other community members.

Making a large poster of the results can make it easier to discuss this information in group settings, and the poster can be amended or added to during the course of the research period. It is always critical to make sure your respondents understand that you are asking for generalizations of the annual calendar, though this is often risky as most respondents, including researchers, have poor memories and tend to give you idealized versions of the seasonal calendar. One strategy to increase accuracy is to ask someone to create a calendar of what they actually saw and experienced, including what they actually did, over the past year. Then you can ask if these were typical of all years. A conversation focused on this type of comparison is also an excellent way to collect information on *'trends'* in the environment (see Box 8.13).

A discussion of current trends visible to community members today can easily be turned toward the subject of the future by asking people to project current trends into the future, for instance by asking them 'What will this be like in twenty years' time?' You might also use a *weighted ranking exercise* to ask people to rank the importance of these trends 20, 40 or even 100 years into the future (see Section 8.2.5). *Scenario building* uses similar exercises to ask communities to consider a variety of different futures and assess their desirability and achievability given current conditions (see Wollenberg et al., 2000).

Finally, a *timeline* or community history can be used to study important events of the community and how they have influenced its development and relationship to the environment. This is a simple exercise. Often the exact year an event took place is not known, so you must find ways to help people to pin it down in history. This is done by referring to more widely known events, such as economic and political changes and natural catastrophes (see Box 8.14). Alternatively, the timing of events may be remembered with reference to some

Box 8.13 Instructions: how to study trends

With individuals or groups:

1 On large piece of paper create a matrix of columns, one for each theme to be analyzed.
   a Themes could include rainfall, droughts, diseases, population (people as well as animals), in-migration, agricultural productivity, hunting yields, cash income from forest products or wage labour, forest cover, and any other topics raised by respondents.
   b You may ask key informants or groups to work on one theme at a time, or on several at once.
2 The number of rows is determined by the number of years or time segments to be discussed, such as 'today', '20 years ago', '40 years ago', '60 years ago'.
3 To investigate local perceptions of the future, add rows for time periods in the future, such as '20 years from now' or '100 years from now'.
4 Respondents may use graphs, draw diagrams or maps or profiles of the landscape, or just write down text in the space for the theme.
5 Interviews using results can address what respondents consider the cause(s) of these trends, and the consequences for both environment and community life.

Box 8.14 Instructions: how to make a timeline

With individuals or groups:

1 On large piece of paper, create a margin and use it to list dates from the present going back in time as far back as you or the respondents wish to go. The main space to the right is used to write down events that occurred.
   a You can add additional columns to write down events, impacts on the community, impacts on the environment and local responses.
   b The timeline may be correlated to genealogies or outside events if dates are difficult to remember.
2 Respondents may draw diagrams to represent key events.
3 Timelines can also be displayed publicly and updated during the research period.
4 Interviews based on the results can address what respondents consider to be the consequences of events for both environment and community life.

other historical event or circumstance. In Bornean farming societies, history is often described in reference to a long lineage of village chiefs, while in many African pastoralist societies historical events are linked to successive generations of men that pass through initiation rituals at the same time.

8.4 Conclusion

This chapter has covered a variety of methods that can be used in structured interviews with individuals, focus groups and communities. They are all directed toward learning about local people's knowledge and use of the environment. Cultural domain analysis can help us to understand local perceptions of important domains, such as animal and plant resources, and how these vary among different people. The same methods can also be used to understand environmentalists' understandings and values with regard to those same domains. They may also serve to focus a community's attention on their own knowledge and environmental circumstances, which may be beneficial to conservationists seeking to establish a dialogue concerning conservation or management policies.

Summary

1 Conservationists need to be aware of local environmental knowledge because it underpins in part the choices people make about what and where to hunt, fish and gather plant resources.
2 Cultural domain analysis (CDA) is used to understand how people in a society think about and define their world. Since all cultures use some system of categories to order experience, the researcher tries to determine what categories are important to people, how they are arranged and what values are attached to them.
3 These important categories are known as *domains*, each of which is composed of a group of elements or *items* organized according to rules or criteria that are culturally determined

and may be culturally specific. In other words, domains are the stuff of culture. Important domains in conservation include food, hunted prey, forest products, soils, medicinal plants, illnesses and so on.

4 Instructions for collecting data and analyzing results for the following methods were described:
   a *Freelisting* is used to elicit the *items* of a domain, to check whether the domain is locally *salient* or meaningful and which of the individual items is most salient. Freelists can also be analyzed to compare informants' different perceptions of what items are important and identify potential experts or knowledgeable amateurs.
   b In *identification exercises* respondents are shown different items of a cultural domain and asked to identify them with local names and, in many cases, answer a series of follow-up questions about them.
   c *Pile sorts* are used to discover the internal structure of a domain by asking respondents to show how items are related to each other, and how they are classified in subgroups. Of particular interest are the criteria used to classify items and the underlying logic governing the arrangement of subcategories.
   d *Paired comparison* exercises are used to examine the criteria underlying the classificatory schemes revealed by pile sorts. A respondent, or group, is asked to compare two items at a time from the same domain according to some predetermined criteria, such as importance, size or value.
   e A *weighted ranking* exercise can be used to identify local preferences among a set of items, including a measure of magnitude. Respondents score (or vote for) each of a set of items on one or more attributes using counters (for example, maize kernels, beads or stones).
5 Conservationists can use these methods alone or in combination to compare individuals and groups' understandings and values associated with landscapes, natural resources and their uses.
6 Local knowledge includes an understanding of natural cycles in the environment, such as weather, phenology and migration, as well as an understanding of the history of land use, conflicts over resources, and the particular circumstances and events that have led to the current situation. Seasonal calendars, trends and timelines are methods used with individuals or with groups to compile local knowledge of change.
7 Scenarios may be used to study local perceptions of the future.

Further reading

Borgatti, S.P. (1996) *ANTHROPAC 4*, Natick, MA: Analytic Technologies. [Steve Borgatti's User's Guide for his software ANTHROPAC (1996), which is available at www.analytictech.com. Contains more detailed description and instructions for analyzing CDA.]

IIRR (International Institute of Rural Reconstruction) (1996) *Recording and Using Indigenous Knowledge: A Manual*, compiled by R.M. Pastores and R.E. San Buenaventura, Silang, Philippines: IIRR. http://www.panasia.org.sg/iirr/ikmanual/layong.htm. [One of several manuals on the methods developed for participatory rural assessment (PRA). This one describes the use of seasonal calendars, timelines and other techniques for documenting local knowledge.]

Martin, G.J. (2004) *Ethnobotany: A Methods Manual*, London: Earthscan. [Gives clear explanations and good examples of the use of CDA in applied ethnobotanical research.]

9 Community workshops and the PRA toolbox

9.1 Introduction

Community workshops have become a standard tool in conservation projects, and if you follow a career in conservation – especially in social aspects of conservation – sooner or later you will need to get to grips with how to run a workshop. Workshops are often used simply for basic consultation and information dissemination, but they can also be a valuable element in a research project. They are used most often in participatory action research (PAR), since they provide a forum where researchers and local people can work together on issues of shared concern (see Section 1.4). Through the use of brainstorming exercises, discussions and more structured exercises, a wealth of information can be generated in a short time based on the perspectives and ideas of the participants, which can then be explored further through the use of other methods.

Several of the techniques that are commonly used in workshops are described elsewhere in this book. Pilesorts and ranking exercises (described in Chapter 8) can be used not only for the theoretical purpose of cultural domain analysis, but also in a more applied setting, to identify people's priorities and the reasoning behind them. Timelines (see Section 8.3) can be used to construct an outline of historical events, which may be important in understanding the current situation – for example in terms of people's relationship to a protected area or in terms of changes in the way that natural resources are used. Seasonal calendars can provide information on seasonal variation in activities such as farming, hunting, fishing, or the availability of natural resources such as fruits or fungi, or more broadly to document seasonal variation in the principal sources of household income. Similarly, trend analysis can map changes in environmental conditions or social factors over longer periods of time (see Section 8.3). Participatory mapping is treated separately in Chapter 10, both because of its importance in applied conservation research and also because it has developed from a simple workshop exercise where people draw a map on the ground to a complex and advanced stand-alone methodology involving GPS technology and the use of Geographical Information Systems (GIS) software. Other methods have been developed for other purposes – for example to map social relations, to trace back causes of problems, or to assess a situation and identify priorities for action. Section 9.2 gives an overview of the different kinds of method that are available and the kinds of question that they can address.

What all of these techniques have in common is that they can be done by groups of people working together, and the process of doing them tends to generate a lot of discussion and reflection. As a group, they are often referred to as PRA techniques, reflecting their use in PAR (although remember that it is not the use of particular methods that defines PAR, but the level of input and motivations of the participants – see Section 1.4). PRA techniques can be used with individual small groups, which function rather like task-oriented focus groups, or in community workshops.

Community workshops differ from other kinds of community meeting in that there should be a substantial level of active participation by all the people who attend, and PRA techniques are designed to accomplish this. Often there is a large number of participants – sometimes well over 100 – and therefore, in order to give everyone a chance to have their say, the participants are divided into several smaller groups for part of the time. The small groups work in parallel and then report back to everyone in *plenary* sessions (sessions with everyone together). However, the overall number of people participating does mean that in terms of research, there are serious theoretical problems with workshops related to group response effects and sampling, and also practical problems in terms of recording what is said. For these reasons some researchers discount workshops as a 'serious' research tool, but if they are used sensibly and with full recognition of their methodological weaknesses, they can be a valuable component in multiple-methods research that provides a wealth of information on local people's knowledge and views.

Section 9.2 gives an overview of different kinds of PRA techniques, and then Section 9.3 discusses how to design a community workshop and some practical issues in running workshops. Section 9.4 describes some of the issues related to recording and analyzing the results, and finally Section 9.5 draws some general conclusions.

9.2 PRA techniques: an overview

There is an enormous variety of PRA techniques and there are many different ways of classifying them. There is no definitive list; once you grasp the principles involved, you can design your own techniques. A lot of them use visual tools – maps and diagrams – and the details can be filled in using simple drawings to represent different items rather than words, making them particularly valuable in societies where literacy rates are low. All of them can be used either in extractive research or in participatory action research, depending on the level and kind of involvement of the participants.

Table 9.1 summarizes some of the most common techniques, using a classification system adapted from Kumar (2002). Methods related to organizing and prioritizing items in a list and methods related to time and space are described in Chapters 8 and 10, respectively. The following text discusses each of the other categories of techniques in more detail.

Brainstorming is a core activity in most workshops, and is in effect the group equivalent of freelisting (see Section 8.2.1). It can be used not only to generate lists of plants and animals but also to generate lists of other things – people, institutions, problems and issues, motivations, ideas, possible courses of action and so on. The principle is that all suggestions are accepted and noted down uncritically, without any change to the wording except insofar as is necessary to clarify the meaning. Accepting every suggestion uncritically means that everyone's contributions are seen to 'count' and encourages people to keep contributing. However people are often hesitant to speak up in front of a large number of others, and therefore *buzz groups* may be used as part of a brainstorming exercise. This involves getting people to brainstorm first in groups of two to four (because they should be more confident to speak in front of fewer people). Each group can then report back to the whole gathering, or if there is a very large number of people involved, they can first join together into successively larger groups (4, 8, 16) and combine their results until there are few enough groups to make it practical for each group to feed back to the whole workshop in a plenary session.

Table 9.1 Some common PRA techniques

Method	*What it involves*	*Use*
I. Methods for generating suggestions, ideas, or items in a list		
Brainstorming	Ask a question or introduce an issue and ask for people's responses. Every response should be accepted uncritically, so that a long list is created.	To generate a list of everything that occurs to people on a given topic. Can be used either in plenary or small group sessions.
Buzz groups	Discussion in pairs or small groups. The groups can then come together into larger and larger groups so that ideas are fed forward.	Used as an alternative to a brainstorm in a plenary session in order to capture everyone's suggestions.
II. Methods for organizing and prioritizing suggestions, ideas, or items in a list		
Pile sorts	A list of items is sorted into piles of 'similar' items (see Section 8.2.3).	Often used after a brainstorm or buzz group activity in order to group the items and come up with a smaller number of categories.
Ranking	A list of items is put in order according to specified criteria.	Used to put items in order – for example of preference or importance.
Paired comparisons	Pairs of items in a list are compared according to specified criteria (see Section 8.2.4).	Used to put items in order – for example of preference or importance. Better than ranking if there is a large number of items.
Weighted ranking	Each item is 'scored' on one or more criteria (see Section 8.2.5).	As above, but also gives an indication of the 'strength' of preference or ordering.
III. Relational methods		
Venn diagrams	Items – usually different organizations, stakeholders or social institutions – are represented as circles and placed on a diagram. Their size and position represent their relative importance and degree of interaction.	Shows the relationships between different actors or stakeholders.
Problem tree	A simple diagram of a tree is drawn. The trunk represents a problem or an issue. Immediate causes are identified by brainstorming and added to the diagram by drawing roots; further items are then added further down each root in order to trace the 'causes of the causes'. Similarly, chains of effects are represented by branches.	Analysis of chains of cause and effect.

Table 9.1 (Cont'd)

Method	*What it involves*	*Use*
IV. Participatory evaluation and planning		
SWOT analysis	A box is drawn and divided into four cells representing strengths, weaknesses, opportunities and threats. The cells are filled in through brainstorming.	Used to evaluate a project, programme, institution or activity.
Action planning matrix	A table is drawn up and one or more specific problems or opportunities are listed in the left hand column. Through small group discussion, the remaining columns are filled in to indicate possible actions, who should be involved and what the timescale should be.	Planning of specific actions to address one or more problems or to build on opportunities.
V. Time-related methods		
Timelines	A line is drawn representing the passage of time, and key events are marked in at different points along the timeline (see Section 8.3).	Used to construct an outline of historical events.
Seasonal calendars	The characteristics of different seasons or months of the year are identified in relation to specific themes (see Section 8.3).	Used to document seasonal changes.
Trend analysis	Changes over longer periods of time are identified in relation to specific themes (see Section 8.3).	Used to document change over time.
VI. Spatial methods		
Participatory mapping	Varies from the production of a simple hand-drawn map on the ground to the production of a very detailed and sophisticated map based on a combination of local people's knowledge and other sources (see Chapter 10).	Used to map features of the landscape or distribution of different natural resources.
Transect walks	Involves walking together along a line that traverses major aspects of the landscape and recording types of vegetation, land and resource use or any other features of interest (see Chapter 10).	Can be used to ground-truth the results of participatory mapping (see Section 10.5) or as a technique in its own right to document local perspectives on features of the landscape, natural resources and land use.

The result is usually a long list of items, many of which may overlap and a few of which may be barely relevant. In a separate step, irrelevant items may be discarded (if their original proponent now wishes to do so) and like items can be grouped together in order to come up with a smaller number of categories. The latter can be done either by participants in full session, or by the workshop facilitators in between sessions. Alternatively more structured exercises can be used, including pile sorts, ranking exercises and paired comparisons.

Brainstorming is used in many of the other techniques listed in Table 9.1 and where this is the case, the results that are recorded may simply be the final list of categories or consolidated items that emerge from the whole process. However, where brainstorming is used in its own right it is useful to retain the original list – the raw data – as well. Showing the full list also sends a message to people that their suggestions have stayed in the picture, which can be important in maintaining a high level of participation.

Relational methods are methods for uncovering relationships between different items. In a sense, pilesorts and ranking exercises are relational methods, but I am using the term here for a set of more complex diagrammatic methods that do more than simply organize items in a list. There are several types of relational method, but probably the most commonly used is the *Venn diagram*. Venn diagrams are usually used to map relationships between different stakeholders – individuals, organizations and other social institutions – but they can also be used for other kinds of items. Box 9.1 outlines the process for producing a Venn diagram to explore stakeholder relations (adapted from Kumar [2002: 236] and Govan et al. [2008]).

Box 9.1 How to construct a Venn diagram: stakeholder relationships

Ask people to:

1 Produce a list of stakeholders with an interest in the local area and its natural resources, through brainstorming.
2 Rank the stakeholders in terms of one specified dimension related to the management of local resources (for example, influence, power, importance), either through open discussion or through a paired comparison or formal ranking exercise.
3 Cut out different-sized circles of paper or card – one for each stakeholder. The size of the circles should represent the relative importance (or influence or power) of the stakeholders. Alternatively, existing objects of different sizes such as stones or seeds can be used rather than paper. The use of cut-outs or objects rather than drawing directly on the sheet of paper makes it easier for people to make changes as they discuss their ideas.
4 Write the name of the stakeholder on each of piece of paper or card (or make sure everyone is clear about what each of a set of objects represents).
5 Draw a circle in the middle of a large sheet of paper to represent the local community. Then place the paper circles (or objects) representing the stakeholders on the paper, positioning them so that their distance from the centre represents their level of interaction with the community. Circles that represent stakeholders who are closely associated with each other can be placed so that they overlap; for other circles, the distance between them should represent the degree of interaction or association.
6 Once the Venn diagram is finished, ask someone from the group to present it to the plenary, explaining the size and positioning of the circles. Make notes or record their explanations.
7 The easiest way to record the diagram itself is to take a digital photograph. Alternatively, if you ask the group to draw around the circles on the paper and write the names of the stakeholders in the appropriate circles, the diagram itself can be preserved.

Like many PRA techniques, the value of making a Venn diagram is not only in the diagram that is produced, but also in the discussion stimulated by the process of producing it. An exercise like this will often stimulate lively debate in a group of people as they exchange their knowledge and impressions of the different stakeholders and attempt to justify their opinions. In a focus group setting, the whole process can be audio or video recorded to provide a wealth of qualitative data. In a workshop setting it is usually carried out by small groups working in parallel and therefore it may not be possible to catch all the discussion, but if each small group is asked to feed back to the plenary, then a substantial amount of information on their reasoning should emerge. If the number of small groups is not large, there is much to be gained from carrying out an interview with each group at a later time in order to ask them about their reasoning.

A second common relational method is a *problem tree*, which is used to identify chains of cause and effect (Kumar, 2002: 194–201). A simple picture of a tree trunk is drawn, leaving plenty of space above and below it. The tree trunk represents a specific problem, such as a shortage of a particular natural resource. Immediate causes are produced by brainstorming ('why does this problem happen?') and a root is drawn to represent each cause that is identified. More indirect causes can then be identified by again asking 'why?' for each cause that has been identified, so that you construct a chain of causes, each chain represented by one root. Once this is finished, a second round of brainstorming is used to identify the effects of the problem ('What is happening as a result of the problem?'), and a branch is drawn to represent each immediate effect. Indirect effects can then be added further up each branch.

As with Venn diagrams, the items that are produced by brainstorming can either be written directly on the problem tree diagram, or written on separate pieces of paper or card that are then positioned on the diagram. The latter gives more flexibility to move causes and effects around as you go along and as more ideas are generated – particularly in terms of where they are placed in a particular chain. Also as with Venn diagrams, the diagram itself is only one part of the output; the discussion generated can provide a rich source of information and analysis, and should be recorded by note taking or recording.

The results of a problem tree exercise could be used in a further exercise to brainstorm about solutions, and come up with a plan for action. This leads us on to the next group of PRA techniques shown in Table 9.1 – techniques for *participatory planning and evaluation*. This group of methods is most closely associated with action research and implementation activities such as project planning, since the direct aim is to facilitate a process of reflection, planning and problem-solving by the participants. Again, there is a wide variety of methods, but one of the best-known and most frequently used is the *SWOT analysis* – an analysis of the strengths, weaknesses, opportunities and threats of a particular project, programme, institution or activity. One advantage of a SWOT analysis is that it is not necessary for everyone to reach a consensus – all the ideas generated can be recorded, even if they are contradictory. The contradictions can be discussed later on in a separate exercise. Box 9.2 gives step-by-step instructions on how to carry out a SWOT analysis.

Conceptually, a SWOT analysis is a simple exercise and most people participate readily. However, it does need careful facilitation. People often find it difficult to distinguish between strengths and opportunities, and between weaknesses and threats. It helps not to introduce all four concepts at once, but to work on strengths and weaknesses first and then move on to the slightly more difficult concepts of opportunities and threats, emphasizing that these are to do with external influences that might affect the project rather than with the internal characteristics of the project itself. However, if people do make suggestions that you think are not

Box 9.2 SWOT analysis (analysis of strengths, weaknesses, opportunities and threats)

1 Introduce the project, programme or activity on which the analysis is to be performed and explain that the purpose of the exercise is to think about how well it is working and why, and how things could be improved.
2 Draw a large box and draw lines splitting it in two both vertically and horizontally, so that there is a table with four equal sized cells. Write titles in the two boxes in the first column, thus:

Strengths	
Weaknesses	

3 Explain that for now, you want people to think about the way the project itself is working; later on the second column will be filled in to represent other things – the external conditions – that might affect it. Ask people to brainstorm on the strengths and write the results in the first box in the table.
4 Repeat for weaknesses.
5 Once this is done, label the two remaining boxes – opportunities in the top box and threats in the lower box. Explain that these are to do with external factors that might affect the project rather than with the project itself. Repeating the above process, ask people to brainstorm first on opportunities and then on threats, and write the results in the appropriate boxes.

appropriate for that particular box, then in keeping with the spirit of brainstorming, you should accept them uncritically rather than get bogged down in arguments about what is or is not a 'correct' suggestion. If they are written on cards and stuck on the diagram, then later on you can ask people whether any items should be moved to a different box. And as with other methods, the diagram that is produced by a SWOT analysis is only one part of the results; the discussion generated can also provide a lot of useful information.

It should be clear by now that individual PRA techniques can build on one another in order to explore a particular issue in more and more depth. For example, the results of either a problem tree exercise or a SWOT analysis can form the basis for working on an *action planning matrix* (see Table 9.2).

An action planning matrix can be constructed either to address one or more problems or weaknesses (or causes or effects of a problem), or more positively, to build on identified strengths and opportunities. Several columns may be completed for a single problem. The second column in the table is important in order to learn from past successes and failures. The fourth column identifies who *of the people present* will take a lead on action; it is no good allocating actions to a person or institution who is not there, because they may not

Table 9.2 An action planning matrix

Opportunity or problem (or strength or weakness or threat or cause or effect)	What has been done so far to try to address it or build on it	Actions: what we could do	Which of us will start doing it	Who else do we need to involve	By when we should aim to have done it

Source: Adapted from Govan et al. (2008: 32).

agree to do anything. 'Taking a lead' may simply involve going and talking to someone else who is in a better position to take the action forward, but if one of the people present has been allocated responsibility, they must seek other alternatives if this does not work. The fifth column, then, should identify potential collaborators. The sixth column should suggest a deadline for each action to make sure that time does not slip by without any progress.

An action planning matrix can work very well with a group of people who are used to the kind of formal planning and organizational management that is common in 'western' institutions, but it may not work well in other groups of people. This kind of rational, linear approach to planning and decision making may simply be inappropriate in societies where actions are directed by more informal processes of discussion and interaction. Also, if an activity of this kind is carried out in a workshop with little previous discussion, people will not have very much time to think and consider what is best, in which case they may not follow up on the responsibilities that have been allocated to them.

This highlights some of the serious limitations of PRA techniques in terms of action research. It is easy to get carried away in designing complex, imaginative exercises, but different kinds of exercise and levels of complexity are appropriate in different social and cultural settings. Also, the more structured the exercises, the more you are framing the results according to your own preconceptions and constraining the way people can express their views (just as occurs in the more structured conventional research methods such as questionnaires). Moreover no single technique or workshop alone can bring people to formulate a well-considered plan for action; that takes time. PRA techniques have been heavily criticized on these grounds (for example, see Cooke and Kothari, 2001; Mosse, 2005). However, when used appropriately as part of an ongoing collaborative process, taking the social and cultural context into account, they can be extremely valuable.

9.3 Designing and running a workshop

If you are planning a community workshop, the first thing you need to do is to think carefully about what you want to accomplish and how it will fit into an overall research design or a longer process of practical collaboration with the community concerned (see Section 9.5). The more participatory you intend it to be, the earlier in the planning process you should involve community representatives. If the workshop is about an urgent problem faced by the community they may want to work with you on the design, especially if they have prior experience of workshops; in other cases they may be involved in defining the aims and leave the workshop design up to you. You should also think about how long the workshop should be and how many people it is likely to involve. A community workshop may involve anything from about 15 to over 100 participants, and can be anything from a three hour data gathering exercise to a week-long evaluation and planning session.

Once some clear aims have been defined, the next step is to design the workshop itself – to decide on the specific activities to be included, and then to put them together into a programme with a timetable. All workshops should consist of a mixture of 'plenary' sessions (with everyone together) and small group work. Activities in the plenaries can include formal presentations, open discussion and brainstorming, and feedback from the small groups; small group activities can involve open-ended discussions or more structured activities of the types described in Section 9.2.

Box 9.3 gives an example of a simple programme for a short workshop on the potential for a community's involvement in tourism. It consists of an opening plenary session in which a speaker introduces the subject and aim of the workshop and presents some background information, followed by a brainstorming exercise in small groups, followed by feedback from the small groups and a concluding discussion in another plenary session, and then some closing remarks. I would expect such a workshop to take between two and four hours, depending on the logistical challenges and on how much people had to say. In a longer workshop, there may be several groups of sessions following this pattern – an introductory plenary, a small group exercise, then a feedback to the plenary – interspersed with breaks for food and refreshments, and possibly longer outings and field-based exercises.

Typically, each small group has between six and eight members and operates somewhat like a focus group. People can be left to divide themselves, or the facilitator can allocate them to groups – sometimes divided by age, gender or other attributes. As with focus groups, the results from contrasting groups of distinct types of people may reveal differences in their perspectives that would have remained hidden if the participants had all been mixed in together. Also, certain types of people may feel freer to express their views in a group of similar people (for example, women in a society where they are subordinate to men). The different groups may all be asked to work on the same task, or they may be allocated different things to do – for example, to work on different issues generated in a previous session, such as different problems or causes of problems or opportunities. Feedback from small groups can take any of several forms: a short verbal statement by one person from each small group, a set of bullet points on a flipchart sheet, or a more detailed set of written notes that is handed to the facilitators.

Ideally, a series of sessions should build from one stage to the next; it should be clear from some of the examples in Section 9.2 how this can be done. The breaks between sessions give time for some initial processing of the results from one session for use in the next

Box 9.3 Example: possible programme for a community workshop—the potential for community involvement in tourism

Section I.	Plenary: introduction to the workshop and presentation on tourism. Followed by questions.
Section II.	Small groups: brainstorming on ways the community could become involved in tourism.
Section III.	Feedback from small groups, discussion and closing remarks.

session, so do not expect to have much time off during facilitation of a workshop. While everyone else goes off for a relaxed lunch or (in a workshop stretching over several days) an evening social, the facilitators are often working frantically to process the results up to that point in order to feed into the following sessions. You should consider what processing will be needed between sessions when you put together the workshop timetable; try to make sure that sessions that will need a lot of immediate processing are placed before the lunch break or, in a workshop of several days, at the end of a day.

You should also think about the general dynamics of different types of session. A plenary session at the beginning and the end of the workshop gives a feeling of a communal event that may be lacking if you start or end in small groups. For the rest of the workshop, interspersing plenary sessions with practical activities in small groups helps to keep people actively engaged. In a workshop that lasts for several days it is common to vary the activities further by showing videos, taking people on field trips or sending them out into the countryside to carry out practical exercises. Finally, bear in mind that people will be least alert straight after lunch, and everyone will probably have lost some degree of concentration by the last session of the afternoon, so it is best to put substantial presentations and other 'heavy' material in the morning if possible.

Finally, the design of a workshop should also take account of the social and cultural context in which you are working. Are the people used to discussing things openly together? Will women feel able to talk freely in front of men, or young people in front of their elders? Is everyone literate? Are people likely to have very different levels of knowledge about the topic to be discussed? There are also important social and cultural differences in what level of structure is acceptable. The workshop should be sufficiently structured to generate discussion and allow for active participation of everyone involved, but not so structured that people feel bullied, community leaders feel that their authority is being usurped or the structuring actually gets in the way of in-depth discussion that would happen if you took a more flexible approach. If you will be visiting a community for the first time, you cannot make assumptions about what will be appropriate. I once tried to run the same workshop in two communities only some 20 kilometres apart, to find that in one, a high degree of structure was expected and in the other, any attempt to break out of plenary – or to include women – was resisted strongly. The main reason for the difference appeared to be that the first community had a lot of previous experience of workshops whereas the second, which was much less accessible, had practically none. The only way to be sure about what is appropriate is to get to know the community in advance. If this is not possible, you should make a short preliminary visit, or at the very least, take local advice from someone who is familiar with the community. You also need to be flexible once the workshop gets under way and adjust the activities if necessary. Therefore it is a good idea to prepare a few alternatives.

Box 9.4 provides a checklist of things you should do and consider when designing a community workshop. Most novice workshop designers (and a high proportion of experienced ones, too) are far too ambitious in what they set out to do in a workshop of a given length – partly because of a fear that they will run out of things to do. It is tempting to design a timetable that is crammed full of imaginative activities, but if you do, you probably will not get through them all. There will always be delays and distractions, and most workshops fall well behind the planned schedule. In rural workshops it is not uncommon for people to arrive an hour or more late – especially in remote areas where travel is difficult, or in parts of the world that have a relaxed approach to timekeeping – and they will take time to settle down. The community leaders may wish to open and close the proceedings in their own way; for example, they may make a series of long, formal speeches, or say a prayer, or sing the

Box 9.4 Checklist: designing a workshop

- Define your aims and make a provisional 'long list' of activities that could address them.
- Are you familiar with the community? Do you know what would be socially and culturally acceptable to them in terms of kinds of activity, degree of structure and participation of different kinds of people? (If not, visit them, or at the least, take local advice.)
- Based on the above, revise your list of activities.
- If you will be dividing people into small groups, think about whether the small groups should all be similar or whether they should be different in some way. Should everyone be mixed together, or should different kinds of people (such as men and women) work separately? Will all groups work on the same task, or will they work on different tasks?
- Decide on the length of the workshop and think about how many people might participate.
- Review the draft list of activities again, cut them down to fit the available time, and put them into a draft timetable.
- Review the timetable and adjust it, cutting it down further if necessary to make sure there will be plenty of time for delays, for feedback, and, if appropriate, for processing the results of one session for use in the next session.
- If you are worried about running out of things to do, keep a couple of extra activities in reserve that are relevant but not essential.
- Now review the timetable again and think about the different levels of activity. Is there a mix of plenaries and small group sessions and of different kinds of activity? Are the 'heavy' sessions in the mornings? Is there a 'lighter' session after lunch?
- Finally, review everything once more and make sure that the design you have ended up with will provide the information necessary to address your original aims.

national anthem, or perform some other kind of ceremony. This may take up a lot of time, but it would be completely inappropriate for you to interrupt them or cut them short.

Another cause of delay is that people are usually slow returning to each session after a break, and again, it would be inappropriate for you to take a hard line and scold them for being slow. In terms of the sessions themselves, even the simplest small group exercise takes at least half an hour, because it takes time for each group to get organized at the start and to agree on what they want to present back at the end; a more typical length for a small group exercise is between one and three hours. Feedback from small groups is also very time consuming. A verbal report from one representative of each small group to the plenary is the simplest and most immediate form of feedback, but if each group representative takes five minutes and there are six small groups, the feedback session will take 30 to 40 minutes in all, during which most of the participants will be sitting doing nothing and may therefore lose concentration. If there are more than about six groups then some form of written feedback may be better – either in the form of lists, diagrams or bullet points on flipchart sheets that are put up where everyone can look at them during the breaks, or as longer written statements. It is then usually up to the facilitators to process these and produce whatever is needed for subsequent sessions.

Whatever you plan to do to, it is quite likely that you will fall behind the timetable. One way to deal with the uncertainties of timekeeping is to limit the publicized schedule to a basic outline, so that you can more easily adjust the detail as you go along according to how much time is left. You can always have extra exercises prepared that you could use in the unlikely event that you have extra time; alternatively, you could think in advance about which of the scheduled activities are really essential, and which you could cut out if time gets short.

In addition to the above issues to do with the workshop design, there are also many logistical factors that must be considered when planning a workshop (see Box 9.5). A workshop with a single community is usually held in the community's own meeting space, or if representatives are invited from several communities – for example, those neighbouring a particular area – in a community or other venue that is centrally located. If people will come together from different communities to a central location, you may need to arrange or fund their transport, and you should have refreshments ready for them when they arrive. People may arrive with all their children, so you need to think about whether it will be possible to run the workshop with the children in attendance or whether it would be better to

Box 9.5 Logistical tasks involved in organising a workshop

Preparation

In participatory action research, some or all of these tasks should be done in collaboration with community leaders or representatives:

- Define the aims and the outline structure and content of the workshop.
- Agree the length of the workshop, a location and date, and the rough number of participants that can be expected.
- If necessary, arrange transport to and from the workshop location for the participants.
- Arrange for light refreshments or full meals to be provided, as appropriate.
- Arrange for enough assistants to help with logistics, facilitation and note taking.
- Work on the more detailed design of the workshop and the specific activities.
- Prepare any presentations, information sheets, or materials to be distributed to small groups for specific tasks.

Supplies

- Stationery for small group work – pens and pencils, flipchart sheets (and something with which to hang them up), notebooks and so on.
- If appropriate, technical equipment for making presentations, showing videos and so on.
- Notebooks and recording equipment.
- Food, drinks and possibly cutlery and crockery (and one or more people to organize them during the workshop itself).
- Basic first aid supplies.

arrange for someone to look after them separately in a crèche. For a full-day workshop you must also provide lunch. Whether you are working with rural villagers, conservation professionals or academics, the quality of the food is often a major talking point after a workshop, so it is important to allow sufficient budget to make sure it is acceptable.

Like focus groups, workshops need several people to assist with the different aspects of *facilitation* – making logistical arrangements, dealing with enquiries and problems, facilitating group discussions, and taking notes or operating recording equipment. Researchers acting as facilitators usually present some information at the start of a series of sessions to set the context and provide some background, but otherwise, just as in qualitative interviews and focus groups, they should stay out of the way of discussion as much as possible and remain neutral; they should not interrupt any more than is necessary to keep the discussion on track. Facilitators are inevitably in a position of great power – they not only define the terms of the debate, but may also be responsible for deciding how it will be debated, who will debate with whom, the form in which the results will be presented and so on. However they should make every effort not to influence the actual process any more than they can help.

At the end of a workshop, the facilitator should always thank the participants for taking part and explain what will happen next. The latter could include a series of action points; a commitment to run another workshop a few weeks or months later to follow up on some of the issues raised or report back on aspects of your research; or the distribution of a draft research report to the participants. Finally, after a community workshop, try to stay around. In a remote community, a workshop is a significant event, and people often talk about what happened for days – both about the quality of the organization and the food, but also about

Figure 9.1 Community workshop in Buayan, Sabah, Malaysia on the potential for community-based tourism.

the subject matter that was discussed. This is an excellent time to explore people's views and ideas informally, simply by hanging around, listening to what is said, and probing gently. People may even seek you out to explain something they said in greater detail, to tell you that what somebody else said was wrong and explain why, or to give you further information on a specific point.

Running a workshop is hard work, and there *will* be logistical glitches. People will arrive late (sometimes hours late), go over time in their presentations and fail to come back promptly after refreshment breaks. Equipment may fail, or the food may be late in coming. However, as long as the facilitator comes across as relaxed and friendly, and as long as people are genuinely interested in what is being discussed, it really does not matter. The most important rule is to expect the unexpected, be flexible, and never to get flustered or show annoyance. You must be hospitable, encouraging and enthusiastic throughout. I have been in workshops that were two hours or more behind schedule by lunch-time, yet by the end of the day had accomplished their goals, generated a lot of information, and motivated people.

9.4 Recording the results

The most straightforward kind of data to be generated in workshops are the lists, rankings, diagrams, drawings and maps that are created during different kinds of structured group exercises. In many consultation and planning workshops, these kinds of output may be all that is recorded, and they may appear in the workshop report in their original form, with little processing and analysis other than the construction of an overall summary. In research workshops too, data of these kinds may be all that is needed. They may be presented and summarized qualitatively, or they may be used in more technical quantitative analyses (for example, see Chapter 8).

However, most workshops also generate an enormous amount of qualitative information, particularly during small group discussions. If you wish to record this information, the options available are the same as for focus groups and other qualitative methods: note taking, audio or video recording, or a mixture of these (see Section 6.5). However, the big challenge in comparison to focus groups is simply that there is so much going on that recording the detail is a major undertaking.

Discussions can be documented either by a note taker (who should not try to facilitate at the same time), or else by audio or video recording. Note takers may either be trained research assistants (best for methodological rigour) or volunteers from among the participants themselves (best for a participatory process). Trained research assistants are often used to take notes in plenary discussions, but more rarely with small groups, both because you would need a large team to cover all the small groups and also because their presence may inhibit discussion. Participant note takers from small groups may also act as 'rapporteurs' – the people who report back to the plenary – in which case it may be their spoken report, rather than their actual notes, that is used as a source of data.

Audio (or video) recordings will of course provide a more detailed record of the discussion, but only if you can get them to work well. Sound recordings of a large meeting may fail to pick up the voices of people on the far side of the group clearly, especially if the workshop is in the open air and there is a lot of background noise. During small group work, several groups may be in the same room, and therefore recordings of a small group discussion may suffer from the noise from the other groups working nearby. A video camera can of course provide the fullest record of all, but the filming process may be very obtrusive, especially if someone is moving around with the camera in order to capture different speakers or

Box 9.6 Types of data and outputs from workshops

Raw data

- Lists of items, diagrams, drawings and maps.
- Written notes by rapporteurs or research assistants.
- Audio or video recordings.

Outputs – levels of information that may be included in reports:

- The final conclusions that were reached during each discussion session or set of sessions. Often reported in the form of a narrative summary plus copies of lists of bullet points or diagrams, attached in appendices (common for consultation workshops; also sufficient in research projects where workshop conclusions will be used simply to inform the following stages of data collection).
- The above plus further details of what was said in discussion – arguments made for and against different points, and so on. The output may consist of a simple summary report as above, or it may be a more technical research paper based on formal qualitative analysis procedures such as coding.
- The above plus details of who said what and non-verbal aspects of interactions (uncommon).

activities on film. Whether or not this is acceptable depends on whether people are comfortable being filmed and whether the subject under discussion is sensitive.

Box 9.6 summarizes the different kinds of data that may be gathered at workshops and the levels of information that may be presented in reports after data processing and analysis. Obviously what you need to record and at what level of detail depends very much on the original aims of the workshop and of your larger research project. Bullet-pointed lists may be sufficient for very broad comparisons between several workshops (for example, in different communities), or to inform subsequent stages of research. However if you are interested in the process as well as the outcomes, much more detail is needed. If you are struggling to find a way to record enough detail in a workshop setting to address your research needs, then it may be more appropriate to run a series of focus groups – in effect, to run the small groups separately as different events rather than in parallel in a single event.

9.5 Conclusion

Running a workshop is hard work and can be expensive. There are a lot of logistical matters to arrange in advance, and during the workshop itself you will need one or more assistants to help with logistics and facilitation. There are also ethical questions about running a workshop purely for the sake of research: why should people give up their valuable time, what existing tensions and conflicts might you bring to the surface, and will there be any follow-up to address them? What expectations will be raised, and again, will there be any follow-up?

If you have some experience of workshops and have a chance to design a workshop that fits in with a larger process (for example, an ongoing conservation project), then this can be a very productive part of a multiple-methods study. Conversely, if you are working by yourself and do not have experience of workshops, it would be unwise to take on the planning and implementation of a full workshop from scratch, but if you are given the opportunity to contribute to a workshop being run by someone else – perhaps a conservation organization or protected areas authority – then it is a very valuable opportunity. Structured PRA techniques can generate a wealth of data in themselves. More broadly, workshops are good at giving an overview of the participants' perspectives at the start of a research project, and therefore they can inform the design of methodological tools to explore the issues that emerge in more depth. Similarly, presenting some initial findings to a workshop near the end of your research can generate useful comments, clarifications and corrections that can then be worked into your analysis and interpretation. Finally, workshops can be an excellent way to build a bridge between research and practice. If people learn about your research through a workshop and feel they have been involved, then they are more likely to consider the issues that were discussed or even read your final report when deciding on future courses of action.

Summary

1 Community workshops are often used simply for the purposes of basic consultation and information dissemination, but they can also be a valuable element in a research project.
2 Through the use of brainstorming exercises, discussions and more structured PRA techniques, a wealth of information can be generated in a short time based on the perspectives and ideas of the participants, which can then be explored further through the use of other methods.
3 There is an enormous variety of PRA techniques. A lot of them use visual representations, and are therefore particularly useful where literacy levels are low.
4 They include methods for generating, organizing and prioritizing suggestions, ideas and lists of items; relational methods; methods for participatory planning and evaluation, and methods related to time and space.
5 The information that is generated includes both the specific products of different kinds of exercise, and also a wealth of information from the discussion along the way. If the latter is important to your research aims it can be documented through note taking or audio and video recording.
6 Using PRA techniques does not mean that you are 'doing' participatory action research. Action research is defined by the motivation and involvement of the people concerned, not by a particular set of methods.
7 Designing and running a workshop is a major undertaking and should not be attempted until you have some experience of workshops. However, you do have experience, or if you are asked to contribute to a workshop being organized by others, then a workshop offers a very valuable opportunity, both in terms of gathering data and also in terms of linking research to practice.

Further reading

Chambers, R. (1992) *Rural Appraisal: Rapid, Relaxed and Participatory*, Brighton: University of Sussex. [A classic text from the 'guru' of PRA.]

Cooke, B., and U. Kothari (eds) (2001) *Participation: The New Tyranny?*, London: Zed Books. [A landmark volume, making an important critique of PRA approaches to development.]

Kumar, S. (2002) *Methods for Community Participation: A Complete Guide for Practitioners*, Bourton on Dunsmore: Practical Action Publishing. [The best manual I have seen describing when and exactly how different PRA techniques should be used. For each technique there are step-by-step instructions, and also a discussion of applications, requirements in terms of time and materials, use with other methods, and advantages and limitations.]

Pomeroy, R.S. and Rivera-Guieb, R. (2006) *Fishery Co-management: A Practical Handbook*. Wallingford: CAB International/International Development Research Centre. [An excellent, practical and detailed case study of participatory techniques and processes in action.]

Pretty, J., Guijt, I., Thompson, J. and Scoones, I. (1995) *Participatory Learning and Action: A Trainer's Guide*, London: IIED. [A manual for training workshop facilitators. Consists of two sections: one on training, facilitation, and organization, and the other giving a run-down of specific exercises for trainers.]

10 Participatory mapping

R.K. Puri

The mapping of indigenous lands has been one of ACT's most powerful tools in conserving the Amazon rainforest in partnership with indigenous peoples.

(The Amazon Conservation Team 2010)

10.1 Introduction

Participatory mapping is a means for gathering information about natural resources, special sites and local perceptions within a shared geographical framework. It is participatory because researchers work with local informants to identify both the purpose of the mapping exercise and those aspects of the environment that are to be mapped. For outside researchers, and sometimes for community members as well, participatory mapping offers a chance to explore local understanding and uses of the natural and social environment, which may be important for land-use planning and development as well as designation and management of protected areas and 'use zones' within or around protected areas (Eghenter, 2000; Sirait et al., 1994; see Box 10.1). Local people may themselves request assistance in mapping their territorial boundaries and important resources as well as sites of historical and cultural significance. When these maps are used to challenge official maps this is known as *counter-mapping* (Peluso, 1995; Rambaldi et al., 2006). Conservation scientists may want to learn about the location of special sites, hidden habitats, new species, and ranges of endangered and vulnerable habitats and species (Sheil et al., 2003; Stockdale and Ambrose, 1996). Therefore, like other 'participatory' methods, participatory mapping can be used either for extractive research or for participatory action research, depending upon whether local participants regard the research as valuable to them and the extent to which they have a role in its design and implementation. In the literature participatory mapping may be referred to as *community mapping*, though it need not involve the whole or even subsections of one or more communities. Individuals acting as key informants may be sufficient to get an initial idea of the lay of the land and important locales.

When conducted with communities in workshops, mapping exercises can serve as *ice-breakers* (exercises at the start of the workshop, designed to put people at their ease). They may also identify local experts and potential research assistants and provide information on patterns of variation in people's local knowledge and the use of land and resources. Even though maps typically portray the distribution of resources for a particular time, multiple maps can be made to represent past and even predicted future distributions, thereby adding a temporal dimension and local understanding of environmental change to the exercise.

Recent innovations to the basic methods of PM include *participatory GIS* and *participatory 3D modelling*, which have greatly increased the ability of communities to analyze and

Box 10.1 Mapping for community use zones in Crocker Range Park, Sabah, Malaysia

The Global Diversity Foundation (GDF) conducted a three-year (2004–7) ethnobiological assessment of key resources and anthropogenic landscapes that are important for the indigenous Dusun communities living inside and adjacent to the Crocker Range Park in Sabah, Malaysian Borneo. The Crocker Range Park Management Plan allows for community use zones that are specifically set aside for continued local community use. GDF, PACOS (a local NGO) and Sabah Parks worked with a team of eight community field assistants plus community leaders, key informants and local researchers to obtain baseline data and develop methodologies for the future monitoring of natural resource use in and around the community use zone.

Participatory mapping exercises, along with other ethnobiological methods, demonstrated the extent and intensity of non-timber forest product gathering, swidden agriculture, hunting and freshwater fishing by the local communities. In follow up projects, the GDF teams have pioneered the use of participatory GIS and participatory three-dimensional modelling (P3DM) to increase the ability of communities to present their research results in a technologically sophisticated, powerful manner. The results will guide the formulation of rules and regulations for the community use zone management agreement that will govern the joint management of the community use zone by the local communities and Sabah Parks.

(Source: Global Diversity Foundation [GDF] – www.globaldiversity.org.uk (last accessed 16 July 2010))

present their land use patterns and knowledge in a powerful and technologically advanced way (see IAPAD, 2008, Public Participation on GIS at www.ppgis.net).).

There are several steps to participatory mapping: consulting with the community or group in question; preparation of a base map; mapping exercises with individuals or groups; ground truthing maps; and the preparation of final results. The following sections describe each of these in turn, using the examples in Boxes 10.1 (from Sabah, Malaysia) and 9.3 (from Malinau, Indonesia) to illustrate different points.

10.2 Getting started: consulting with communities

Since this is a participatory method, by definition consulting with the people whose land is to be mapped is a critical first step. As mentioned above, it may be that leaders of a community have already contacted researchers to request that their lands be mapped. It would then be the researcher's responsibility to sit down with the community to determine the objectives of the mapping exercise. They may wish to prepare a very informal map to be used in the community itself, perhaps for school children or visitors. Alternatively, they may want a professional map made with much greater precision and presented using modern cartography, to be used in legal disputes with neighbours over the location of boundaries or with government officials about boundaries, resource use, protected areas and development planning. Researchers should also be aware of government regulations regarding officially recognized maps, for they may require professional or government licensed surveyors and cartographers in order to be accepted in legal cases. In either case, the communities should decide what they want mapped and give consent to the researchers to proceed (Chapin et al., 2005; Eghenter, 2000).

For conservationists wanting to better understand the spatial (or even temporal) dimensions of local land and resource use patterns in communities in or around parks and other protected areas, a process of *free and prior informed consent* (see Box 11.3 and Section 13.4.2) must be followed, which involves agreeing or explaining the objectives and asking for local approval, support and assistance. Community leaders may request that certain resources are not mapped and thereby made public, often because they are valuable and outsiders may poach them. Boundaries in dispute may also be left unmarked. Researchers should make it clear that participatory mapping may require a representative sample of informants to capture all the different ways that residents make use of their environment. You may have to request assistance from leaders and others to get a variety of people to come to a community mapping workshop.

Mapping workshops are often good places to find knowledgeable assistants for ground truthing and future research. You may also need some local assistance in preparing base maps and running the workshops. Ask a local leader for recommendations for knowledgeable people that can lead you around the territory (you may need transport), who are familiar with place names (rivers, mountains), possibly property owners, and have some experience working with researchers or other outsiders. However be aware of the danger that they may put someone forward for political reasons rather than for their skills; you can reduce this risk by asking them to suggest three or four people who you could try out and then choose between (see also Sections 12.2.2 and 5.3.1).

10.3 Preparation of a base map

Base maps should be provided to local people to draw on (Figure 10.1). They provide orientation features so that the map makers can locate themselves and get a sense of the scale at which they are drawing. Sometimes the base map may have just a few lines to represent major waterways or roads and the location of the village or house currently being used for the exercise. In other cases you may be able to use an official geological or vegetation survey map or even a remotely sensed image, though many of these maps may be at too large a scale to be useful for small-scale community mapping. Researchers from other disciplines, such as cognitive psychologists, may just give people a blank piece of paper to draw a map on, but these are probably more useful for studies of local conceptions of space than for collecting information to produce an accurate and useable map of the land and its resources.

Box 10.2 lists some guidelines for preparing a base map. If they are available, you should first compile and consult any existing maps (for example, geological or geographic surveys) for important landmarks and features, especially roads, rivers, villages and peaks. With the help of field assistants you should check the location of place names on these maps. If possible you should also create a database of important locations using a global positioning system (GPS). Eventually this database can be expanded to include the locations of resources or land use activities, and you may also be able to correct or update previous maps though this initial exercise. Finally, on large sheets of paper, draw a simple map of the main tributaries or road intersections, present villages and landmarks. Make multiple copies for use in mapping exercises with the community or your informants.

Figure 10.1 Residents of Buayan, Sabah, Malaysia prepare a large base map to transfer information collected during mapping exercises.

Box 10.2 Instructions: how to prepare a base map

1 Collect and compile suitable information from all available maps of the area (major features, particularly rivers, roads, villages, logging camps and peaks).
2 With local informants and a basic map, begin to collect and check location names around the village, at forks of main tributaries and at road intersections. If possible create a global positioning system (GPS) database of these points. Add these to the base map. Note the scale. If you can add latitude and longitude markers on the edges of the map, these will be helpful when ground truthing.
3 Leaving a corner or edge of the paper blank for a legend, prepare a simple map of the main rivers, tributaries, location of present villages and landmarks, with the local names as provided by informants.
4 Make sufficient copies for mapping exercises on large paper (A1 or A0).

(Source: Sheil et al. [2003])

10.4 Mapping exercises

There are several ways to work with people to draw maps of their areas. You can visit individual households or hold community or focus group meetings. You may even be able to hand out base maps and ask key informants to complete them in their own time. The desired

Box 10.3 Community mapping of important resources in Malinau, Indonesia

CIFOR's Multidisciplinary Landscape Assessment (MLA) project in rainforest dependent communities in East Kalimantan, Indonesian Borneo (1999–2001) used community mapping workshops to identify local land cover and land use categories that could then be assessed using standard biodiversity inventory techniques. The maps also contained lists of what each group (differentiated by age, gender and ethnicity) considered to be their most important plant and animal species. Once a base map had been constructed, community members were invited to an evening meeting in a large community centre. They undertook a mapping exercise which usually lasted around two hours, but some groups would take considerably longer to perfect their drawings.

For the exercise the participants were divided into groups, and a facilitator/secretary was chosen by each group to write things down. Each group also had someone who spoke the local language as well as Indonesian, and was willing to help explain and answer questions as they arose. Members of the MLA team circulated and would sit with groups and help them as needed.

The mapping exercise usually started with the community members finding their orientation with respect to the map, naming and drawing in numerous tributaries, and identifying their direction of flow. This often took a lot of time. The groups were then asked to draw additional reference sites (such as old village locations and hill tops) and to start locating positions associated with specific land cover types, resources, features or activities, including special or unusual sites. A key of specific symbols or colourings was developed. Many elements of this key became standardized across the villages, as examples from previous work often served as templates in subsequent villages.

The resulting maps were generally pinned up on a wall where they could be viewed by community and team members and updated as needed. They then served as a basis for further discussions between the community and the village based research team, and for selecting sites for biodiversity assessment and inventory by the field team. In all, 200 plots were inventoried for plants, animals and human uses, 600 soil samples were taken and 15,000 plant voucher specimens were collected. Community members continually revised the maps over the course of the following weeks. These further refinements required the combined efforts of both the field and village based research teams, as discussions or field observations during the day often led to minor changes or additions. Before the research teams left each village, a master map, combining data from all the initial maps, revisions and additions, was neatly drawn and copies were left with the village leaders. CIFOR organized a workshop for local government officials where the results of all seven of the community mapping exercises were presented by community leaders.

Note: Naming was complex when dealing with communities of mixed ethnicity as in some cases, especially for tributaries, a feature could have multiple names (depending on language). However, usually community members were aware of alternative names, and so it was only the researchers who were confused.

(Source: Sheil et al. [2003]. See also the MLA website at: www.cifor.cgiar.org/mla/)

content of the maps varies considerably, depending on your research needs. For instance, you may ask household members to draw a map of their land holdings and areas they use for resource extraction, such as orchards and forests. Additionally, you could ask them to specify, using different colours or labels, who in the household uses each type of area and when. In another case, you might be looking for local knowledge on specific resources such as the location of birds, and so ask individuals to map where they have seen, heard or even captured all or certain bird species. These same tasks can be performed by individuals or by groups – for example, older men, younger women, wild plant gatherers, herders – in meetings where through a process of discussion, knowledge sharing and consensus building they decide what to put on the map. This was done in the example from Indonesia presented in Box 10.3.

It is important to be clear whether you are looking for information based on people's own personal experiences (fruits they have collected this year) or what they know or think they know about their community (fruits collected by village members in general) or what they know about their environment (the locations of fruit trees). Some community members may have little experience with maps, so careful explanations and examples may be needed.

One helpful tip is to leave a space on the edge of the map (by drawing a margin or a box in the corner) to be used as a key to list resources, or land cover/use types, and then assign symbols and/or colours to represent each item in the list on the map. It may be simplest to ask informants to list the fruits, birds or other resources first before they start mapping them. These lists can be analyzed as freelists (see Chapter 8) and compared between different individuals and groups. If many items are listed then it is probably best to use an additional sheet of paper as a key, or make several maps for different domains of information (for example, tree palms, mammals, black soils, water sources). As in the Indonesian example (Box 10.3), maps can be continually updated and corrected by mappers in the course of a project. A compiled map may also be made from separate group maps, though you should be aware that some groups (such as women) may want their maps to preserve important distinctions in resource use and knowledge.

10.5 Ground truthing

Ground truthing involves going out onto the ground with map in hand and checking to see whether the information on the map matches up with what you find. Often neglected by researchers, ground truthing is essential to validate participatory maps. It can increase their accuracy and precision and thereby their acceptance by all parties as legitimate representations of local knowledge. Not all aspects of a map can be verified (for example, the location of highly mobile animals or animals which are only seasonally present would present obvious difficulties), but land cover/use types are easily verified through visual inspection.

Ideally you should use a GPS unit for ground truthing. This involves visiting a location or boundary noted on the map, taking a GPS reading of it (latitude, longitude and altitude), and then comparing the reading to the map location. In the Malaysian example, community field assistants checked locations of reported extraction of non-timber forest products (NTFPs) by visiting them and collecting GPS points (Figure 10.2). Following the initial mapping exercise, field assistants have been adding new points to their map each time community members engage in hunting, fishing or NTFP collection. As long as the base maps are produced from geo-referenced maps, it should be fairly straightforward to use the GPS database to confirm reported locations on the map (Figure 10.3).

Figure 10.2 Buayan, Malaysia, research assistants collect GPS points in the process of ground truthing locations on their community map.

If you do not have a GPS unit, you can ground truth maps by using a *transect walk*. This is particularly effective for smaller scale maps, such as a map of a small settlement, and involves walking with some key informants though the territory on a line that traverses major aspects of the landscape (for instance, the dominant land use types or the contour lines). Taking the map or a copy of it with you, you pace out distances and record the types of vegetation, land and resource uses and any other features of the landscape you come across. You can draw a *profile* diagram – a cross-sectional view of the transect line showing elevation, vegetation, features and so on – as you go, or even video record the walk to compile information that can be added to the community map. These tools also provide interesting material for interviews and group discussions.

Ground truthing is essential because even the most knowledgeable person may not remember everything about their landscape; mistakes are to be expected and should not be taken as evidence that local people are not 'knowledgeable' about their own territories.

10.6 Data analysis and conveying results

With regard to data analysis, all maps can be photographed, digitized and examined by hand or with the help of specialist qualitative analysis software such as Atlas.ti or NVivo (see Box 14.3 in Chapter 14). Simple 'eye-ball' comparison between maps made by different people can produce very interesting findings. For example, by comparing the resources mapped by different subgroups in buffer zone communities in Pu Luong Nature Reserve in northern

Figure 10.3 Buayan, Malaysia, research assistants update community maps with new information on location of resources.

Vietnam, researchers were able to identify that young boys were spending nights hunting small reptiles and amphibians with flashlights, to sell to traders for money that paid for school fees, clothes and books. It was a group of young women who identified the locations of these practices and the species collected, whereas maps produced by men and the youths themselves made no mention of them. In terms of quantitative analysis, the frequency of mention of each resource mapped was calculated to see which resources were widely mentioned and which were mentioned only by certain groups (Puri and Maxwell, 2002).

Digitized maps can be imported into GIS software (for example, the freeware QUANTUM GIS) and the area of designated land uses (such as, rice paddy) or vegetation types (secondary forest) can be calculated, and, for example, then compared with other maps from other communities. Distances to hunting grounds and collection areas for NTFPs can also be calculated (by hand as well as on the computer) and can help to estimate the effort expended and the range of people's activities. If maps are made of the same area over several seasons or years (or historical maps and old aerial photographs are discovered) then a GIS can be used to quantify changes in the area of landscape units. These maps may also be used for a preliminary ground truthing of remote images, by correlating the information contained in local maps with satellite images or aerial photographs. When combined with on-the-ground biodiversity assessments, participatory maps and remote sensing can provide very useful maps for understanding human landscapes and use patterns over much larger scales (see Fox et al., 2003).

In the Indonesian example described in Box 10.3, analysis consisted of identifying all the different land cover/use types drawn on maps, which then guided the field assessments, as described above. The resources mentioned were compiled and formed the basis for follow up interviews with households, but no further analysis was conducted on the maps. However, they did play an important role in the communities as tangible evidence of their participation in the research and as important evidence of their knowledge and use of their lands that could be used in discussions with government officials and logging and mining companies.

In the Malaysian example (see Box 10.1), all the maps have been digitized and imported into a GIS database. Analysis has involved delineating the extent of resource use in time and space. As this GIS database grows over the years, community leaders and Sabah Park officials will be able to analyze how resource use patterns are changing; for instance examining seasonal differences in hunting and NTFP collection and the changes in these patterns during periods of unusual climatic variability. As described above, this information will be critical in adaptively managing the community use zone into the future. Using the updated maps, the community has now constructed a three-dimensional model of the community lands that straddle the Crocker Range Park boundary, and labelled rivers, settlements and other important sites (Figure 10.4). The model is now on public display in the local Sabah Parks office, and will eventually be placed in a community centre. The centre will also have computer terminals at which residents will be able to access copies of the original participatory maps as well as the growing GIS database.

10.7 Conclusion

Participatory mapping exercises have been conducted around the world for almost two decades now, and have proven to be relatively easy to set up, often enthusiastically received and extremely productive tools for both researchers and the communities that make them. There are of course important ethical issues that arise in helping communities to represent themselves and their knowledge (Eghenter, 2000; Peluso, 1995), and you should be aware

Figure 10.4 Residents of Buayan finalize details on a 3-dimensional community map.

that proper consultation and consent are essential starting points in conducting mapping exercises. As we have seen in the Indonesian, Malaysian and Vietnamese studies described above, mapping can be the starting point for important conservation research and management decisions concerning particular resources as well as whole areas in and around protected areas. Primarily, they help communities compile and consolidate their own knowledge about their territories, their important resources and their own resource use patterns. Maps can also show how people from the same community can view and use their environment in different ways, and, when mapped repeatedly, how resource distributions and use patterns are changing over time. Whether presented in simple paper formats or in more advanced technological platforms, participatory maps give local people a universally understood and attractive way to convey often vital information to a variety of interested parties, including extractive enterprises, government officials, non-governmental organizations and research organizations. Even if such goals are beyond the scope of your study, mapping exercises are a fun and interactive way to get oriented in a new location, get to know the people and even find research assistants. Map making is a great way to bring local people together to share information about their environment and the way they use it.

Summary

1. *Participatory mapping*, or *community mapping*, is a means of gathering information about natural resources, special sites and local perceptions within a shared geographical framework. It is used both in extractive research and in participatory action research.
2. Recent innovations include *participatory GIS* and *participatory 3D modelling*, which have greatly increased the ability of communities to analyze and present their land use patterns and knowledge in a powerful and technologically advanced way.
3. Conservation scientists can learn about the location of special sites, hidden habitats, new species, and ranges of endangered and vulnerable habitats and species.
4. Mapping allows researchers and community members to explore local understandings and uses of the natural and social environment, which may be important for land use planning as well as designation and management of protected areas and 'use zones' within or around protected areas.
5. If maps are made of the same area over several seasons or years (or historical maps and old aerial photographs are discovered) then a GIS can be used to quantify changes in the area of landscape units.
6. There are five basic steps to participatory mapping: consulting with the community or group in question, preparation of a base map, mapping exercises with individuals or groups, ground truthing maps and preparation of final results.
7. A process of *free and prior informed consent* (see Box 11.3 and Section 13.4.2) must be followed, which involves agreeing or explaining the objectives and asking for local approval, support and assistance.
8. Base maps are provided to local people to draw on. They provide orientation features so that the map makers can locate themselves and get a sense of the scale at which they are drawing.
9. Map making can take place with individuals, focus groups or in large workshop settings. It is always important to try and capture variation in local knowledge and use of a landscape by ensuring that many people participate in these exercises.
10. Ground truthing – checking to see whether the information on the map matches up with what you find – is essential to validate participatory maps. It can increase their accuracy

and precision and thereby their acceptance by all parties as legitimate representations of local knowledge.

11 Completed maps can be photographed, digitized and examined by hand or with the help of specialist qualitative analysis software. When combined with on-the-ground biodiversity assessments, participatory maps and remote sensing can provide very useful tools for understanding human landscapes and use patterns over large scales.

Further reading

Corbett, J. (2009) *Good Practices in Participatory Mapping: A Review Prepared for the International Fund for Agricultural Development (IFAD),* Rome: IFAD. Online at http://www.ifad.org/pub/map/PM_web.pdf. [This is the latest from one of the pioneers of participatory mapping, prepared in consultation with scientific experts at IFAD, with a particular focus on agriculture.]

Eghenter, C. (2000) *Mapping People's Forests: The Role of Mapping in Planning Community-Based Management of Conservation Areas in Indonesia*, Washington, DC: Biodiversity Support Program. [Demonstrates the role of community mapping in the establishment of Kayan Mentarang National Park in Indonesia.]

Rambaldi, G., Corbett, J., McCall, M., Olson, R., Muchemi, J., Kyem, P.K., Wiener, D. and Chambers, R. (2006) *Mapping for Change: Practice, Technologies and Communication*, Participatory Learning and Action No. 54, London: IIED. [An excellent manual describing mapping in greater detail.]

Section III

Fieldwork with local communities

11 Preparing for fieldwork and collecting and managing data in the field

11.1 Introduction

By the time you start fieldwork, except in the case of participatory action research you should have developed a well-defined research design as outlined in the previous chapters. Unless your study is entirely inductive you should also have designed some specific methodological tools such as interview guides, draft questionnaires, workshop activities and so on – although you may not be able to finalize them until you start consulting, interviewing or piloting. Finally, you should have read the relevant chapters in this book on ethics and data analysis so that you can make sure you behave and collect the data appropriately. This chapter is concerned with final preparations for fieldwork – including many practical ones – and the actual process of collecting and managing the data.

Obviously the options for organizing fieldwork depend on whether you will be collecting data within reach of your home base or travelling to a distant field site for a single extended period. If you are working from home you can visit the field site repeatedly throughout the whole research process from design to write-up, which has considerable advantages in terms of setting up the logistical arrangements, consulting with collaborators during the research design and carrying out periodic analyses of data along the way in order to check what more you need to do. In contrast, if you will be travelling away from home then you have no choice but to collect data in intensive blocks of time, and you need to do as much preparation as possible before you leave. However many of the issues that you need to consider are the same; the main difference is simply in the timing.

There is also a big difference depending on whether you can use a computer at the field site or whether you will be relying on handwritten notes and records. The latter is relatively uncommon nowadays but may apply if you are working in a remote site with no power supply. If the field period is long enough to justify it and if you have funding available it may be worth investing in one or more solar panels, but they need to be substantial to provide enough power to run a laptop.

Taking these variations into account, Section 11.2 discusses some issues that should be addressed before you begin fieldwork. Some of this material is specific to research overseas, but even if you will be working at home there are many relevant aspects, especially in relation to documents and equipment. Section 11.3 describes what you need to do to get up and running once you arrive, both in practical terms and also in order to finalize the details of your research design. Finally, no matter where you are working, you need to be extremely organized in terms of how you will manage your data in the field. This is the subject of Section 11.4. Section 11.5 draws some conclusions.

11.2 Preparation for the field

Whether your research site is distant or within reach of your home, you should have been in contact with people on the ground since at least part way through the research design process. As the start of fieldwork approaches you need to sort out the final details with them. This includes checking the details of any logistical support or introductions that they have offered to help with, letting them know when you will arrive and, if appropriate, setting a time and place for a first meeting.

Before you set the dates for fieldwork you need to think about seasonal constraints. For example, if you intend to carry out a visitor survey, then obviously fieldwork should be at a time of year when there are significant numbers of visitors. If you want to describe the harvesting of a wild fruit, then you can only do it during the harvesting season. If you intend to track a particular management process, then you should try to arrange the fieldwork to coincide with important meetings or decisions. And if you are working at a remote site, it may not be accessible all year round.

Whether you are working from your home base or travelling to a distant field site, you also need to map out how much time you will need. Draw up a draft timetable, estimating how many days it will take you to collect each data set and adding days for sorting out logistics and for travel between different locations. If possible, add a couple of blank days for the unexpected. As you do this you may find that you are being too ambitious, in which case you need to adjust your research design by cutting out some research questions or reducing your target sample sizes. More rarely you may find that you will have time to spare, in which case you can add some extra activities or increase the sample sizes.

11.2.1 Paperwork and permits

Probably the next issue to be considered is whether you will need a visa, research permit or some other kind of formal authorization. In many countries foreigners need a research permit for any kind of research, and even nationals must have a permit to work inside state protected areas or visit remote indigenous communities. Further permits are likely to be necessary for work with listed species or to collect plant samples – especially if you intend to export them. It can take months for a permit to come through, and some countries regularly refuse them, so you should start this process as early as possible. It often helps if an influential local person or local institution is prepared to liaise with the government office involved on your behalf; this is something that you could reasonably ask of a non-academic collaborating organization such as an NGO. If time is getting short and the permit has still not been issued, it is sensible to have a backup plan – either a less controversial project at the same site (or outside instead of inside a protected area) or a similar project in another country. Box 11.1 provides a useful checklist.

Vaccinations for some diseases such as yellow fever are a legal requirement in many countries and you may be required to produce a certificate of vaccination as proof. Extra vaccinations may be needed if you will be handling animals. Again, these can take some months to complete, so it is important to get the process under way early on. Preparing yourself for adequate health care at remote sites where doctors may not be available is beyond the scope of this book; you should seek advice both from your doctor and from someone who has been there. Apart from medical supplies you may need to take a mosquito net, water filters and other equipment.

Box 11.1 Checklist: paperwork and permits

Documents to take to the field:

- Passport, visa and permits.
- Proof of insurance.
- Letters of introduction.
- Vaccination certificates (legal requirement for some countries).
- Your full research design.
- Any extra reference sources or notes that you may need.
- Details of all contacts and other information that may be useful.

Documents that may be required by your institution:

- Notification of your travel dates and contact details while you are away.
- Risk assessment.
- Ethics statement.

Your institution may also require you to complete a risk assessment form (see Box 11.2) and provide an ethics statement, and you should leave them with details of your travel arrangements and, if possible, your contact details.

Whether fieldwork is at home or abroad, it is wise to have insurance that will cover any unexpected emergencies. Insurance may be covered by your university or institution, but it is important to check this. Finally, it can be useful to have a formal letter of introduction

Box 11.2 Risk assessments

Carrying out a risk assessment before any period of fieldwork serves two purposes: it makes you think carefully about what the risks will be and how you can minimize them, and in many countries it is also a legal institutional requirement. Some factors are relevant to most projects, such as transport (both to the field site and locally) and health. Other factors will be specific for each project. A risk assessment should include both risks connected to the nature of the field site and also risks connected to the nature of the work itself (for example, climbing trees, diving, handling animals or using powered tools). The complexity of the assessment should reflect the nature of the fieldwork, and therefore it needs most attention when travelling abroad or to remote areas.

Your institution will probably have its own format for a risk assessment. One common format is to list risk factors in the left hand column of a table, and then in the subsequent columns to categorize each one by the level of probability (high, medium, low) and the level of severity (high, medium, low). If any risk is marked as 'high' for both factors, then in effect you are saying that something serious will go wrong, so you need to change your plans. The last column should state what measures you can take to minimize the risk.

from your institution confirming your position there, the subject and time period of your research, and that you have the institution's backing. If you will be working in a country where people speak a different language, also take a translated version of the letter (with signatures and official institutional stamps). All documents should be photocopied in case anything goes missing. I take the originals and one photocopy set with me and leave another set at home. If I have to travel in remote rural areas, I take local advice about whether I need to take the original documents with me, or can leave them safely in town and take a set of photocopies in case they are damaged or lost along the way.

In addition to permits issued by the state and documents required by your home institution, in work with local communities you may also need the formal consent of community leaders or representative community organizations (see Box 11.3). If possible, you should arrange this before you leave for the field, or you may arrive and then find that you are unable to do what you had planned. However, this is unlikely to be possible unless you already know the people concerned or are working through an intermediary on the ground. You may have to leave it until you get there and can talk to people directly – in which case you should keep an open mind about which community you will work with and allow some time to find one that is happy to host you. Chapter 12 discusses how you are likely to be regarded by the host community and a range of other issues to do with your relationship with them.

Box 11.3 Key term: free, prior and informed consent (FPIC)

Free, prior and informed consent (FPIC) means that people are fully informed of a particular action that may affect them in advance, understand the implications of that action, and are free to give or withhold their consent. It is a standard term in international law, especially with respect to major development projects, impact assessments and intellectual property rights over local or traditional knowledge. It has also become a standard term in research ethics.

When carrying out research with people, the level at which FPIC should be sought and the form it should take varies according to the circumstances. When working with officials in different organizations, verbal consent at the start of an interview may be sufficient, or their organizations may require a written agreement addressing issues of confidentiality. When working with communities, FPIC is often negotiated at the level of community leaders or representative organizations and recorded in the form of a formal exchange of letters or an entry in the written minutes of a community meeting. Every individual who is interviewed should be aware of the issues in general terms and has the right to refuse to participate, but written consent at this level may be impractical. Even if people are literate, asking them to sign a consent form before an interview may formalize the situation and create levels of anxiety or suspicion that make data collection impossible. At this level, FPIC is also a matter of building and respect and trust between all the parties involved. However if the researcher is concerned that FPIC is documented with every participant – for example, if there is commercial value to the data being collected – then FPIC should be obtained either in writing or, if necessary, by audio or video recording.

Box 11.4 Practical tips: equipment

- Equipment may include dictaphone, camera, laptop, binoculars, mobile phone or extra USB stick; and also more minor items that rely on batteries, such as a torch or travelling alarm clock.
- If you are buying equipment for one specific lengthy field trip, then using a well known brand makes it more likely that you will be able to buy accessories or have repairs done in the field.
- Make sure you have all the computer software you will need and that you know how to use it. This may include a word-processing package, a spreadsheet such as Excel (for entering questionnaire data) and SPSS for initial data analysis; or a qualitative data analysis package such as NVivo.
- In the case of audio or video equipment, think about the type and quality of output needed for the planned analyses and use.
- As soon as you receive equipment, check to see it is working correctly, that all the connecting parts/accessories are compatible, and that you know how to use it.
- Keep all receipts and warranties.
- Valuable items may need to be named individually on your insurance policy or insured separately. Write down a list of serial numbers and give a copy to your institution, another to the insurance company and keep a copy for yourself.
- Can the equipment run off the mains or off a car battery? Check the voltage and amperage in the country you are travelling to – will you need an adaptor?
- Will you be able to buy spare batteries, bulbs and so on locally? If in doubt, take enough with you for the whole trip – or else consider buying a solar-powered battery charger pack.
- For digital cameras and recorders consider buying an extra battery and memory card – especially if you will only be able to download photos infrequently.

(With thanks to Laura Robson for many of these suggestions)

11.2.2 Equipment

Box 11.4 lists some of the most common items of equipment that are needed in the field. The list is necessarily quite general because conditions and needs will be specific to each piece of research and to each study site; on the latter, the only really effective way to find out about local conditions is to ask someone who has worked there before, so if you will be working abroad, try to find someone who has worked in the same country before as a foreigner.

You will almost certainly have problems with equipment at some stage. Be as prepared as you can: back up your data, and when equipment breaks down, do not panic. Find a repair shop, buy cheap alternatives locally or think about what low-tech alternatives you could use.

11.2.3 Travel arrangements

Box 11.5 lists some points that may be useful as you make your travel arrangements. Many of these are relevant to any travel abroad, but field research is usually in more remote areas that other forms of travel, so they need to be considered carefully. This is especially the case for health care. Wherever your field site is you should take a basic first aid kit, make sure you

Box 11.5 Checklist: travel arrangements

- Booking flights
  - Which airlines fly there? Do certain days of the week or time periods make the tickets cheaper?
  - What time do you arrive at your destination? If in the middle of the night, is it safe to arrive there at that time?
  - What are the baggage weight restrictions? If not enough for everything you need to take, consider sending in advance by freight rather than paying very high excess baggage charges.
  - There may be some items you need that will not be allowed through customs – for example, bottles of liquid, mercury thermometers or certain drugs. If you are in doubt, contact the airline in advance – they may be permitted with special packaging or they may have to be sent separately.
    - Consider wrapping luggage in plastic at the airport so that you can see if it has been tampered with.
    - Fragile items can be packed separately and marked FRAGILE. You may need to check them in or pick them up at a separate desk; ask the person at check-in.
    - Do you need a ticket that is flexible on the date of the return journey?
- Currency
  - Find out what is safest, most convenient and cheapest to take. Cash? Travellers' cheques? Credit or debit cards?
  - Are travellers' cheques easy to cash? If so, what currency should they be in (dollars, pounds or euros?)
  - Will there be a bank or cashpoint near the study site?
  - Which credit or debit cards are widely recognized in the country you are going to? (Visa is probably the most widely accepted; American Express and Mastercard are much less so).
  - Debit cards (but not credit cards) can be used at cashpoint machines in major towns throughout the world. However, debit cards are often unusable in places where there is an unreliable telephone line or electronic connection, so if you are going to a remote location, take both.
- Security: If you are travelling to an area where theft is a serious problem
  - Consider taking money-belts/stitched in pockets for long journeys.
  - Keep some money (£40?) and old credit cards or store-cards (of no value) at hand which can be given to muggers to get rid of them.
  - List all valuable belongings and keep track of them.

- Health care
  - First aid kit.
  - Additional medicines as appropriate such as antihistamines, malaria prophylactics, and for very remote sites, antibiotics, treatments for diarrhoea and constipation, vermox (for intestinal worms), and so on.
  - Vaccinations.
  - Sunscreen, a hat and insect repellent.
  - Decent shoes to protect your feet against ants, snakes and spiders.
  - Water purification tablets or water filters.

(With particular thanks to Laura Robson and Olivia Swinscow-Hall for their suggestions)

know how to get to a hospital quickly, and give contact details of your family and insurance company to someone on site who can call these people in an emergency. For more remote sites you need to take health issues extremely seriously. If you are going to the tropics you can get anti-malarial prophylactics from your doctor before you go (although they may not provide them for more than a maximum period of time). It is essential to take really good care of scratches, insect bites and minor wounds because in the tropics they can quickly become infected. Keep them clean, use an antibiotic or antihistamine cream, and try very hard not to scratch them. There may be no medical assistance immediately available so be ready to treat common ailments yourself – not only your own but also possibly those of your hosts. If you know in advance that this will be the case, you may be able to get broad-spectrum antibiotics from your doctor before you go (but you must be sure you are not allergic to them or you could become very dangerously ill). Many medicines that need a prescription in the UK may be freely available over the counter in other countries, but always check the packaging and especially the expiry date before buying them, and consider taking one of the books listed in the further reading at the end of this chapter.

Before you finally leave for the field you need to develop your time schedule in greater detail. You must have a clear plan for where you will be based during at least the initial stages of fieldwork, what you will do there, and in what order. Book some accommodation in advance for at least the first couple of nights, and let people know where they can find you. If you intend to interview people in town when you arrive before travelling on to a rural location, make as many appointments as you can by email before you arrive, or you may find that they are unavailable. Communicate with them well in advance in case they are away or not checking their emails. Similarly, unless someone in-country has offered to do it for you, book any internal flights from home before you travel – otherwise you may waste several days waiting when you arrive and find out that they are full.

11.3 On arrival

When you first arrive at the field site, you need to allow some time to meet with collaborators, finalize the details of your project with them if necessary, and sort out the logistics of your stay. Non-academic collaborators may want quite a limited role – they may expect you just to 'get on with it' after an initial discussion – or they may wish to be involved in every step of the process, so it is important to follow their lead rather than making assumptions. The more collaborative or participatory the project, the more time you need to allow to discuss it with them. They may also be willing to accompany you and introduce you as you visit government offices to pick up permits; visit the proposed study community for the first time; arrange transport if you will be working in a rural location; exchange money, and buy local supplies such as stationery or spare batteries or a local SIM card for your mobile phone. Introductions are particularly useful – turning up unexpected in a remote community may not go down very well – but this must be balanced against the expectations and preconceptions it will establish (see also Sections 12.2 and 13.3.2). If a conservation organization introduces you to a community that is hostile to them, it may be a drawback rather than a benefit in establishing open relations with community members. More broadly, it is important to be as self-sufficient as possible rather than relying on collaborators too heavily and risking being let down. As one of my PhD students recently put it, 'be wary of people who say "yes" but never say "when"'.

In terms of finalizing your research design, this is the point at which you need to make any remaining decisions on sampling. If you are planning to use probability sampling then you should review your definition of the study population and sampling units, see if it is possible to construct a sampling frame, and then select a sample, balancing your original aims with the logistical challenges on the ground. For example, are all communities accessible and willing to cooperate? How many schools exist in the study area, and which ones fit the profile you are looking for? In which government offices do you have contacts, and who is available? Which tour operators are currently at the field site?

During the first few days at the field site it may also be necessary to finalize arrangements for field assistants. This should be relatively straightforward if you just need a driver or guide, but if you will be working with research assistants (for example, to administer questionnaires or carry out interviews) then it is important to spend some time with them and make sure they understand exactly how you want them to carry out their work; you may even need to arrange some structured training sessions. First, you need to be confident that anyone working with you will behave professionally and courteously at all times. Second, if they will be carrying out interviews, then they should be aware of basic interview techniques, including prompting and how to minimize response biases. You will also need to show them exactly how you want them to record the data (by audio recording unless you already know that they are proficient note takers). Ideally you should find people with some training in social science methods – perhaps through a local university – and it may be valuable to sit in on their first interviews or meet with them afterwards to review any problems that have arisen. It may also be important to consider the identity of the research assistants in terms of ethnic group, gender and so on; in most multicultural societies, people will talk more openly to others of the same ethnic group; and in many places it would be unacceptable, for example, for a male interviewer to interview women by themselves. These issues are more important in more qualitative forms of interview, because the interviewer has a greater responsibility for constructing and directing the interview and noting down what is said, but even in questionnaires there is plenty of scope for getting it wrong.

These challenges are compounded if you will be working through interpreters (see also Section 12.4.2). Even trained and highly experienced interpreters can cause a major response bias in the participants – particularly if they are powerful members of the host community, or, alternatively, if they belong to a community or institution or ethnic group that is on poor terms with the host community. Apart from this, untrained interpreters will often go far beyond simple translation; it is by no means unheard of for them to actually misrepresent

Box 11.6 Checklist: issues to address on arrival in the field

- Final discussions with collaborators about logistical issues, about the details of the research design, and possibly about the extent to which they wish to be involved.
- Clarify arrangements for local accommodation and transport.
- Find out where you can exchange money, buy supplies and get access to the Internet.
- Make sure you know where to go for emergency medical assistance.
- Pick up permits and, if appropriate, pay courtesy visits to authorising institutions.
- Review population and sampling issues.
- Confirm precise study sites and introductions to host communities.
- Meet with potential field assistants and, if necessary, provide orientation and training.

what people say in order to give you the answers they think you want. Even if they do not do this they may – with the best of intentions – fill in gaps for you as they go along or add their own interpretations about what people mean. There is a balance to be struck here: insisting on as literal a translation as possible minimizes the influence of the interpreter's own preconceptions and prejudices, but may introduce other problems. Translation is never an exact science because words and phrases in different languages do not map onto one another with absolute equivalence. In order to make sense, your questions may need rephrasing in translation in order to be appropriate to the language and culture of the interviewees. Good translation that gets this balance right is extremely difficult and to some extent you have to trust your interpreter's judgement. However, you can learn a lot about local perspectives by discussing these issues with the interpreter, getting them to translate back in literal terms, and asking them about how certain things have been rephrased and why.

Box 11.6 provides a checklist of issues you may need to address when you arrive in the field. In undergraduate studies, most of these points should be arranged in advance by your university or its field partners, but even then it is wise to be aware of them so that you can be sure that everything is in place.

11.4 Collecting and managing the data

Once you are ready to start collecting data, the first stage will probably be to carry out some piloting and refining of the methodological instruments you intend to use. Section 7.7 described piloting for questionnaires. Piloting for qualitative methods is not such a formal process, but standardized semi-structured interview guides or exercises for use in focus groups or workshops should be tested on one or two people or groups if at all possible and adjusted if necessary before you use them again. In more inductive studies, 'piloting' does not really apply; instead you constantly make small adjustments to your methodological tools, and the fine focus evolves as you follow up lines of enquiry and explore different subject areas as they emerge.

Whatever methods you use, once you start to collect data it is essential to keep a record of what you have done and what you have collected so far. If you need to collect a lot of written documents such as local publications, unpublished reports, archival documents or policy papers, keep a log of what you have collected and where, including a properly formatted reference list (Section 17.4 describes how references should be formatted) and preferably an indication of the broad issues that are covered in each document. Sets of interviews or questionnaires should also be tracked. If you keep a tally of how many completed questionnaires you have collected or how many semi-structured interviews you have conducted, then you can see at a glance how much data you have collected so far – which is particularly important if you are aiming at a target sample size.

There are several additional types of material that you may assemble during fieldwork. These include maps, correspondence, your own photographs and audio or video recordings, cumulative lists of contacts or potential information sources, and a list of vocabulary (whether in a foreign language or simply local or specialist terms in your own language). For each set of materials you need to develop a cataloguing system. In this way, when you have finished fieldwork you should have an ordered and easy-to-search set of materials. If you do not catalogue them you will end up with a big pile of paper (or sets of computer files) with little idea where to find any one item. Leaving the cataloguing until you get home invites other problems and frustrations. First, you are bound to find that some key elements are missing – and it may be too late to do anything about it. Second, you will have forgotten where and when you collected some materials or what they were meant to represent. This is particularly common with photographs; if you want to make full use of them you need to note down the date, circumstances and subject matter of each photo or set of photos as soon as possible after you take them.

Each of these should be backed up regularly. If you are able to use a computer at your field site this is relatively straightforward – electronic data can be copied onto a flash drive or external hard disk, or in addition, emailed to yourself or a colleague at your home institution. If you are reliant on hand-written notes then you must photocopy them regularly – even if it means extra trips to the nearest town – or photograph them with a digital camera, or fall back on the tried and trusted if now rather antiquated method of using carbon paper. Store the backups away from the originals in case of fire, theft or some other disaster, and protect them from damage. In the tropics this means keeping them in a sealed, airtight box with a drying agent, to protect them from the damp, mould or insect attack. If they contain confidential information they must also be safe from prying eyes, so put a lock on the box.

In terms of data processing, in qualitative studies you will need to dedicate some time every day to writing up notes, checking them through, making sure recordings have worked, and doing some initial processing in the form of annotations, memo-writing and coding. If you are writing in notebooks, each notebook and each page should be numbered so that periodically you can index what they contain – both in terms of the date, subject and participants, and also (later on) by the codes you enter on each page.

Keeping up to date with notes in this way can take two or three hours a day, and therefore you have to be realistic about how much processing you do in the field and how much you leave until later. The more inductive the study, the more processing you will need to do as you go along, but even for sets of standardized interviews or focus groups some initial processing is important – both to make sure the notes are legible and make sense, and also to help you to think about what issues you should look out for in further data collection and whether you are nearing saturation (see Section 4.3.1). A second issue to take into account is that if you are living in a community – especially one where many people cannot read or write – spending a lot of time writing and processing notes can cause a barrier between you and your hosts. I have found two (not mutually exclusive) ways to address this. First, be open about why you are there and ask people to accept that you need to spend time writing, even if this seems strange to them; in most cases they will get used to it. Second, leave the study site occasionally and spend a few days catching up on your notes, thinking through what you have found out so far, and redefining what you need to do next. This is especially valuable when you begin to feel swamped by the detail in the data you have collected and need time out to look through it and think about where you have got to. It is also valuable in maintaining your personal sanity, allowing you to escape from the round-the-clock gaze of your hosts and have a little privacy.

In questionnaire surveys the initial piloting phase may be quite intensive (see Section 7.7) but once you finalize the questionnaire format, further processing in the field can be kept to a minimum. You need to keep track of the number of completed questionnaires that have been received so that you know when you are approaching your target sample size. It is also important to check completed questionnaires as they come in for missing or invalid data or for illegible handwriting. You cannot always go back to the respondent to correct this – you may have to simply discard that questionnaire – but at least you then know you need to collect an extra questionnaire in its place. Further, if you are able to enter data onto a computer spreadsheet while in the field it will not only save you a lot of time during writing up but will also allow you to carry out some initial analyses in order to check whether the sample size is enough to address your research questions. However, if this is not possible, you can leave data entry until after the end of fieldwork.

Bernard (2006: 389–91) suggests keeping a small pocket notebook with you at all times for 'daily jottings' – notes recording interesting observations or statements or ideas as they come up during the day. This is most important in inductive studies, but I find it a useful practice regardless of the methods you are using; in addition to the above you can also jot down contacts, appointments and reminders to yourself as they come up.

The following two chapters go on to discuss particular aspects of fieldwork in more detail: first, the role of the researcher in relation to the host community, and, second, ethical issues that should be taken into consideration.

11.5 Conclusion

This chapter has gone into some detail about the amount of preparation that is needed for fieldwork and some of the risks involved, but do not be put off. Fieldwork should be the fun part of research – both in intellectual terms as you see the theoretical issues you have been reading about unfold in real life, and also in much broader terms. Fieldwork can be a life-changing experience; it is one of the few kinds of work that allows you to immerse yourself in a different place and a different culture. Fieldwork in conservation can also take you to some of the world's great wilderness areas. The better prepared you are for fieldwork, the more enjoyable the whole experience should be – and the more smoothly the data collection should go.

Summary

1 The options for organizing fieldwork depend on whether you will be collecting data within reach of your home base or travelling to a distant field site for a single extended period. However, many of the issues you need to consider are the same.
2 Before departing for the field, these include:
   a Deciding the length and dates of the fieldwork.
   b Getting paperwork in order – both that which is required at the field site and that which is required by your home institution.

   c Collecting together the equipment you will need and making sure you know how to use it.
   d Making travel arrangements.
   e Preparing a time schedule.
3 Once you are in the field you must:
   a Make final arrangements with collaborators.
   b Collect permits.
   c Orientate yourself in terms of where to buy things and how to get things done.
   d Gain an introduction to potential study communities and gain their free, prior and informed consent.
   e Finalize the research design – especially in terms of sampling and piloting.
   f If necessary, organize and train field assistants.
4 In relation to data collection:
   a It is very important to catalogue or otherwise keep records of the data you have collected – including photographs, documents, maps and so on.
   b All data should be backed up regularly.
   c In qualitative studies you should allow a significant amount of time each day to writing up and reviewing notes.
   d In questionnaire surveys, the initial piloting phase may be quite intensive, but after that, further processing in the field can be kept to a minimum.

Further reading

Bernard, R. (2006) *Research Methods in Anthropology*, 4th edn, Walnut Creek, CA: Altamira Press. [See Chapter 14 for a more detailed account of collecting and managing data in the field, especially for qualitative studies.]

Borgerhoff Mulder, M. and Logsdon, W. (1996) *I've Been Gone Far Too Long: Field Trip Fiascos and Expedition Disasters*, Oakland, CA: RDR Books. [A compilation of stories from the field by established researchers. Great fun.]

Eddleston, M. Davidson, R. Wilkinson, R. and Pierini, S. (2004) *The Oxford Handbook of Tropical Medicine*, Oxford: Oxford University Press. [A technical medical companion that may be a life-saver if you will be working at a remote location with no accessible medical assistance.]

Werner, D., Thuman, C., Maxwell, J., Pearson, A. and Cary, F. (1994) *Where There is No Doctor: Village Health Care Handbook for Africa*, Oxford: MacMillan Education. [For years, a standard travelling companion for those working in rural locations in tropical countries. It is laid out in the form of a simple instruction book, providing step-by-step guidelines for diagnoses and treatment of yourself or your hosts when professional medical assistance is not available.]

Winser, S. (ed.) (2004) *Royal Geographical Society Expedition Handbook*, London: Profile Books. [From the ultimate institution for advice on expedition planning, full of practical pointers.]

Young, I. and Gherardin, T. (2008) *Africa: Healthy Travel Guide*, London: Lonely Planet. [One of several regional guides produced by the Lonely Planet to act as a reference volume for travellers. Not as technical as Werner et al. (1994) or Eddleston et al. (2004), they contain have useful information on common problems, such as ticks and lice.]

Website

The Royal Geographical Society website has extensive information on expedition planning and fieldwork practicalities, and also advertises regular events, courses and workshops – http://www.rgs.org/OurWork/Fieldwork+and+Expeditions/Fieldwork+Expeditions.htm

12 The role of the researcher

C.W. Watson

> Fieldwork is an intensely personal experience and the quality of that experience is often as important for the final analysis as the data which are gathered.
>
> (H. Russell Bernard)

12.1 Introduction

Wherever the field – an NGO office, a government department, a hamlet on the edge of a forest, the weekly meeting of a women's market traders' association, the playground of a school, the queue outside a polyclinic or a donors' lunch at an expensive hotel in a capital city – the researcher is always a stranger, someone recognized as coming from the outside, an unknown quantity, perhaps preceded by a reputation, whom the collective group will want to observe and sound out. In most circumstances, at least initially, the stranger is given the benefit of being treated as a guest, and consequently is welcomed and privileged but at the same time expected to behave like a guest, not being too intrusive or demanding and alert to the specific circumstances in which she finds herself. There is, however, always a context to these first meetings, and both the researcher and the group are only too aware of this and of the different assumptions that each party is bringing to the occasion and to the subsequent collaboration.

This chapter considers both the position of the researcher in the field: the different expectations that the societies may have of the researcher and the effect of her presence among them, as well as how the researcher must take steps to attune herself to local circumstances in order to make the most of the opportunity to do research in the field, whether of a longer or shorter duration. Many of the examples of encounters and situations that are described have taken place in far flung locations, but the general principles are the same, whether you are doing research abroad or at home. Even apparently familiar environments can turn out to be stranger than you anticipated in terms of the occupations and preoccupations of the people who surround you and the opinions they hold. It is wise therefore to think through in advance what might be the best approach to working in a particular location by running through the possibilities that may arise. And of course this is especially important when making a thorough general risk assessment prior to going into the field (see Box 11.2). Box 12.1 lists some of the points you may need to consider, both before going into the field and when you are there.

12.2 How do they see you?

For the receiving group the researcher, however much she presents herself as an individual, is always a representative – of a development team, of a government initiative, of a foreign

Box 12.1 Practical tips: knowing the local context and its potential influence on how you conduct your research

You should consider the following issues.

Prior to going into the field

- How is it likely you will be regarded?
- Who have been your predecessors?
- What are the gift-giving traditions and notions of hospitality?
- What is the linguistic etiquette?
- What are the conventions of eating together?
- What are the strong religious taboos?
- What has been the recent history of the region?

When in the field

- Who are likely to be key informants?
- Who are the decision makers?
- What are the local alliances and tensions among families and interest groups?
- What are the characteristics of the local sense of humour?
- What are the rules of behaviour governing relations across gender and age boundaries?
- How has recent history been perceived by the people themselves?
- What is being said about the researcher and the research?

company, an aid organization or simply, as in the case of, say, the lone doctoral researcher, of a source of academic knowledge and external privilege. Prior to the arrival in the field, then, there always exists a set of preconceptions, favourable or otherwise, depending on the group's previous experience of such representatives of the various categories. There are few places in the world today that have not been visited – or invaded in the perspective of some – by external observers, all in some way seeking to derive something from being there, even if this is disguised as selflessly wanting to help or render assistance.

12.2.1 As patron

Communities receiving the stranger in their midst know that outsiders want something and are often wary of the demands which they know will be made of them. In many if not all cases these days they want to see some kind of reciprocation for what they will be giving, for their time, for the information, for the disruption to their lives. To the novice researcher this can sometimes come as a shock: 'I wanted to be treated as one of them but they kept on asking me for things'; 'They wanted to be paid for having a conversation with me'; 'I had to bargain for information'. Sometimes, even prior to the research, conditions have to be negotiated: 'Yes we will accept you to live among us but we will expect in return that you will teach us English, or give us the loan of your mobile phone or laptop'. All this can be upsetting if it comes unexpectedly, and even in circumstances where you think you understand traditions of hospitality and gift-giving, unpleasantness can easily arise. There are some classic cases of this in the fieldwork literature. Clifford Geertz (2000: 21–41), probably the most well-known Anglophone anthropologist of the last 50 years, for example, describes how when he was in Java he made the mistake of asking for the return of a typewriter that an informant of his had taken on a semi-permanent loan and how that soured the relationship. In some circumstances it may be difficult to reciprocate friendship without at the same time building up unrealistic expectations but, conversely, taking on the role of a benevolent patron may sometimes be critical for the success of the fieldwork. Box 12.2 gives an example of one issue that frequently confronts researchers working with people who do not have full access to modern health care.

12.2.2 As client

In the normal course of things the researcher, far from being a patron, at least initially, puts herself under the patronage of others. You are introduced into the fieldwork setting by someone in authority who passes you on or entrusts you to the protection of an influential member of the local community (see Section 5.3.1). Very often in a rural setting this will be done through the regional and local government or an NGO community-based organization. The official permissions will have been sought in the offices of central government, the stamped letters will have been issued, and armed with this authority the researcher will make contact with a village head and request, if only implicitly, protection and assistance. Or the local representative of an NGO or a development office will escort the individual to the same village

Box 12.2 The researcher as patron

I'm sure it's very common in the field that local people look to the researcher as a source of Western medicine, which the BaAka usually use alongside forest medicine. A friend of mine, for example, asked if I could buy him injections of penicillin for his young son who had a cut on his foot, although it was not infected. Injections of antibiotics are a very common way of treating anything and everything in Africa. I told him I would help, but explained that I didn't think an injection was necessary, and could even be dangerous. He, of course, had his own views on injections ('like snake bites, they get the medicine into the body quickly'), and we disagreed. This is just one example of an almost daily problem. How do you try to help people when they ask, using your own knowledge of health, without undermining local knowledge and privileging your own? As the source of money for medicine, I found this created uncomfortable power dynamics between me and my friends in the field. Usually I would defer judgement to the local doctors, but sometimes, as in their over-use of injections, I disagreed with them. Not being a doctor myself, this left me in a difficult position. It would be easy to simply hand over the money and let the doctors deal with it, but I felt this was not ethical. So I spent a great deal of time at the hospital with people, helping them to understand the process and get the best treatment.

(quoted with the author's permission from unpublished correspondence by Olivia Swinscow-Hall)

Box 12.3 The well-connected researcher

Emmerson (1976) tells an amusing story about research that he conducted into Indonesian politics in the late 1960s. When he first embarked on his research he found people unwilling to talk to him since in their eyes he was an insignificant doctoral student asking impertinent questions. Puzzled, and responding to the question frequently posed to him about who had recommended he see them, he decided to procure proper letters of introduction. Being well connected through his American contacts he obtained letters from the most senior representatives of the Indonesian political establishment and armed with these he went back to his interviewing. But now the problem was the reverse; precisely because he appeared so well-connected, people were afraid to talk to him for fear of the information going back to his patrons and the careers of his informants being jeopardized.

A cautionary tale, then, of how one must be careful in making use of one's connections and patrons. Those in positions of power, at all levels of the political pyramid from government ministers down to village landlords, might well be respected but association with them may need to be kept at a distance if reliable and representative data are to be collected. The same can apply to relationships with powerful conservation organizations.

chief and the introductions will be made at a more personal level. Usually the system works well; the established networks of hierarchy and patronage facilitate the slotting in of the outsider into a position from which she can take her immediate bearings before striking out on her own. However, it is best to be aware of some of the pitfalls that can occur. One can, for example, sometimes be too well-connected (see Box 12.3).

A similar problem exists at the other end of the social scale. There are those who are only too willing to befriend the lone friendless researcher during the first few weeks of acclimatization. These individuals can, however, often turn out to be millstones round one's neck as the research develops. The researcher becomes, as it were, contaminated by association with them to the extent that others will not want to know or at least become too closely involved with the researcher. I remember a friend, Michael (not his real name), telling me of an experience in the central highlands of Sumatra where an individual had approached him in a tea shop, a conversation had been struck up and an acquaintanceship had developed. Michael had been only too glad to have some company. However, as the weeks passed, the new acquaintance became possessive of him, mediating all his contacts and conversations with others, and, worse, it turned out, was the man's obsession with conversations about sex. Try as he could to shake him off Michael found the man sticking to him like a limpet. Quite apart from the association hampering his research, Michael became quite depressed by the situation in which he found himself. Having been so willing to accept the man's overtures of friendship in those first crucial weeks, he felt a sense of obligation to him. Experiences like this are not isolated, and you should always be on your guard against an over-willing eagerness to be friends while at the same time avoiding the risk of appearing too aloof.

Sometimes, contrary to the experience just described, these acquaintances at the margins can pay dividends. One well-known example is the anthropologist Victor Turner's (1960) friendship in West Africa with Muchona, the Hornet, an individualistic man, as Turner describes him, somewhat marginal to his own society yet extremely well-informed and subsequently the source of many of Turner's subtle interpretations of Ndembu ritual. The best advice in the circumstances would therefore appear to be that you should be open but cautious before committing yourself to binding friendships.

12.3 Working with friends and informants

Inevitably, though, in research settings, whether a village, an office, or a group of activists, the researcher does find herself identified with a particular group of individuals. And in some cases it is difficult to avoid the suspicion that comes with that association. In rural situations again you may be compelled to put yourself under the protection of a village head or a powerful landlord and thus initially find yourself excluded from sources of comment and information which you would otherwise want to know about. You need to recognize this and work around it. It is one of the advantages of fieldwork of reasonably long duration that you have time to seek out alternative accounts and counter-interpretations. I recall from my own experience that during my doctoral fieldwork, when I was living in Kerinci in central Sumatra with a family known in the village for its modernist orthodox interpretation of Islamic practice, it was taken for granted by others that I would share the family's views, and so people were reluctant to talk to me about what they felt I would regard as superstition or perhaps heterodoxy: belief in spirit possession, offerings at harvest time to rice-spirits, the burning of incense to invoke the blessing of ancestors and taboos on the naming of certain animals. In time I did get to hear of these things after gradually making it clear that I was interested in everything that people had to tell me and I had time to listen to them, for, after all, it was indigenous knowledge from all sources that I was after, as I explained. At the same time I had to make it clear that this was not simply an opportunistic strategy of running with the hares and hunting with the hounds, a dilemma that Shore (1999: 37) faced when he did research into small town local Italian politics, and that there were certain principles guiding my research. Again, similar dilemmas may face you if you are working in collaboration with a conservation organization: local people may assume that you share the conservation organization's views and they may therefore be unwilling to talk to you about any views that contradict what they perceive to be the conservation organization's perspectives and interests.

12.4 Learning the language of the field

In explaining openly and honestly what it is that you are interested in and making clear that you welcome information from all quarters, the problem can be, as just described, one of breaking out of associations and networks which are too constricting, but if you have sufficient time this is not insurmountable. A more taxing problem is knowing enough of the local language and idiom – and, as we shall see, this includes, even when you are working in your own country, making an effort to discover local terminologies and the customary local ways of referring to local practice – to be able to put your ideas across as intelligibly as possible.

Most fieldworkers, and certainly all social anthropologists, would agree that the key to acceptance within any group is knowing how to use the local language appropriately. All researchers are of course aware of the need to know something of the language at a superficial level, and everyone knows the usefulness of learning important keywords and phrases, but frequently there is less acknowledgement that you should know the language well. Researchers coming to an area or a region that is unknown to them will often be expecting to rely on the services of an interpreter, especially when their brief is a rapid appraisal or a quick social impact analysis where there is no time for them to develop linguistic competence. Even some experienced researchers question the need to know a language thoroughly – Margaret Mead, for example, was not convinced of the need for fluency (and according to Derek Freeman [1983] this had disastrous consequences for her classic anthropological fieldwork in Samoa). The research to be done and the data to be collected do not, allege Mead and others, depend on knowing the nuances of the language. In some cases indeed this may be correct. Very often, obtaining basic statistical material, doing headcounts, and observing production and consumption patterns will not require comprehensive linguistic skills. But it is surprising how even the collection of apparently transparent data is dependent on being able to phrase questions correctly in a survey document or a questionnaire (see Chapter 7) and be confident that your interpreter has understood the questions which you want to pose. Box 12.4 describes an example of this from my own fieldwork experience of the dangers of interpreting locally recorded information without knowing the local norms.

12.4.1 Knowing the vocabulary

The significance of terminology, special phrases and local ways of expressing concepts needs to be properly recognized, since the issue is not, as is sometimes thought, simply a knowledge of the common language. Researchers frequently find themselves in the position of doing research in environments where their own language is being spoken but where the technical vocabulary is alien to them. If they are to understand the situation they are investigating and obtain the data they need, then it is vital that they master that specialist vocabulary as quickly as possible, so that they will be taken seriously by those to whom they are addressing their enquiries. Similar to barristers mastering a brief so that they can cross-question specialist witnesses in court, researchers need to have done a lot of preparation beforehand. This is as true of researchers studying in office environments as in national parks. One of the major difficulties often faced by academic researchers asked to do consultancy work for national and international companies is the lack of fit between their perceptions of their role and that

Box 12.4 The need to be familiar with local practice

On one occasion in the field I was trying to find out the price of land through a perusal of the land sales register in the village head's office. The figures were there carefully registered, and it should have been relatively easy to calculate the price of various types of land, good rice-field land, upland small-holdings, building plots, but the more I looked at the register the less confident I felt, if only because having lived there for some time I knew that land was changing hands at much higher than the recorded prices. Discreetly I made one or two comments to the clerks working in the office using the parlance of land transactions, and very quickly I learned that in fact the register was totally unreliable, first because many transactions were conducted under the table and never recorded, but also because there was consistent under-reporting of the real prices – without there being any recognized measure of a fixed percentage by which a figure was underreported – in order to avoid the seller's taxation fee. It was knowing not just the vocabulary but the specific terminology and how to phrase the enquiries that elicited the information I needed.

of their clients: both think that they are speaking the same language, give or take a bit of blurring over the precise meanings of specialist words or phrases, but in fact they are frequently talking past each other. In these situations it is incumbent on the researcher to master the appropriate vocabulary, understand the technical terms, and grasp the hierarchies implicit in the nomenclatures and the shorthand phrasing of key terms in order to conduct the research efficiently. In this way one can go some way to avoiding what is frequently a mismatch between the perceptions of the researcher and the employer, whether an NGO, a government department or a business company.

12.4.2 Working with interpreters and informants

If this injunction to learn the language and the relevant vocabulary is so important in an environment where speakers and listeners are, at least ostensibly, employing the same language, or at least the same language rules, then how much more important is it in situations where fundamental translations are taking place in all conversations and exchanges, when, for example, one is making use of an interpreter. It is not simply that so much is 'lost in translation' but that so much is entirely omitted from the realm of the discussion in the first place, because one does not know the follow-up questions to raise, or the right incidental comment to make. And this becomes doubly an issue because it applies both to the information one is giving and to what one is receiving. Reverting for a moment to what we said earlier about the preconceptions and foreknowledge that a group has of the stranger-guest who is descending upon them, all researchers are aware of the ethical obligation to explain the nature of their research projects. However, researchers are equally aware of the potential difficulty of explaining their research aims – whether to civil servants in a ministry of development or to a cattle grazer in the grasslands of the tropics. It is all too easy to slip into a short over-simple explanation which seems to satisfy the enquirer and then move on. But not only is this unethical; it can also be counter productive. Hearing that a researcher's interest is historical, for example, an informant assumes that it is about names and dates, myths and legends, and famous events of the past that the researcher wishes to hear, and may withhold information that might otherwise be relevant. The researcher herself is unable to explain even through an interpreter that the history with which she is concerned is the changing patterns of land usage or the development of inheritance rules, all of which information the informant thinks is boring and uninteresting for an outsider.

Given then the general difficulty of translating ideas and eliciting the desired information, it is very well worth spending time learning the local language. If time constraints make this impossible, it is imperative that you have confidence in your interpreter. Consequently, you are well advised to take great care in selecting the person you will be working with and not simply accept the first one who comes long. And even after you have found someone potentially good you should ensure that you allow time to explain your objectives clearly and if necessary provide some training in techniques of questioning and putting informants at their ease: a sophisticated urban dwelling university graduate may not know how best to approach a knowledgeable old grandmother lovingly tending her jealously guarded herbal garden.

12.5 The researcher as learner and teacher

For practical reasons relating to the immediate research objectives, then, the greater the familiarity with the language and the sooner that familiarity is acquired the better. There are, however, broader and more important reasons for developing language skills, which takes us back to the question of how the researcher positions herself in the field and how the society perceives her as an individual and responds to her. It is now commonplace for expatriates going to live and work abroad, and within this category one can include not only businessmen and government advisers but also short-term consultants and researchers, to receive cultural awareness training and to be issued with booklets on 'culture shock' in which advice is given on how to shake hands and with whom, where to entertain, how to deport oneself or what clothes to wear. Taking in this advice and what they hear from colleagues and mentors who have been to the area, researchers can sometimes be lulled into a false sense of security that they are sufficiently prepared and will know how to avoid making elementary blunders. This false sense of confidence can be especially the case when nationals of a country carry out research in areas unknown to them from personal experience. Several of the contributors to Srinivas' collection of essays by Indian anthropologists describing their fieldwork in the Indian countryside bear out how shocked they were by their own ignorance (Joshi, 1979; Unni, 1979). Similarly, having taught elite urban dwelling students in Indonesia, on one occasion taking them for a week's research in rural west Java, I have seen at first hand the awkwardness that can arise despite goodwill on both sides. It is only through the slow mastery of the language and its local variations, and observing the situations in which certain phrases and locutions are used, that one comes to understand how to behave appropriately in the give and take of formal and informal discussions. Note, incidentally, that in this context the giving side of the discussions needs to be constantly borne in mind. Informants also like to be listeners and are frequently eager to hear about impressions in worlds of which they have no direct experience (see also Section 6.3). Unni (1979: 70), for example, evocatively describes how when responding to questions, he used to hold his village audiences spell-bound with simple descriptions of urban ways of life. Reluctance to answer simple queries and an appearance of simply wanting to extract information rather than participate in a two-way conversation can quickly be interpreted as aloofness and so militate against a free flow of conversation.

12.6 Finding your cultural bearings

It would take too long to describe in detail how an understanding of the field of socio-linguistics – how language meshes with social behaviour – is critical to how you are received by a community, to the richness of the personal experience you acquire and to the quantity and quality of the information you obtain. (For a good introduction to socio-linguistics, see Trudgill [1995] and for an interesting volume of readings see Pride and Holmes [1972].) With a knowledge of the language comes also a knowledge of behaviour, a sense of place and occasion and, perhaps most important of all, a capacity to hold in balance different ways of relating to others and an ability to make appropriate comparisons not predicated, at least not consciously so, on the assumed superiority of your own social upbringing. In terms of the theme we are pursuing here, how the researcher is perceived, it is above all the understanding of the language that sharpens our sense of self and allows us to alter our behaviour appropriately. However, we should note that there are other common factors that have particular significance for the relationship of researchers with their hosts – eating together, humour, reciprocity (giving and receiving), and finally knowing the limits of your understanding of local behavioural norms.

12.6.1 The significance of eating together

One of the best ways to illustrate the principle of openness to difference is through the conventions of commensality, the often unspoken rules that govern eating together. None of the books on research methods or the experience of fieldwork that I have come across seem to me to give sufficient attention to this dimension of living in another society. True, the culture shock books may give advice about how to eat with your hand and how to position yourself on the floor or at a table and when to accept and decline food, but these tips do not come anywhere near indicating the significance of the occasions of meals. As far as anthropologists are concerned there is something of a paradox here, since there is a great deal of ethnographic writing now about meals – in particular ritual meals – and how they are prepared (see for a short bibliography Bourque [2001: 98–100]), and the work of eminent anthropologists such as Lévi-Strauss and Mary Douglas in their structuralist analysis of food and meals is very well-known throughout the discipline. There are also, however, one or two well-described examples of a more general kind that bring out the social significance of eating and drinking, and which all fieldworkers might usefully read.

In a well-known article entitled 'How to ask for a drink in Subanun' Charles Frake (1964) describes how a thorough knowledge of drinking conventions is essential in any endeavour to elicit conversation among the Subanum in the Philippines. As the concluding sentence of his article puts it, 'In instructing our stranger to Subanun society how to ask for a drink, we have at the same time instructed him how to get ahead socially' (Frake, 1964: 131). And Nicole Bourque (2001), writing of meals in the Andes and the significance of serving guinea-pig, perceptively describes how the perception and evaluation of the social and political background of the guest determines what food will be served, and how the meal thus becomes a forum for indirect communication. Frake's and Bourque's remarks on the social function of commensality would, I am sure, find ready corroboration from the common observation of researchers in other regions that it is through eating and drinking with others in the field and understanding culinary etiquette that you can most easily and quickly win acceptance in a society.

Eating together means eating most of what others are eating, appreciating the food and demonstrating that appreciation by deed – taking, when the occasion warrants it, second helpings – and word – with favourable comments on taste and enquiries about recipes – and understanding the circumstances in which it is being offered to you. It does not necessarily mean liking everything that it is put in front of you. Strangers are not expected to like everything, nor indeed are they expected to be able to consume some of the culinary extremes of the host group, such as extremely hot food or certain kinds of offal. Indeed, if they do consume such food with relish they can sometimes be considered slightly odd. There is well-known cautionary (apocryphal?) story frequently told to graduate fieldworkers about to go into the field about a researcher in the Amazon who nearly died of malnutrition through always eating what was put in front of him when he ate by himself in his hut. He was only saved by the chance observation one day that the other members of the community were eating a very different diet from his own. On enquiring why he was being served different food he was told that they had been experimenting with him, and since he said that he enjoyed whatever they put in front of him they had decided to try him on food that they themselves would never eat and they had been doing this for some time.

12.6.2 Humour and conviviality in the field

Humour is probably the most difficult of the skills that the fieldworker has to master: play it wrong and you risk jeopardizing your status altogether. You need to observe very closely whom you can joke with, and in what circumstances and with whom joking is not in any circumstances permitted, for example, across gender or age or class barriers. Most societies

Box 12.5 Food preferences and turning things to your advantage

Not liking certain kinds of food can sometimes be turned to advantage, both because it leads to a discussion of food preferences and sometimes an invitation to prepare and serve a meal from your own national or local cuisine, but also because it can frequently become a hoary topic of banter. For example, I like hot food, but the expectation of friends in Sumatra is that being European I cannot tolerate it, so whenever we are eating together they will always advise me that I should not take too much of particular dishes. At first I used to resist this advice and say I enjoyed hot food, but eventually I capitulated to their preconceptions, and indeed I began to see that they knew very well that I could eat and indeed liked hot food but they enjoyed the little conceit that my stomach was not up to it and would have been disappointed if I had spoiled the joke. I was quite happy to sacrifice my taste for piquant dishes for the sake of adding to the humour of the occasion, since of course shared humour is another of the quicker routes to acceptance in whatever field setting one finds oneself.

have very clear prescriptions about joke telling: a risqué joke in front of the wife of the managing director of a multinational corporation at a dinner party – something to which I was once an embarrassed witness – is as unacceptable as a piece of slapstick at a solemn religious festival, though both may have their place on other occasions. Once, however, you have mastered the give and take of humorous banter, and even though you are likely to come off worse in any exchange, you feel a certain comfortable security in having found your level within the everyday exchanges in which you participate. In all societies repartee, wit, wordplay, satire, amused scepticism and teasing provocation all play their role and to have those skills in your repertoire can greatly ease acceptance into a society.

12.6.3 Knowing the limits

Nonetheless, to return to what we said earlier, despite the excellence of an individual's interpersonal skills there frequently remain obstacles to successful integration and acceptance within the society, as researchers have often found to their cost. Sometimes these obstacles arise from the bitter experiences that individuals and communities have had at the hands of other outsiders. Sometimes, however, they will arise from a long tradition of hostility or at least wariness towards outsiders (and sometimes, specifically towards conservationists). The anthropologist of Borneo, Peter Metcalf (2001), in a book tellingly entitled *They Lie, We Lie. Getting on with Anthropology*, amusingly describes his frustration at being unable to win round a woman who was potentially a key informant among the Iban of Sarawak, who simply refused to entrust her knowledge to him, for reasons that the title of his book suggests he well understood. In other circumstances informants, while recognizing the sincerity of researchers, will be appalled at their political naivete and rather than risk jeopardizing their own safety will prefer to keep their distance. A further dimension to bear in mind here is that the establishment of friendships and working relationships needs to be kept constantly under review and friendship, as we all know, can be a 'high maintenance' activity. Several tales from the field indicate how easy it is for threads to be broken. One of the most disturbing is *Okubo Diary* in which Brian Moeran (1985), as integrated into Japanese society as any outsider could hope to be, finds himself confronting an intractable situation after an accident to his daughter in public swimming baths in Japan when those in authority refused to accept liability for what had happened. Jean Briggs (1970) in a classic account describes how she was cold-shouldered and ostracized after a display of temper among the Inuit – something that is completely unacceptable to them. All these are cautionary tales to remind us to take nothing for granted and that fieldwork never proceeds in a smooth linear fashion. Knowing this in advance should help keep the researcher alert to those small changes in circumstances that affect your everyday relationships and so can have such an impact on your research agenda.

12.7 Working as a member of a team

One final point to be made here relates to how researchers working in a team can assist or hinder each other, and how as an individual in a team you need to be constantly aware of the impact the team as a whole is having on a society. Clearly there are similarities with the position in which a researcher finds herself as a consequence of impressions left by her predecessors or her association with marginal informants or difficult landlords or officious government officials or indeed conservation professionals whose presence may be unwelcome in the local context in which research is being conducted. You should always try to distance yourself from associations that are burdensome, provided of course that you are not breaking any moral or social commitment that you have entered into. That is, however, not always possible, especially when, for example, you are identified with a team of consultants living together in a rest-house or a hostel and members of the team are regarded as indistinguishable in their attitudes and behaviour. This can be difficult, and the solution of trying to move out of the collective group may not always be possible or desirable since, apart from anything else, it may lead to hostility from your co-researchers. Again simply being aware of the situation helps both in terms of helping you to adjust your behaviour, but also in terms of systematically feeding back to the team information on the dynamics of their interactions with the community. Ultimately, however, the longer the research proceeds the more that the people you are coming into daily contact with will be able to recognize the differences between individuals, their different research interests and their different personalities, in just the same way as the researchers too learn that the society is not homogenous nor its culture easy to describe, but consists of individuals with different characters and temperaments who sometimes act collectively and sometimes prefer to plough their own individual furrows.

12.8 Conclusion

There is no such thing as a typical situation in the field which the researcher in conservation can use as a yardstick or template by which to gauge the right way to conduct research. Each situation will be unique in terms of the problems and opportunities that arise and the types of people you will be interacting with. Most of the examples given above have been drawn from fieldwork of longish duration in tropical countries, in a political and social context which is for the most part probably outside the direct experience of the reader. But the conservationist might equally find herself in a temperate zone in an industrialized country working on a short contract in a context with which she is reasonably familiar. Does the advice and do the examples then cease to be relevant? No, because the principle still holds: the researcher will inevitably, whatever the situation, be regarded as an outsider and consequently will be kept at a distance. The sensitive researcher will recognize this and will try to mitigate the consequences of this situation by positioning herself to win and keep the trust of the people among whom she is working. There is clearly a very strong ethical dimension to the work of research and that is dealt with at length in Chapter 13. Here we have been concerned with more practical issues of the best ways of adapting to a variety of different circumstances in the field: avoiding the pitfalls, making creative use of opportunities and feeling at ease in the company of those with whom you are working. Understanding traditions of reciprocity, that is, knowing what is expected in terms of giving and receiving, mastering as much of the local language as time permits, knowing how to partake fully in conventions of commensality, and, finally, learning how to share in everyday episodes of humour and amusement, are all roads which potentially lead to a warm reception and a positive welcome.

Many of the points made above can in fact be seen as simply extensions of the way we should act in any everyday situation, not just in the field but at home. The only difference may be that it will take longer in the former situation to understand the social context in which we are operating and consequently make adjustments in how we position ourselves. In such situations we would be foolish not to learn from the experience of others. The trick, if there is one, is to be self-aware without being self-conscious, and then through that self-awareness should flow that special empathy which we all need to possess if we hope to collaborate successfully with others.

Summary

1. Before you go into the field learn as much as you can about the communities in which you will be living and working.
2. Remember that people will already have established preconceptions about you, some negative, some positive, based on their previous experiences. Find out what these are and try to dispel negative or unhelpful impressions.
3. As a potential patron you may be expected to give assistance to informants. Beware of over-committing yourself while nonetheless recognizing that you have obligations.
4. Choose your friends and informants carefully and take care that you do not allow one individual or group to monopolize you.
5. The importance of familiarizing yourself with the local language and idiom cannot be over-emphasized.
6. If you have to use interpreters assure yourself that they are reliable.
7. Learn the art of commensality, eating together and enjoying the meal in accordance with local conventions and practices.
8. Participate when you can in humorous give and take.
9. Never assume that you have been fully integrated in to the society.
10. Being part of a team has its advantages and disadvantages. It is important that you establish yourself as an individual.
11. The social skills of adaptation you use at home when you move from one environment to another – home to office, sports club to university classroom – are exactly the same as those that should come into play in more unfamiliar settings. The sensitivity is all.

Further reading

Lassiter, L.E. (2005) *The Chicago Guide to Collaborative Ethnography*, Chicago, IL: University of Chicago Press. [A book that recognizes many of the charges of exploitation made against the old anthropology but argues while acknowledging postmodern criticisms that there is an important role to be played by working together with individuals and communities in the field and suggests ways of doing just that.]

Robben, A.C.G.M. and Sluka, J.A. (eds) (2007) *Ethnographic Fieldwork: An Anthropological Reader*, Oxford: Blackwell. [An excellent selection of readings from a wide variety of fieldwork practitioners covering many of the issues dealt with in this chapter and many more besides. Usefully divided into parts such as Part III: Fieldwork Relations and Rapport and Part V: Fieldwork Conflicts, Hazards, and Dangers.]

Spindler, G.D. (ed.) (1970) *Being an Anthropologist. Fieldwork in Eleven Cultures*, New York: Holt, Rinehart and Winston. [Although slightly dated now this collection of essays beautifully describes the tribulations and joys of fieldwork in places ranging from the jungles of Malaysia to the streets of Chicago.]

13 Ethical issues in research

C.W. Watson

> In the field ethical research relationships must be actively, if not creatively, negotiated and adapted to the specifics of the situation or context.
>
> (Jeffrey A. Sluka)

13.1 Introduction

Trying to ensure that we act appropriately and with a due awareness of moral obligations when we undertake research has increasingly been recognized as not only a proper and socially responsible way to behave, but also, quite pragmatically, as offering a better chance of success than acting as though morals and scientific research were two unrelated things. What we need to bear in mind is that, though as conservationists our goals are to understand the natural environment, its flora and fauna and general ecology and, where it is necessary and where we can, to propose action to conserve that environment, we will never be successful if we neglect the participation of people and societies within that ecology we are studying (in this respect see also Section 5.2). We must constantly be aware of the social, political and cultural context in which we are working. This is especially the case when we are outsiders coming in to assist at a local level; and we should remember that we can also be outsiders in our own societies where we speak the language and share the nationality of the communities among whom we are working. The record of intervention from outside over the last 150 years, even well-intentioned intervention, has not always been a happy one. In this respect, especially in the last two decades, we have made progress by reflecting on and learning from the mistakes of our predecessors, but we must never drop our guard against what might be described as an assumption of arrogance. This section begins with some comments on colonial practice and the rest of the chapter then takes the reader through some of the recent debates about ethics; it goes on to discuss the range of responsibilities that researchers should bear in mind, and finally it suggests ways to address the ethical dimensions of research at the stages of preparation, fieldwork and writing up and publishing.

In the fields of research and policymaking there is always value to be gained from considering the recent past and looking at the similarities and differences in relation to our situation today. This is especially the case when we observe how initiatives in the field of conservation and environmental sustainability have been undertaken in the last 150 years. In the European colonial period of the nineteenth century the primary motive of imperial governments was to acquire territory, the resources of which could be economically exploited for the benefit of the metropole, the mother country back in Europe. To exploit those resources meant constructing an infrastructure in transport, engineering and agriculture, and this required a trained, fit and cooperative labour force which meant in turn providing educational and health services for the indigenous population. It also meant planning for conservation and sustainability. Concern for the welfare of the people of the country was, according to critics of colonial policy, nothing to do with a sense of ethical responsibility and everything to do with optimizing economic production for the present and future. Even missionaries in darkest London – as William Booth of the Salvation Army called it – or darkest Africa, could be seen to have ulterior motives in offering charity. Be that as it may, and even though the larger picture may conform to this characterization, it needs to be recognized that many of those who worked as researchers or in philanthropic organizations of the time started not from a premise of wanting to contribute to exploitation but in order, as they saw it, to bring civilization to savage societies and to raise standards of general welfare. Their intention was summed up by the phrase *mission civilisatrice*, and to pursue this mission they were prepared to ignore the wishes of local people, dismiss their religious beliefs and outlaw their practices and institutions, which indeed they saw as inimical to the establishment of civilization. We of course know better in the twenty-first century, or should that be: we are no better? After all we frequently ride roughshod over local sentiment in the interests, not of civilization this time, but of modernization, public welfare – and conservation. And we frequently think that we know best. There is always a potential danger here that we run the risk of repeating the mistakes of the past and ignoring local values and world-views that people hold with a passion but which are different from our own. At the very least a consideration of the range of ethical issues that may arise in connection with our research will prompt us to be better aware of what the implications of our actions are and how we might pre-empt potential problems.

13.2 Professional guidelines on ethics and responsibility

Dilemmas and concerns arising from our position as outsiders, whether as representatives of conservation or government agencies, or as researchers, still continue to worry us: what right do we have to intervene or even to be somewhere where we do not subscribe to the local norms? When can we ignore those norms and the wishes of individuals in the interests of the wider society? It is the frequency with which social science researchers have been coming up against such questions that has led to major changes in our perception of research ethics. We can see this in renewed debates about ethical responsibility, in the attempt to draw up frameworks or codes of ethics and ethical guidelines and, of fundamental importance these days, in the importance attached to ethics forms included in research proposals to help monitor, though they can never guarantee, compliance with accepted moral responsibilities. Nowadays all professional associations publish easily accessible guidelines with which all members of a profession engaged in research are expected to be familiar.

The websites in Box 13.1 give the codes of ethics for several different professional bodies of social science researchers. Conservationists, in addition to consulting the code of the Society for Conservation Biology, and the guidelines of sociologists and geographers who also do fieldwork, will find the codes of social anthropologists especially useful since social anthropologists, like conservationists, spend especially long periods in the field and do a lot of work with different groups of people from grassroots to government offices. Most websites also give further material and links on ethical issues, including examples of dilemmas from the field.

All funding agencies and research units within organizations also require evidence of completion of ethics forms, and all university departments require evidence of some recognition of the ethical dimensions of research. Many have ethics committees to monitor good practice within universities. In my own department we have a research ethics committee that checks the ethics statements on all research proposals before approval is given for research, and all students are given advice by supervisors on the potential ethical dimensions of what they intend to do.

Box 13.1 Codes of ethics for research in the field

American Anthropological Association
http://www.aaanet.org/committees/ethics/ethics.htm
[Website also includes briefing papers on common dilemmas faced by anthropologists conducting research in field situations, including: health emergencies of subjects; payment of subjects; impacts of material assistance to study population; negative impacts of publication; informed consent; and sexual relations with a member of the study population.]

American Sociological Association (ASA)
http://www.asanet.org/cs/root/leftnav/ethics/ethics

APSA (The American Political Science Association) Committee on Professional Ethics, Rights and Freedoms
http://www.apsanet.org/imgtest/ethicsguideweb.pdf

Association of American Geographers
http://www.aag.org/Info/ethics.htm

Association of Social Anthropologists of Aotearoa New Zealand Incorporated
http://www.asaanz.rsnz.org/codeofethics.html

Association of Social Anthropologists of the UK and Commonwealth (ASA-UK)
http://www.theasa.org/ethics/guidelines.htm

International Sociological Association (ISA)
http://www.isa-sociology.org/about/isa_code_of_ethics.htm

International Society for Ethnobiology
http://www.ethnobiology.net/global_coalition/CoE-Eng.php

National Association for the Practice of Anthropology
http://www.practicinganthropology.org/about/?section=ethical_guidelines

Society for Applied Anthropology
http://www.sfaa.net/sfaaethic.html

Society for Conservation Biology (SCB)
http://www.conbio.org/Publications/Newsletter/Archives/2004-8-August/v11n3014.cfm

Society for Economic Botany
http://www.econbot.org/ethics/

Royal Geographical Society (RGS)
www.rgs.org/grants-ethics

13.3 Responsibilities

In classes on research methods, ethics constitute, then, a major topic of discussion and debate, but it is surprising how limited in their scope some of these discussions are. When, for example, asked in a class to name whom one has a responsibility to, all will respond 'To the human subjects of the research (stupid!).' But then when pressed about whom they mean by the 'subjects' their answers become vague: our respondents, the disempowered, the disadvantaged, the society, the community. So although we are all aware that there are ethical implications about the way we should behave, and although we know we have responsibilities, it is not at all straightforward where these obligations lie. At this point it is useful to suggest a list of those individuals and agents we should at least consider, even though in particular instances not all the categories may be relevant.

Most of the list should be obvious. One clearly has a responsibility to those among whom one works, and it is easy to see that that responsibility at a minimum lies in ensuring that no harm comes to those involved as a result of your research. This is sometimes known as

Box 13.2 Responsibilities to whom?

1 To the host community and its gatekeepers

- Community in which the research is being carried out.
- Specific groups within that community, e.g. specific age groups, men, women, minorities.
- Gatekeepers, those responsible for issuing permissions and guaranteeing the bona fides of the researcher.
- Individual informants.

2 Sponsors, employers and academic institutions

- Employers, university departments, supervisors.
- National and local governments.
- Funding agencies.
- Sponsors and guarantors.
- Fellow researchers in the field.
- The NGO community.

3 The global research community

- Research community in general.
- Academic profession.

4 Personal

- Family.
- Friends in the field and at home.
- Yourself.
- The world at large which may benefit from or be disadvantaged by your work.

guarding against *malfeasance*. On the question of harm the guidelines of the Association of Social Anthropologists of the UK and Commonwealth, for example, have this to say:

> (2) Anticipating harms: Anthropologists (and for our purposes this applies to conservationists) should be sensitive to the possible consequences of their work and should endeavour to guard against predictably harmful effects. Consent from subjects does not absolve anthropologists from their obligation to protect research participants as far as possible against the potentially harmful effects of research:
>
> (a) The researcher should try to minimize disturbances both to subjects themselves and to the subjects' relationships with their environment. Even though research participants may be immediately protected by the device of anonymity, the researcher should try to anticipate the long-term effects on individuals or groups as a result of the research.
>
> (http://www.theasa.org/ethics/guidelines.htm, accessed 16 November 2009)

Others would put the responsibility in stronger, more positive, terms arguing that not only should no harm be done, but the outcome of the research should do some good, the principle of *beneficence*. The Code of the International Society for Ethnobiology says about this:

> 12. Principle of Reciprocity, Mutual Benefit and Equitable Sharing.
> This principle recognizes that Indigenous peoples, traditional societies and local communities are entitled to share in and benefit from tangible and intangible processes, results and outcomes that accrue directly or indirectly and over the shorter and longer term from ethnobiological research and related activities that involve their knowledge and resources. Mutual benefit and equitable sharing will occur in ways that are culturally appropriate and consistent with the wishes of the community involved.
>
> (http://www.ethnobiology.net/global_coalition/CoE-Eng.php, accessed 20 July 2010)

However, what can we say about others in the list besides the immediate community? There is not the space to go over all of these here, but many of the available ethical guidelines referred to in the box do a good job of drawing attention to what is involved in relation to all of these categories. In addition, the case studies of ethical dilemmas (see Box 13.3) – all modified from real dilemmas faced by researchers – provide material for reflection on some of the ways in which people and institutions within the different categories can be caught up in the decisions which the researcher makes. It may, however, be instructive to discuss here three of the less often mentioned categories: different groups within a society, gatekeepers and sponsors.

13.3.1 Different groups within a society

The very real danger to which novice researchers doing fieldwork in unfamiliar places or environments fall prey is to regard the people within that environment as being a homogeneous community: the villagers, the workforce, the farmers, the military. Making such an assumption can appear to facilitate research. When for example conducting surveys, if the whole population is taken to be the same, then sampling is so much easier. The results of such a survey may, however, be invalid. We all know that societies are not homogeneous, that within them there are groups with different interests and differential access to knowledge and privilege. It is imperative therefore that in every fieldwork situation we are able to

Box 13.3 Ethical dilemmas

1 *The unscrupulous landlord (the gatekeeper)*
You are working in South Asia and you have been introduced to a local landlord very influential in the area. He has befriended you and as the local gatekeeper has approved your research and spread the word that people should respond positively towards you. He has also found you somewhere to stay, in fact one of his own houses, which is comfortable and ideal for your needs. You then discover that in fact he is engaged in illegal logging and frequently goes on hunting expeditions in local forest reserves. What do you do in these circumstances?

2 *The potentials of bio-piracy (the host community)*
Working in a small community in the Himalayas you discover that the people there make use of an endemic plant as a natural contraceptive and it works effectively and efficiently with no side effects. If you make known your research findings it will inevitably mean that pharmaceutical companies will descend on the village and make off with this indigenous knowledge. True, they will pay the village royalties but in fact this sudden access to large supplies of cash will destroy the integrity of the community and will lead to tension, political conflict and damaging divisions. What are the issues which you must consider and what ultimately do you decide to do?

3 *Conflicts of opinion within the community (specific groups in the community)*
You are employed by a local green NGO in Sumatra as a consultant anthropologist/regional specialist. The NGO wants to put money into the community to raise household incomes by persuading women to develop a small handicraft industry, which would supplement household income. Some logging concession-holders seeing the potential of the timber in the area want to build a road from the community area to the nearest major town. The provincial and local government support this idea. Within the community there are mixed feelings about the potential benefits. Opening up the area to greater markets and the stimulus of better communications with the outside world, and better and quicker access to health and educational facilities appears very attractive. On the other hand, some people including some of your co-workers in the NGO fear that the construction of the road will mean not only the destruction of the environment but also the breakdown of the community itself. What position do you take?

Consider also this dimension. You decide to lobby the government to prevent further logging. This becomes known to the timber companies. You then begin to receive a series of text messages on your phone threatening your life and saying that the senders of the messages know where you and wife and children live and that you should beware of the consequences to them.

4 *The chained monkey (host community, gatekeeper and self)*
You arrive in a rural community in Pakistan at the start of fieldwork and ask to be shown to the house of the family that is hosting you (who know in advance that you are coming). You are led to a relatively wealthy household and greeted by the village head who welcomes you. You all sit down and talk. As you talk, you soon

notice that there is a young monkey chained to a post at the side of the house that is obviously in distress. The chain has worn sores on its leg and it looks starving. At one point a group of men who are passing by start tormenting the monkey and throwing sticks and stones at it; the owner does nothing to stop them, and it appears that this is clearly acceptable behaviour here. You are shocked by the way the monkey is treated, but you don't want to start an argument and don't think people would listen to you anyway. Also, you don't want to lose your chance to work in the community. What should you do? (Thanks to Mike Fischer for this example.)

5 *Sleeping with the enemy? (Funding agencies)*
An international nickel mining company asks you to advise them on their operations in Borneo. Your brief is to assess the impact of the operations on the local community. An NGO website has stated that the company is guilty of gross environmental damage, and the result of dumping toxic waste in the mouth of the river has been the contamination of the sea offshore. The consumption of fish caught by local fishermen has led to malformations among the children of the community. The company supported by local government claims it is anxious to establish the veracity of the accusations about which it is sceptical and hence it wants to employ you. You are anxious that your name and the research might be used to cover up the company's malpractice, but on the other hand you see this as an opportunity to help the local community by your work. Should you accept the commission to do the work?

seek out those differences in the interests of both justice and accuracy, and that consequently we can make the necessary discriminations between, for example, the wealthy and the poor, the single and the married, the young and the old. In short, we have a responsibility to ensure that we have elicited the views of each group and do not allow one to speak for the other.

The besetting fault of early colonial research, to glance back at the past once more, was to assume that articulate chiefs and community leaders spoke for all within a society, and we are rightly appalled by the glib way in which a superficial knowledge of a people was thus obtained. We now regard such cursory data collection as methodologically and ethically irresponsible. Nonetheless, the risk of taking the part for the whole still exists. It is especially visible in societies where men dominate the public forum and act as intermediaries between the local society and outsiders. In such circumstances researchers have a difficult time eliciting the voice of women, especially when the latter may be unwilling to express their opinions. However, the researcher has a responsibility here, one that should not be ignored whatever the time constraints on the research.

However, one should also be aware that pushing issues too far in the direction of the advocacy of disadvantaged groups – whether, say, women or ethnic minorities – during the course of research may lead to harmful consequences. The latter possibility is beautifully illustrated in Vikram Seth's wonderful novel *A Suitable Boy* where the philanthropist son of an Indian landlord makes things worse for his father's tenants by his championing of them.

No hard and fast rules are going to solve this kind of dilemma, but in general terms one can say that before taking up any political position with regard to members of a society one needs to be sure of one's ground and to have a deep and broad understanding of not only the present but also the past of the society. This applies whether one is, for example, conducting research on how to change fishing practices in a community in Southeast England or the consequences of introducing a minimal school leaving age in rural Bengal.

13.3.2 Gatekeepers

Researchers sometimes have a tendency to regard *gatekeepers*, the term used to describe those who are in the position of opening doors and giving permissions for research, to be a nuisance, or at least a hindrance to their research. They are perceived to hamper access to one's respondents, perhaps because they have something to hide or sometimes simply because they want to exercise their authority. Gatekeepers can be the manager of an organization, the head of a government department, the director of a project, a village head or the organizer of an NGO or a sponsoring government national research agency. When such individuals, who can often be powerful and influential, choose to be of help in signing letters of permission, making introductions, acting as a guarantor, providing access to archives and institutions, this can be a welcome boon. In some cases, however, that helpfulness may come at a high price, namely, researchers becoming beholden to these individuals, and accepting the implicit terms on which that patronage has been extended to them. However things ultimately work out in practice during the course of the research, it is always best at the outset to assume good rather than bad intentions and to recognize that the gatekeepers are often working on behalf of the community of people to which they are allowing you access. Their fear is that the work that you are doing may be disruptive and may bring few advantages, and at worst may cause tension and friction within the community whether that community is a group of villagers, or schoolchildren, or workers in a government office, or contending parties to a land dispute, or a party of hunters and gatherers. Consequently, as gatekeepers they have to monitor your activities and will regard you as having an obligation regularly to inform them of the progress of your research. As researchers and gatekeepers become more familiar with each other's practices during the course of a project, relations between them inevitably change, sometimes for the better, sometimes for the worse and sometimes simply with a greater understanding on both sides of what the implications of the research are going to be.

13.3.3 Sponsors and funding organizations

Increasingly today in addition to funding from academic and quasi-academic institutions, researchers are having to look elsewhere for support, and in doing so, need to scrutinize carefully the source of their funds and the conditions which might be attached to accepting employment or receiving an award. Receiving a grant from a philanthropic institution funded by the profits of a multinational mining company might be seen as endorsing the company's global exploitation strategy and condoning in some regions its damaging effects on the natural environment. A joint research programme sponsored by a university and a department within the ministry of agriculture might require you to keep your research findings, or some of them, confidential, or subject to a veto on their disclosure. In taking the money how far are you obliged to keep to those conditions imposed on the research, especially in cases where disclosure could be of immediate short-term benefit to your primary respondents? May the broader agenda of the sponsor, unknown to you, have greater beneficial consequences in the long run for society at large? Bear in mind that taking funding from a conservation organization may raise similar ethical issues, especially where the interests of conservation and the interests of local people do not coincide.

Within this category of sponsors you might also consider academic supervisors or senior collaborators in a research project. It has been my experience that researchers frequently fail to recognize an obligation to sponsors of this kind, often indeed exploiting them and regarding them as a means to an end, access to funding, patronage, or a mediator with institutions. For the researcher it is the research and the publications and reports which in these circumstances remain paramount, with sometimes a recognition of the responsibilities to respondents but rarely with a corresponding recognition of the need to reciprocate the sponsor's help and hard work by remaining in regular correspondence, providing interim reports of the progress of the research and respecting the generosity of the agencies and individuals to whom the sponsor has provided introductions. This does not mean that all advice and instruction has to be unquestioningly accepted, but it does mean making an effort to observe common standards of professional courtesy such as making due acknowledgement of help given, responding to letters and enquiries, and keeping to agreements.

Sponsors, too, have obligations, not to put unnecessary constraints on the research, to respect the judgement of the researcher, even when possibly disagreeing with it, and to bear in mind the shared desire that the research will be carried out with due regard to possible malfeasance and with the hope of benefit to all. Furthermore, and this seems to be an increasing issue within academic circles, senior collaborators have to be always aware that they must give full and proper credit in the public domain – in publications and reports – to researchers and research assistants for the contributions they have made to projects.

13.4 General principles to bear in mind

13.4.1 Preparation

Although researchers do not always encounter ethical dilemmas in the course of their research, they do occur frequently and often unexpectedly, and consequently it is important to be as well prepared as possible for what might arise. Conscientious preparation will require regular consultation, especially at the early stages, with supervisors and project managers in order to understand not only the scientific and academic nature of the research but also the social and political context in which it is to be carried out. People who are knowledgeable about the area and the potential problems of research need to be sought out and consulted. To supplement this information you need to do library and perhaps archival research, reading what has been written about a subject and a region. In order to get a general feel for the ethical dimensions of a research project, besides looking at ethical guidelines, it is also useful to look at some of the books which deal with ethics in research and provide illustrative case studies of problems encountered in the field. Some of these books are mentioned at the end of this chapter.

You should, however, remember that each situation, although it may resemble what you have heard or read about, is unique, and that when it comes to decisions to be made on the spot, these will have to be based on general rules and principles of the kind that are indeed applicable in our everyday lives. Kent (in Burton, 2000: 64–6) has set out what some of these principles are, and how they should guide research: *veracity* – telling the truth and presenting yourself and your intentions honestly; *privacy* – respecting the rights of respondents not to disclose information about themselves and indeed when sometimes they have second thoughts about what they have said to withdraw information from the record; *confidentiality* – abiding by the wishes of respondents not to disclose information given to you in confidence, and understanding the limits to this provision when harm and danger to others may be at stake;

fidelity – representing honestly what has been said and observed and not distorting information to fit hypotheses. For a good sense of what the immediate issues here are you should look again codes and guidelines. Those of the Association of Social Anthropologists of the UK and Commonwealth and the Society of Ethnobiology are especially helpful.

At some point you may be tempted to do *covert research*, that is, not telling those with whom you are interacting that you are actively seeking information, perhaps by secretly filming or recording, which you intend to disclose to others and place in the public domain. In some very rare cases this practice may be justified if, for example, you are working with the police or the government to expose illegal activity – trafficking in endangered species or exploitation of child labour, for example. But unless this is your explicit task – and very few researchers are engaged to do this type of undercover investigation – then you should keep clear of such work. It is certainly not something that the novice fieldworker should do despite the temptations that may arise when you see the opportunity to blow the whistle. Let someone else with more professional experience deal with the situation, and concentrate on your own immediate objectives.

13.4.2 Reflecting on ethical issues

It should be clear by now that ethical issues arise at all stages of research from the original conceiving of the research through fieldwork up to publication and post-publication. Some of the most significant of these issues are set out as a checklist of points which researchers should bear in mind (see Box 13.4). At the *writing the research proposal stage*, for example, you should think carefully of what the consequences of the research might be for the people of the area where the study is being carried out and what their attitudes to it will be. At the *pre-field work* stage, identifying potential problems beforehand (as far as possible) and considering pro-active ways of mitigating them are required. Doing the actual *fieldwork* means being alert and sensitive to issues in the field. In particular that means maintaining those principles of *veracity* and *privacy* mentioned above and ensuring that the consent one obtains is indeed fully informed (see Box 11.3) and that respondents and participants in the research who agree to be interviewed and observed know of the short, medium and long-term consequences of their willingness to disclose information. At the *writing-up* stage decisions have to be made whether to anonymize data and how much needs to be anonymized if one is to respect *confidentiality*. In addition to the simple change of personal names should you for example try to disguise the locality of the research? Or some of the non-governmental organizations involved? And with regard to *publication* and its aftermath, in addition to reviewing the above issues relating to writing-up, once again you have to obtain necessary permissions for use of material and make due acknowledgements, bearing in mind the issue of *fidelity*.

13.5 Conclusion

The various professional guidelines set out in more detail what the researcher has to be mindful of at each stage, and researchers should consult them regularly, not just at the outset of the research. In the final analysis, however, even when the research is ostensibly limited to ecological aspects of conservation, researchers should recall that people live in the natural environment and have their own claims on it. The researcher must constantly reflect on their interests and welfare, especially if the research is policy orientated and is intended to improve the quality of life for people on the planet as a whole. We noted how during the colonial period administrators and researchers were often concerned with the issues of conservation

Box 13.4 Checklist of ethical dimensions of research practice

Writing the proposal

- Have the consequences of the research for local populations been addressed?
- Have the interests (and potential conflicts of interests) of the sponsors and gatekeepers been considered?
- Does the proposal description comprehensively cover the research which will be conducted?
- Has the ethics form dealt as fully as is possible at the pre-fieldwork stage with all the ethical contingencies which may arise?
- Does the form explain how and from whom informed consent should be obtained?
- What ethical guidelines have been consulted and what sections are most likely to apply to the proposed research?

Pre-fieldwork

- What research has been done to identify beforehand ethical dilemmas which may arise in the field?
- What are the potential areas of conflict within the community which one should be aware of: religious, ethnic, political?
- What is the history of recent research by outsiders in this area?
- What precautionary measures can be identified in order to mitigate the risk of giving offence?

Fieldwork

- Draw up a list of the conventions governing everyday behaviour, distinguishing what is considered appropriate and inappropriate behaviour for the researcher.
- Differentiate between the mildly inappropriate and the grossly insulting.
- Be aware of social taboos especially in relation to gender and age-related behaviour.
- Note and negotiate the social and class divisions within the society.
- How have payments and recompense to informants been arranged?

Writing-up

- Should names be anonymized and if so to what extent? This applies to the names of individuals and organizations and also the names of places.
- Have confidences been respected?
- Is there any potential harm which may arise from the information given?
- Has due acknowledgement been made of the sources of the information?

Publication and post-publication

- Have relevant consents been obtained with regard to material to be published?
- Have due acknowledgements been made?
- Have copies of the publication been sent to informants and where necessary gatekeepers, for example, government research agencies?
- Has consideration been given to the division of royalties which may follow publication or post-publication profits?

and sustainability and indeed their good intentions and practical policies led to the establishment of reserves, the gazetting of forest land and the establishment of scientific institutes. At the same time, however, it was clearly the case that despite their good intentions they failed to see the human consequences of many of their actions. Subsequent generations have had to pay the price of their short-sightedness in, for example, the extreme poverty caused by forced displacement from protected areas. Conservationists and development specialists today are concerned with many of the same issues as administrators of the past and are also invariably well-intentioned, but that colonial experience shows us that good intentions often pave the road to hell. Keeping ethical dimensions constantly under review will, if nothing else, make us reflect on the direction in which we are travelling.

Summary

1. Although we acknowledge that an earlier generation of fieldworkers behaved arrogantly in colonial times we often fail to perceive that we ourselves sometimes act similarly.
2. Professional guidelines of academic bodies usefully remind us of obligations and responsibilities and should be carefully consulted.
3. Ethics forms should be completed and signed before going into the field.
4. Being aware of our responsibility to local communities means recognizing that within these communities there are various different interest groups in terms of gender, age, class/caste, wealth, poverty, education and profession.
5. Ethical dilemmas frequently arise in the field and should be thought through from every angle.
6. Your behaviour in the field has consequences for many people besides yourself.
7. Ethical considerations need to be borne in mind at all stages of research from preparation through to writing up and disseminating results.

Further reading

Barnes, J. (1979) *Who Should Know What? Social Science, Privacy and Ethics*, Cambridge: Cambridge University Press. [Barnes has written quite extensively on ethical problems in fieldwork and although in some respects his work is dated and does not consider many of the kinds of ethical dilemmas which have surfaced in the last two decades he sets out a discussion of ethical principles clearly and straightforwardly.]

Caplan, P. (ed.) (2003) *The Ethics of Anthropology: Debates and Dilemmas*, London: Routledge. [This book contains a number of case studies arising from contemporary fieldwork experiences and includes a discussion of the recent controversy concerning the the work of anthropologists among the Yanomami of the Amazon. The introductory chapter by Caplan has an excellent bibliography and covers a lot of useful ground.]

Current Anthropology 10(5): 505–22 (Problems of Role Conflicts in Social Studies). [Two articles, 'The Problem of Ethical Integrity in Participant Observation' by I.C. Jarvie and 'Role Conflict in Social Fieldwork' by Peter Kloos are followed by thoughtful comments from several well-known anthropologists.]

Current Anthropology 20(2): 101–36 (Professions of Duplexity: A Prehistory of Ethical Codes in Anthropology). [A more recent issue of the journal that contains an article by Peter Pels followed by comments from others, including John Barnes, which touches on much more than ethical codes and considers the roles which anthropologists play and have played in the field.]

Robben, A.C.G.M. and Sluka, J.A. (eds) (2007) *Ethnographic Fieldwork: An Anthropological Reader*, Oxford: Blackwell. [Besides containing an important section of readings entitled 'fieldwork ethics' this reader also includes the code of ethics of the American Anthropological Association.]

Wade, Peter (ed.) (1995) *Advocacy in Anthropology: the GDAT Debate*, Manchester: Department of Anthropology, Manchester University. [This is one of an excellent series of pamphlets documenting annual debates in anthropological theory organized by the department of Anthropology of the University of Manchester. This particular lively debate sets out the perils and privileges when researchers engage in practice in the public arena.]

Section IV

Data processing and analysis

14 Processing and analysis of qualitative data

When one tugs at a single thing in nature, he finds it attached to the rest of the world.
(John Muir)

14.1 Introduction

This chapter considers the analysis of qualitative data – data that do not take the form of numbers and therefore cannot be analyzed directly using statistics. The most common forms of qualitative data are words (field notes, interview transcripts and recordings, documents and so on) or images (photos, drawings, video). Analysis of images such as photographs, video footage or participatory maps is beyond the scope of this book; for more advanced texts on this subject see the list of further reading at the end of this chapter. Data in the form of words are the focus of this chapter.

If you are used to working with numbers, the idea of treating words as data can seem rather puzzling. How can you be rigorous and objective, above all in testing hypotheses, given that you cannot use statistics on words? How can you decide which words or voices are important? Given that much of what people say may be inaccurate or even downright wrong, how can you assess the validity of their statements – especially if different people tell you different things? How can you prove or disprove anything?

Some of the above questions are addressed in Chapter 6 – for example, techniques to reduce inaccuracies and to double-check information for consistency (see Sections 6.4 and 6.6). It is not possible to eliminate all inaccuracies, but the same could be said for numerical data. The important thing is to be able to minimize them and assess what level of uncertainty you are dealing with. The question about rigour and objectivity is more complex; there are tools and techniques to deal with these issues but it is undeniable that qualitative analysis is weaker than quantitative analysis in these respects (though it is stronger in other ways).

Before addressing these questions further, it is important to make a distinction between quantitative and qualitative *data* – raw data in the form of numbers and non-numbers, respectively – and quantitative and qualitative *analysis*. Quantitative analysis involves counting things and using statistics to describe patterns of variation, to make inferences from a sample to a larger population, or to check for relationships between different variables (often in order to test pre-defined hypotheses). Qualitative analysis, on the other hand, involves building in-depth description and interpretation of a situation or topic. There are many ways in which this can be done, but it usually involves summarizing and discussing what the data tell you about different themes and illustrating your summary with direct quotes or with narratives recounting particularly relevant incidents. Thus quantitative analysis and qualitative analysis perform complementary roles. One gives statistical rigour, which is important for

Box 14.1 The relationship between quantitative and qualitative *data* and quantitative and qualitative *analysis*

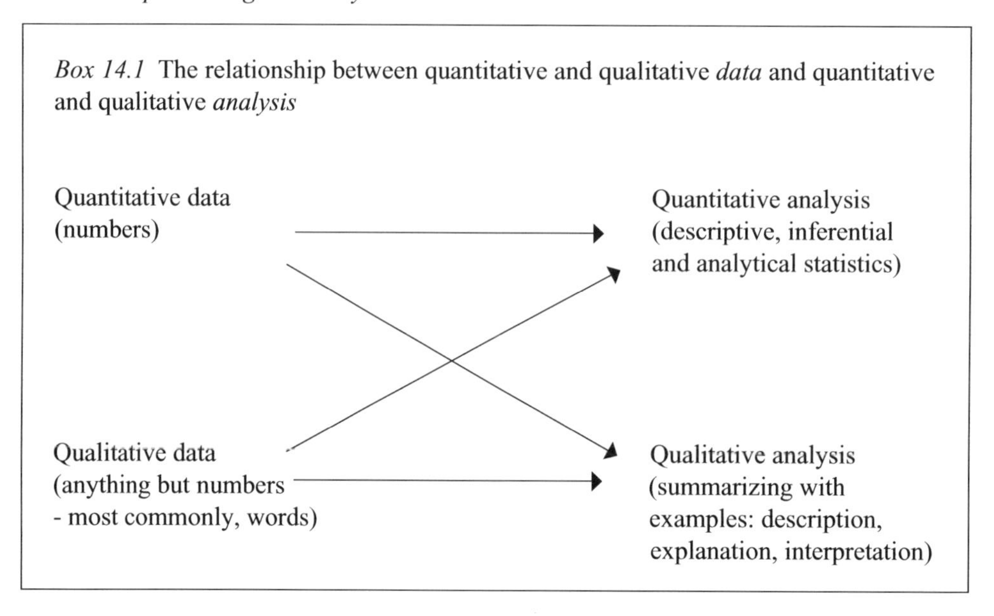

testing hypotheses, and the other gives detail and depth, which is valuable in gaining understanding of complex situations. Some of the best studies use a combination of the two.

It is important to recognize that in theory, either form of data can be analyzed either quantitatively or qualitatively (see Box 14.1). Numerical data can be used to put together a textual description of a situation, and textual data can be used to generate numbers for statistical analysis. In practice quantitative data are almost always analyzed quantitatively, but both qualitative and quantitative analyses are common for qualitative data. Chapter 7 describes how textual data from questionnaire surveys can be processed for quantitative analysis, which involves assigning numerical values to different answers. A more advanced form of quantitative analysis of texts is *content analysis*, which involves counting up the frequency with which different words, phrases or concepts appear (see Section 14.3.2 and the suggestions for further reading at the end of the chapter).

This chapter focuses on some common techniques for the analysis of qualitative data from less structured methods – participant observation, qualitative interviews and focus groups. Section 14.2 describes some initial steps of data processing and analysis that should be started during fieldwork and Section 14.3 describes some techniques for further analysis after data collection is finished. The main focus throughout this chapter is on qualitative analysis. Some methods for processing textual data for quantitative analysis are also introduced briefly, and Chapters 15 and 16 will focus on quantitative analysis in its own right. For more advanced analysis techniques, see the suggestions for further reading.

14.2 Analysis during data collection: annotations, memos and coding

Section 6.5 described some initial, relatively mechanical steps in processing qualitative interview data during fieldwork – 'cleaning up' your notes and cataloguing or transcribing your audio recordings. However, you would be wise to do more than this before you return from the field or you will come back with a big pile of notes with little idea of whether you

Box 14.2 Initial analysis during data collection

Annotations (= marginal notes, footnotes): comments written on your notes, usually either in the margins or as footnotes. They may simply 'mark' sections on a particular subject or they may include comments regarding interpretations, possible links with other sources, or points for further action.

Memos: more substantial notes to yourself that are written separately. Examples include summaries of what you think you have found out from a particular interview; ideas about emerging themes, interpretations or hypotheses, or outlines of issues you want to follow up on.

Coding: standardized codes (either abbreviations or numbers) are written in the margins to mark where a particular topic or issue is raised. Codes can be used simply as an indexing system, or to focus successive stages of the research as you go along, or to build a conceptual framework from scratch that is 'grounded' in the data (an approach known as *grounded theory:* see Strauss and Corbin [1990]).

have enough to address your research questions and no way to find data on a particular theme, other than reading through the whole lot. To avoid these problems three additional things you should consider doing during fieldwork are annotating your notes, writing memos and coding the data.

Annotations, memos and codes are simply your own comments and markers to help you organize and index the qualitative data you gather, reflect on what you have found, and plan what to do next. Annotations are scribbles on the page of data itself; memos are more lengthy, standalone written reflections, and coding is a more systematic form of annotation, involving writing standardized 'codes' on the page (usually abbreviations or numbers) to mark sections of notes dealing with different themes. These three processes are summarized in Box 14.2 and described in more detail below.

There is no strict format for annotations, memos and codes and usually no-one but you need see them, so you should feel free to develop whatever system you find most useful. However, the following descriptions offer some suggestions. I have described these processes as they have traditionally been practised in fieldwork – using notebooks and handwritten notes. However for more complex studies it is worth getting to grips with one of the many qualitative data analysis software packages available (see Box 14.3).

14.2.1 Annotations

Writing annotations is an intuitively obvious process, as is evident by the frequency with which people write comments in books and other texts they are reading (much to the disgust of librarians). Annotations usually take the form either of notes in the margins or of footnotes. They may be used to mark sections you want to be able to find again later on, or to cross-reference to other sections or sources of data, or during fieldwork, to mark something that you should follow up.

Box 14.4 shows an extract of an interview transcription from my own research on collaborative wildlife management and social change in Amazonian Peru (this research example

Box 14.3 Qualitative data analysis software

What it is and what it does
Qualitative data analysis software is specialist database software designed for organising and processing qualitative data – interview transcripts, field notes, documents, and in some cases also recordings, images and video.

Common functions

- Annotations, memo writing and coding.
- Searches for material with a specific code in order to find and extract material on a particular theme.
- Compound searches using several criteria (codes, words, attributes of participants).
- Word searches and counts.
- Some programs have the capacity to carry out very basic quantitative data analysis – e.g. to summarize the socio-demographic characteristics of participants.

There are several different software packages on the market. A discussion of the strengths and weaknesses of each of them is beyond the scope of this book, but some of the leading ones are as follows:

Atlas.ti http://www.atlasti.com (PC)
MAXqda http://www.maxqda.com/ (PC)
QSR NVivo http://www.qsrinternational.com (PC)
HyperRESEARCH http://www.researchware.com (MAC and PC)

In the UK, the University of Surrey offers excellent training workshops introducing the different packages (see http://caqdas.soc.surrey.ac.uk/).

Box 14.4 Example: interview notes with annotations

Check with PM Reserve creation C.f. int with A.R.	In 1973 professional biologists arrived at our settlement, Chino, to carry out their studies and they gave talks to us on communal reserves, who manages them and how. So we reached an agreement between two communities: Chino and Buena Vista. For its management a resolution was attained in 1991 and we started to care for lakes in the Tahuayo and the Blanco river for the resources that they had inside the forests of the reserve, we no longer let loggers, hunters and fishermen enter. We asked support from the National Police to give more strength to the care, a (guard) post so that people from outside or from elsewhere, who're not from Buena Vista — Chino, weren't allowed to go past.
San Pedro creation; flooding	In 1982 in search of high ground to make fields we came, three families, to San Pedro:.. The 'crecientes' [*annual rising of the rivers*] flooded the ground in the settlement of Chino, for this reason we entered the Blanco river to make our home thinking of the future of our children of tomorrow.

will be used throughout much of this chapter). The only change that has been made to the notes is that names and initials have been changed in order to preserve anonymity. There are four annotations. The first is a reminder to myself to check back with 'P.M.' – a biologist who was present at the time of the study. The second and fourth are labels showing what each paragraph is about. The third is a cross-reference to an interview with 'A.R.', which I remembered contained a slightly different account of the same thing. There are no rules about how many annotations you should use or what form they should take; just make sure that they are written in a format that you will still be able to understand when you come back to your notes during the writing-up phase, which may be several months later.

14.2.2 Memos

In contrast to annotations, memos are free-standing, lengthier reflections on what you have learnt up to that point (see Box 14.5 for an example). They may be connected to one particular data source such as a specific interview, or they may record your reflections on broader aspects of the research up to that point. A memo may consist of a brief summary of what you think you have learnt about a particular issue or situation, questions arising and points to follow up, or brainstorms about possible interpretations and explanations. Each memo should be marked with the date.

Memos are invaluable in providing a focus with which to 'interrogate' the raw data (and also to further explore the published literature) once you return from the field. As a simple quantitative example, the "x'' 's in Box 14.5 are exactly as written in the field; during data analysis I counted up the relevant frequencies and replaced them, and some related frequencies are included in a published paper resulting from this fieldwork (Newing, 2009).

14.2.3 Coding

Coding is a systematic form of annotation that involves marking sections of the text with standardized 'codes' (abbreviations or numbers written in the margins) that indicate the themes that they touch upon. Codes are usually hierarchical, including a small number of top-level codes and additional codes for subcategories within each of them. The codes can be defined in advance – from the literature, from your research objectives or from a

Box 14.5 Example: a memo

RCTT

10 July 2003

Discovered so far:

The mobile nature of many Amazonian families makes it difficult to see how a fixed community protected area approach to natural resource management can work. Of xxx families in the Qbda Blanco, X% had moved location on average at least once every x years throughout their adult lives [also calculate this for two communities; and for groups before / after 1989 settlement]. Though it must be recognized that, by definition, these are not long-term settled families, since the Qbda Blanco was only colonized in the 1950s.

Box 14.6 Different uses of coding

- *Indexing:* so that you can find information on a particular issue later on.
- *Iterative research:* coding is used to identify key themes and patterns as they emerge. The next stage of data collection can then be designed to explore them in more depth (see also Chapter 6).
- *Building theory:* theoretical or conceptual frameworks are constructed from the coding system with minimal use of pre-defined categories.

standardized list (see Bernard, 2006: 400) – or they can be developed as you start collecting and reviewing the data. Pre-defining the *full* set of codes in advance is quite rare; it is most appropriate in focused studies where data collection is narrowly targeted on specific issues. It is more common either to develop the codes completely from scratch, or – probably most the most common approach of all – to define some broad codes in advance in line with your aims and research objectives, and then to develop the more detailed codes and subcodes later on, according to what emerges from the data.

Box 14.6 summarizes the different purposes for which coding can be used. At its simplest, coding is a method of indexing your data so that you can find sections on different topics easily. If you do some initial coding during data collection, it can also help you to develop your line of enquiry as you go along. Coding allows you to identify key themes and patterns as they emerge that might be worth exploring further, and also to find all the material on a particular topic very easily, which makes it easier to triangulate between different informants (see section 6.6) and to judge when you have enough data (see section 4.3.1).

At a deeper level of analysis, hierarchical systems of coding categories can be used to develop the basic conceptual framework on which to build theory and through which to interpret results. The most extreme approach of this kind is an approach known as *grounded theory* (Strauss and Corbin, 1990), where the ideal is to start with a completely blank slate rather than identifying categories of interest in advance of data collection. Many researchers regard this as an unattainable ideal in its pure state in that you have to define *some* focus in advance in order to know what data to collect, and coding cannot be *entirely* objective. Nonetheless grounded theory offers a very clear and systematic approach to qualitative data analysis and has been highly influential in the development of coding as an analysis technique.

14.2.3.1 How to carry out coding

Box 14.7 gives a brief description of some common steps in coding. Coding with pre-defined codes is very straightforward: it involves going through the data (notes, transcripts or other documents) and writing codes in the margin to mark sections that deal with different themes and subthemes. If you are working without predefined codes, then you need to read through the data, identify key themes or topics, note them down, review them and when you are happy with them, invent a code for each one. Once you have a list of a few codes, you can go through your data a second time and enter codes in the margins as appropriate.

Sometimes it is obvious from the first interview what the major themes are, whereas at other times it takes longer to decide what themes you should focus on and how to define them.

Box 14.7 Instructions: how to carry out coding

With a set of predefined codes

1 Go through your 'clean' notes and write the relevant code in the margin each time a particular theme comes up.

With no predefined codes

1 Read all the way through your annotated notes from your first few interviews or first few days of fieldwork, and try to note down the main themes (those that recur and / or are especially relevant to your research) on a separate piece of paper.
2 Review the list of themes, group similar themes together, and place subthemes under broader themes, using as many levels as is useful. Try to limit the number of top-level themes to a maximum of about ten. More than this and it is much more difficult to keep track of codes during the coding process.
3 Edit them to minimize ambiguities and cut out overlap.
4 Then invent a code (abbreviation) and subcodes for each theme and subtheme. Where necessary, include a brief definition so that you will know in future exactly how you should interpret each code.
5 Go through the notes again and write the relevant code in the margin each time a particular theme comes up.
6 Use the list of codes you have developed to code further notes periodically during fieldwork. Add new codes as necessary when you come across new themes.
7 You may wish to make some alterations to the coding system part way through the study. If so, go back through all the notes you have coded so far and make sure the alterations are reflected in the coding.
8 Periodically – once every few weeks, or for a short study, after the end of fieldwork – make an index by listing the notebook and page numbers (for written notes) or documents (for computer files) where each code appears.

In a project with no predefined codes, I find it useful initially to use annotations on the field notes to mark the subject matter of each section (and also to mark anything else of interest). I then develop codes based on the annotations once I get a feel for which themes are most relevant and come up most frequently.

Box 14.8 shows again the annotated interview extract that was used in Section 14.2.1, but now with two codes entered in the right-hand margin. Note the format that is used for hierarchical codes. 'RC-cr' refers to the communal reserve (RC: a top-level code) and more specifically to its creation (cr: a subcode). The advantage of this format is that later on it will be easy to search your notes for codes at either of the levels. In this small extract there is no real difference in the function of the codes and of the corresponding annotations, but over a large set of documents there is an important difference in that annotations are completely flexible, whereas codes are (eventually) standardized. Some researchers use annotations alone because they value their flexibility; you can label each section as seems most appropriate in that particular case rather than trying to fit everything into standardized categories. However, in practical terms the great advantage of codes is that you can easily search through your notebooks of data for a particular code and thus identify all material that touches on a

Box 14.8 Example: interview extract with annotations and initial codes

Annotations	Text	Codes
Check with PM Reserve creation C.f. int with A.R.	In 1973 professional biologists arrived at our settlement, Chino, to carry out their studies and they gave talks to us on communal reserves, who manages them and how. So we reached an agreement between two communities: Chino and Buena Vista. For its management a resolution was attained in 1991 and we started to care for lakes in the Tahuayo and the Blanco river for the resources that they had inside the forests of the reserve, we no longer let loggers, hunters and fishermen enter. We asked support from the National Police to give more strength to the care, a (guard) post so that people from outside or from elsewhere, who're not from Buena Vista Chino, weren't allowed to go past.	RC-cr
San Pedro creation; flooding	In 1982 in search of high ground to make fields we came, three families, to San Pedro:.. The 'crecientes' [*annual rising of the rivers*] flooded the ground in the settlement of Chino, for this reason we entered the Blanco river to make our home thinking of the future of our children of tomorrow.	C-SP-cr

Codes: RC-cr = creation of communal reserve; C-SP-cr = creation of community of San Pedro.

particular theme. Some researchers argue that because of this, the use of codes contributes towards greater objectivity during data analysis and interpretation – it is easier to make sure that you check every reference to a particular theme, including those that do not support your tentative interpretations up to that point. To my mind, the use of coding at the broader levels is immensely valuable, but once you get down to the fine detail, annotations may be more appropriate because the detail does not fit easily into fixed categories.

There are no set rules about how detailed the coding should be. Like annotations and memos, codes are mainly for your benefit and you should develop the system that you find most helpful. You may end up with one code per paragraph in some sections and two or three codes per line elsewhere. A single section of text can have multiple codes or it can have no codes at all (for example, this may be the case for parts of interview transcripts where the interview went off subject). However, even if you aim eventually to code in detail, it is best to start at a broad level. If you attempt to go into too much detail too soon it is easy to become bogged down in dozens of very specific codes and lose sight of the big picture.

As you collect more material, you may find you need to change some of the codes you have listed because, on further use, they don't work very well. They may prove hard to apply consistently (in which case you probably need to define them more precisely) or you may find that you're always using two of them together and are not really sure of the difference (in which case either redefine them to make them more distinctive, or if they really are the same, delete one of them). Alternatively, you may want to elevate certain themes to the top level and relegate others to the status of subthemes. To start with, extensive code-editing is part of the normal process, as you try codes out and see what fits. Later on in the study you should keep these kinds of changes to a minimum because every time you delete codes, change their meaning or change their groupings, you have to go back through all the data

Box 14.9 Example: interview extract with further coding

Interview extract with annotations:

Annotations	Extract	Codes
Check with PM	In 1973 professional biologists arrived at our settlement, Chino, to carry out their studies and they gave talks to us on com-	EXT-BIOL
Inter-comm agreement	munal reserves, who manages them and how. So we reached an agreement between two communities: Chino and Buena Vista. For its management a resolution was	RC-cr
Reserve creation	attained in 1991 and we started to care for lakes in the Tahuayo and the Blanco river for the resources that they had inside the	RRNN-mgmt
Cf.int with A.R.	forests of the reserve, we no longer let loggers, hunters and fishermen enter. We asked support from the National Police to	RC-mgmt
Def. insiders	give more strength to the care, a (guard) post so that people from outside or from	EXT-POL
	elsewhere, who're not from Buena Vista – Chino, weren't allowed to go past.	RC-Mgmt-ENF

you have coded up to that point, cross the codes out and enter the new ones – which is both time-consuming and messy.

All further coding needs to be done with reference to the list of codes you have built up so far so that you don't create new codes unnecessarily. However, the list of codes should continue to expand. You can add new codes as new themes come up, invent extra subcodes in order to break down themes on which you have collected a lot of information or that you wish to investigate in more detail, and so on. This is often an ongoing process that continues throughout data collection and well into the writing-up phase. If you are unsure what code to use for a particular section of data or whether a new code is needed, then it is useful to mark the place initially with an annotation and come back to it later.

Box 14.9 shows more detailed coding of the extract used in the previous examples. The additional codes were developed in the course of data collection and are listed in Box 14.10. There are also two new annotations: 'inter-comm agreement' (inter-community agreement) and 'def. insiders' (definition of insiders: the accompanying text touches on who is permitted access to the reserve and who is excluded, which relates to the literature on what we mean by a 'community' and the importance in community conservation of defining access rights in terms of insiders and outsiders). You can decide later on whether to develop these, too, into fixed codes, or to keep them as isolated annotations. If the themes recurred several times a code would be useful in allowing you to pinpoint all the occurrences when you wish to do so; if, on the other hand, they do not come up again then there is nothing to be gained from adding a code. Note also that I have put a line through the two annotations that were reminding me to carry out further actions; this tells me that I have done what I needed to do (and the results should be written up elsewhere in my notes or memos).

Box 14.10 gives a partial list of the codes developed in this study. The terms in this particular list are all straightforward and the meanings are mostly self-evident; the only

Box 14.10 Example: extract from coding list

Extract from coding list developed during a study of collaborative wildlife management in Peru

RC	Communal Reserve
RC-cr	Creation of Reserve
RC-mgmt	Management
RC-mgmt-enf	enforcement
RRNN	Natural resources
RRNN-u	Natural resource use
RRNN-mgmt	Natural resource management
C	Community
C-SP	Community of San Pedro
C-SP-cr	Creation of community of San Pedro
EXT	External actors
EXT-BIOL	"biologists"
EXT-POL	police
EXT-MUN	municipality
EXT-NGO	"NGO"

exception is the way in which the terms 'biologist' and 'NGO' are applied. Local people had clear views on which outsiders were biologists and which were NGO workers – even though many biologists were also NGO workers (and vice versa). I have put these terms in inverted commas in the coding list, to indicate that the coding follows local usage rather than any external judgement. With more complex codes it is important to include a definition or description of the related theme, to make it clear exactly what the code stands for and how you should apply it.

14.2.3.2 Summary: coding

Coding can be rather daunting when you first try it, but it is a surprisingly powerful and rewarding process once you are used to it. Box 14.11 gives some practical tips that you should bear in mind. One issue of major concern to novice coders is 'How do I know which are the right codes to use?' The answer is very simple: there are no right or wrong codes, only more useful and less useful ones. If ten people were asked to code a set of interview transcripts from scratch, they would almost certainly all come up with different sets of codes. However, this is more a question of the focus of their analysis than of a lack of objectivity; the objectivity comes at the stage of applying a set of codes once they have been defined.

An easy way to check the internal coherence of a defined set of codes is to ask several researchers or colleagues to use them to code the same interviews. If there is a big discrepancy in the way they apply some codes, it usually means that the codes involved are not defined precisely enough. If, on the other hand, they all code the material in the same way, you can

Box 14.11 Practical tips: coding

- Do not rush to develop new codes too soon – wait until the underlying themes have appeared several times and you're reasonably sure what you want to code for.
- Initially, do not try to code all the detail. You can go back and do more later. Use annotations as an intermediate step.
- You can code a single phrase with more than one code, and you can leave sections uncoded.
- Do not get too worried about finding the 'right' codes. There is no such thing; only more useful codes and less useful codes.
- Do try to develop a coherent set of codes that are nested in groups. They are much easier to work with than a long list of freestanding codes.
- If you find you are not applying some of them consistently, then they are not very useful, so go back and define them more precisely or delete them and create new ones.
- If you do change some of your initial codes, you must go back over all the data you have coded until that point and make sure the codes reflect the changes.

be confident that you have a coherent, unambiguous set of codes. The extent to which different researchers use the (pre-defined) codes in the same way is known as *inter-coder reliability*, and it is particularly important in studies involving several research assistants.

14.3 Further analysis of qualitative data

There are many possible approaches to further analysis of qualitative data, including both qualitative and quantitative elements. Whatever type of further analysis you use, annotations, memos and coding can act as an invaluable starting point. The following sections look, first, at more advanced qualitative approaches and, second, briefly, at some techniques for transforming textual data into numbers – a prerequisite for quantitative analysis. Quantitative analysis itself is the subject of the following two chapters.

14.3.1 Qualitative analysis

The core activity in qualitative analysis is to build a narrative account describing and interpreting what you have found. This process involves an intensive 'interrogation' of the data – you should read and re-read the material you have gathered, triangulating between different sources (see Section 6.6) and examining them for common threads and patterns, then write about what you think you have found, then go back to the data and check systematically to see whether it supports what you have written. There is thus a constant interplay between thinking and writing and the data itself; unlike quantitative analysis, qualitative analysis cannot be separated from the writing-up process but is inextricably intertwined with it. The purpose is to create a storyline linking the different aspects together into a coherent account. This is described further in Section 17.3 in the context of writing the first draft of the thesis.

Each researcher develops their own technique for building a narrative. Some people start by writing a summary and then checking their field notes or other forms of data for relevant extracts such as quotes and accounts of events, whereas others start by extracting all the data on a particular theme and then writing a summary of that theme based closely on the data in front of them. Whatever method you use, annotations and especially coding are invaluable in locating the relevant material and organizing your thoughts on different topics. In fact it is common to do further coding during the write-up, as you read and re-read the material you have gathered and think about themes and patterns in the data. You may decide to code certain themes for which you have a large amount of material in even more detail. Alternatively you could add *inferential codes* – codes that do not simply describe the factual subject matter but are related to more theoretical concepts or to patterns of connection between different descriptive themes. The 'definition of insider' annotation in Box 14.9 could become an inferential code. You can even use your codes to guide the structure of the writing; the themes represented by the top-level codes could each end up as a separate section of your results. This is a useful technique to break down the mass of material into manageable chunks, although it has been criticized by some researchers on the basis that it segments the analysis artificially.

Qualitative analysis involves a skill not unlike the skill of essay-writing, in that you need to refer to multiple sources (in this case, different interviews or observations or archival documents), extract useful information or concepts, summarize, and – vitally important – make it clear how far you can back up your statements with 'evidence'. Evidence in qualitative data analysis consists ultimately of the raw data, and is most commonly used either by inserting direct quotes, or by reporting observations or by recounting a particular event or a series of events that illustrates the point you are making.

The correct formatting of quotes when writing up the results will be discussed in Chapter 17 (see Box 17.5). However in order to illustrate their role in backing up points in

Box 14.12 Example of the use of quotes: the effects of forced displacement on medicinal plant knowledge amongst the Batwa, Uganda

The most common explanation for knowledge loss was lack of access to the forest. In particular, certain types of knowledge could only be transmitted in the context of the forest:

> *I know [the songs] but still there is no way I can sing and dance it here now. There is no way I can teach my children because they must be done in the forest. If we were still in the forest I would be teaching them how our grandparents used to dance. (Man aged 40–50)*

Cultural stories and music were important forms of communication through which history was shared. It was, however, considered a taboo to tell these stories outside the forest resulting in a whole generation who may never have heard them (Kidd, 2007). The songs referred to in the quote above praise the ancestors, but cannot be sung outside the forest.

(Source: Swinscow-Hall [2007: 39], reproduced with kind permission of the author.)

Box 14.13 Example of the use of short quotes and observations: the importance of medicinal plants today

Today medicinal plants can only be collected from around the community or other areas outside the forest. The extent to which medicinal plants are used today proved difficult to gauge as I never witnessed anyone actually using them. Specific answers concerning when plants were last used or how often they were used were rarely given. Instead, a common response to '*when was the last time you used medicinal plants?*' was '*we use them whenever we need them*'. It may be that plants are not used as much as informants think they should be. Furthermore, I suspect people were reluctant to admit to a researcher interested in their plant knowledge how little they actually use them. There is a health centre very near the community where people often go for help. A common complaint was the lack of money for malaria treatment from the health centre, never the lack of money to visit a herbalist. On the one occasion I was directly asked for help in this respect, money was requested to take a pregnant woman to hospital. In part this no doubt stems from the illegality of collecting plants from the forest, especially as almost everyone agreed that medicinal herbs were more effective than western medicine.

(Source: Swinscow-Hall [2007: 35], reproduced with kind permission of the author.)

the storyline, Boxes 14.12 and 14.13 give two examples, both from a study of the effects on the Batwa pygmies in Uganda of forced displacement from their ancestral lands to make way for a protected area (Swinscow-Hall, 2007). Box 14.12 gives an example of the use of a substantial quote. Note that the speaker is not identified by name but only described by his role (in this case, age and gender; in other cases the role might be the speaker's job or institutional position). As a general rule, the identity of informants should be kept anonymous in order to preserve confidentiality (see Section 13.4.1); the exception is when they specifically state that they wish their statements to be attributed to them by name. Box 14.13 gives a more complex example that uses shorter quotes together with observations to give a convincing account of how little medicinal plants are used. It comes through clearly how – and to what extent – the author's statements are supported by the data.

14.3.2 Quantitative analysis

Quantitative analysis of qualitative data involves processing texts in some way to produce numbers, so that the numbers can be subjected to statistical analysis techniques. Chapters 15 and 16 describe the statistical techniques themselves. This section is confined to a description of when quantitative analysis is or is not appropriate and briefly introduces some techniques for producing numbers out of data from qualitative methods such as interviews, focus groups and participant observation.

Quantitative analysis is most often used on qualitative data in three ways. First, it is used for analysis of answers to closed questions in questionnaires that take the form of words. This involves a very precise form of numerical coding, which is described in Section 7.4. Second, it can be used to explore the internal composition of particular texts – documents, statements and so on. This process, known as content analysis, involves counting up the frequency with which particular words or phrases occur, or in a slightly more sophisticated approach, defining clusters of words or phrases that are associated with a particular concept and counting up their frequencies instead. Frequent purposes for which content analysis is used include policy analysis or analysis of marketing materials (for example, in the marketing of ecotourism). It can also be used to compare the perspectives of different speakers or institutions on the same subject. A detailed account of how to carry out content analysis is beyond the scope of this book but for further information see the list of further reading at the end of this chapter.

A third use of quantitative analysis is to summarize information on particular topics across a set of interviews – for example, to count up how many people expressed a particular view and how many people expressed an alternative view. This is only really appropriate for standardized sets of semi-structured interviews where probability sampling is used (see Chapter 4) or where data are collected from all members of a small population, because otherwise all you end up with is the proportion of people you happened to talk to who expressed each view – which tells you very little about the population as a whole. It is also problematic with focus groups, because it is impossible to define a watertight sampling structure with focus groups (see Section 4.3.1).

Box 14.14 reproduces an extract from the final text of an article based on my work in Amazonian Peru. Since I conducted semi-structured interviews with practically the whole population – 92 out of 94 adults living in the study area – it is meaningful to report the proportions of people who said different things. The numbers are expressed both as frequencies and also as percentages of the complete sample. The inclusion of percentages is particularly important where comparisons are made between subgroups (in this case, men and women), because the number of men and the number of women interviewed was not equal.

Once you produce numbers in this way, and provided you are dealing with a complete or representative sample, you can go on to the full range of quantitative analysis techniques: extract the numbers and present them in tables or graphs; run statistical tests for significant

Box 14.14 Counting up frequencies from a set of semi-structured interviews

One striking feature of the life histories was that sixty-six individuals (72 per cent of the total sample) had at some time lived in the city of Iquitos, in some cases for extended periods; thus, the characterization of local people as rural subsistence dwellers is over-simplistic. … Twenty-one individuals (23 per cent) had spent a period of time in the city either for their own schooling or, if they could afford it, for that of their children. Fifteen (44 per cent) of the women born in rural locations had gone to Iquitos as teenagers, ostensibly to work – either for income or to assist relatives – but also to seek a husband; all but three of them met their future husbands there. … 51 per cent of men had lived in more than three different rural locations, with a small minority (five men, or 10 per cent) having lived in more than ten different rural locations. Wives and children might accompany their men folk on contract work or stay in either the city or a rural community. As a result, women are less mobile; only six women (15 per cent) had lived in three or more locations and none had lived in more than six.

(Source: Newing [2009])

differences between subgroups or for correlations between variables, and so on. The next two chapters describe specific statistical techniques in more detail.

14.4 Conclusion

Qualitative data analysis is similar in many ways to writing an essay – it involves searching, multiple sources, developing a system to keep track of material on different themes and then synthesizing the material in order to produce a narrative account. It is less objective than quantitative analysis and is often criticized on this basis, but careful coding and cross-checking ensures some measure of objectivity. Moreover its great advantage over quantitative analysis is the richness of information and depth of insight that it can generate. Rather than viewing one analysis technique or another as superior, the two can best be regarded as complementary, offering different strengths and weaknesses.

Summary

1 Qualitative data are data that do not take the form of numbers and therefore cannot be analyzed directly using statistics.
2 The most common form of qualitative data is words, but they also include images.
3 Qualitative data can be analyzed either qualitatively or quantitatively.
4 Initial steps in qualitative analysis include annotations, memos and coding. Annotations are notes to yourself written on your interview notes, memos are more substantial notes and reflections that are written separately, and codes are standardized annotations that can be used as an indexing system, to focus successive stages of research, or to build a conceptual framework out of the data.
5 The most common form of further qualitative analysis involves building a narrative account through intensive 'interrogation' of the data.
6 In order to carry out quantitative analysis, texts must be processed in order to generate numbers.
7 In order to analyze answers to closed questions in questionnaires, a very precise form of numerical coding is used (see Chapter 7).
8 *Content analysis* is a technique used for quantitative analysis of texts and involves counting frequencies of words, phrases or concepts.
9 Finally, numbers can be generated by counting up frequencies across standardized sets of semi-structured interviews. However, this is really only appropriate if probability sampling has been used or all the members of a small population have been interviewed.

Further reading

Banks, M. (2007) *Using Visual Data in Qualitative Research*, London: SAGE Publications. [An authoritative text on the use and analysis of of visual data.]

Bryman, A. (2004) *Social Research Methods*, 2nd edn, Oxford: Oxford University Press. [Chapter 9 gives a useful introduction to content analysis.]

Miles, M. and Huberman, A. (1994) *Qualitative Data Analysis: An Expanded Sourcebook*, 2nd edn, London: SAGE Publications. [An advanced account of different analysis techniques for qualitative data, including explanations of many different forms of diagrammatic representation.]

Ryan, G. and Weisner, T. (1998) 'Content analysis of words in brief descriptions: how fathers and mothers describe their children', pages 57–68 in V. de Munck, and E. Sobo (eds) *Using Methods in the Field: A Practical Introduction and Casebook*, Walnut Creek, CA: Altamira Press. [Provides an informative case study demonstrating the use of content analysis on empirical data.]

Strauss, A. and Corbin, J. (1990) *Basics of Qualitative Research: Grounded Theory Procedures and Techniques*, London: SAGE Publications. [The classic text on grounded theory. It sets out the principals of coding clearly and methodically and is worth reading in order to get to grips with coding even if you do not intend to use the full grounded theory approach.]

Van Leeuwen, T. and Jewitt, C. (eds) (2001) *Handbook of Visual Analysis*, London: SAGE Publications. [An advanced text on analysis techniques for visual data.]

15 Quantitative analysis

Descriptive statistics

C.M. Eagle

Without data, all you are is just another person with an opinion.
(Anon)

15.1 Introduction

statistics *n. pl.* 1. Numerical facts systematically collected. (*statistics of population, crime*; VITAL *statistics*). 2. (usu. Treated as *sing.*) Science of collecting, classifying, and using statistics, esp. in or for large quantities or numbers.
(*Concise OED*, 6th edition)

The first two sections of this book introduced you to research design and the various ways you might go about collecting the data necessary to attempt an answer to your research questions. The purpose of this chapter and the next is to show some of the statistical techniques you need to present data and test hypotheses. Regrettably, the mention of 'statistics' can bring a shudder to the best of students, and undergraduate courses on statistical methods are often endured rather than enjoyed. However, without an understanding of the basic techniques, and when it is appropriate to use them, not only can you fail to analyze and present your own research results properly, but you will also find it difficult to cast a critical eye over other people's output.

It is not sensible to leave this understanding until after you have carried out your research. Earlier chapters have stressed the importance of reading around the chosen topic before planning your project: you need some statistical understanding to be able to get the most from the output of others. Moreover, although you do not need to plan the analysis in detail before you collect the data, you do need to have some idea of what type of analysis you will use. You do not want to arrive home after an expensive field trip, and only then discover that you should have collected the data in a slightly different format, or that you could do a more powerful test if only you had a larger sample.

Statistical techniques fall into two broad categories. *Descriptive statistics*, the subject of this chapter, attempts to summarize data and present them in an easily assimilated form. *Inferential statistics*, to be dealt with in the next chapter, seeks to draw inferences about the population of which your sample group is a part. These two chapters will introduce you to some of the most widely used techniques and suggest where you might take things further if necessary. However, it is most important to remember that 'statistics' is not a methodology in itself: it cannot provide hypotheses, nor compensate for poorly formulated ones, poorly conducted research, inadequate data or poor sampling. A well-done statistical analysis will enhance a well-planned project: it will not rescue a poorly planned one (see Table 15.1).

Table 15.1 Limits of 'statistics'

What 'statistics' can do	*What 'statistics' cannot do*
Reveal data patterns	Design your project
Present results clearly	Create hypotheses
Summarize numerical data	Compensate for a poorly designed questionnaire
Test hypotheses	Compensate for a poorly drawn sample
Make predictions	Turn dross into gold!

The following sections make use of data from the questionnaires in Appendices 1 and 2. The first of these is survey data collected as part of a study on the potential conflicts between kitesurfing and birds in the UK (Gilchrist, 2008). Kitesurfing is a recreational water sport, carried out around the UK coastline, often in areas that are also important for wildlife. The survey attempted to build a profile of kitesurfers and their attitudes to the environment in order to inform conservation management approaches to kitesurfing. The second questionnaire is part of a study on community orchards in England and their significance for conservation as a form of community-led protected area (Johnson, 2008). These two surveys will be referred to as 'the kitesurfing study' and 'the orchards study' throughout.

15.2 Calculating tools and statistical software

Even quite simple statistical tests require more arithmetic than most people would be willing to do with paper and pencil alone. It is no coincidence that the development of statistics, a largely twentieth-century branch of mathematics, has been paralleled by the twentieth-century development of computational tools, from mechanical adding machines through to electronic calculators with statistical functions and the computers with sophisticated statistical software packages that are widely available now. In order of increasing complexity and utility, the tools available to you are: pencil and paper, calculator, scientific calculator with statistical functions, spreadsheet program, statistical package (see Table 15.2). It is unfortunately the case

Table 15.2 Useful tools for data analysis

Tool	*Uses*	*Advantages*	*Disadvantages*
Paper and pencil	Simple calculations	(Nearly) always available	Too difficult for all but the simplest of tasks
Calculator	More complex calculations	Cheap and portable	Need to be well-organized to keep track of data
Scientific calculator with statistical functions	All simpler statistical tests	Cheap and portable	No means of safely storing data and results
Spreadsheet (e.g. Excel)	Basic statistical tests and graphs	More likely to be available than a specialist statistical program	More limited in its functions than a specialist statistical program
Statistical package (e.g. SPSS)	A wide range of statistical tests and graphs Easy data transformation	The only tool for the most complex tests	Too easy to misuse

that the more complex the tool, and the more it will do for you by way of complex calculations, generating graphs and storing and saving editable output, the less real understanding of statistical techniques you need to use it. In my own view (not always shared by students!) it is appropriate and sensible to use a software package to analyze your project data, but only after you have gained the necessary understanding by working through exercises using no more than a calculator.

The most widely used statistical package in the social sciences is SPSS (Statistical Package for the Social Sciences), and all the techniques I mention in these chapters, plus many more, can be performed using this powerful package. There are many good books on how to use SPSS and I do not intend to duplicate them (see the list of 'Further reading' at the end of Chapter 16 for details). However, I must emphasize that a statistical package is simply a computational tool. 'I'm learning SPSS' is not synonymous with 'I'm learning about statistics' and 'SPSS' is not the answer to the question 'What method are you using to test your hypothesis?' When using any piece of software, the old adage GIGO (garbage in, garbage out) very much applies. If you have no real understanding of what you are doing and why, it is all too easy to enter the data, press a few buttons and produce very impressive-looking output, with results calculated to several decimal places, that either has no sensible meaning at all, or whose meaning you cannot interpret correctly.

If you do not have access to a dedicated statistical package, then a suitable spreadsheet program could be used. The most widely used one, Excel (part of the Microsoft Office suite of programs) has a good number of statistical functions available and there is extensive help available on their use, with examples, in Office's own Help system. However the same warning applies as above: a spreadsheet will simply do the calculations you ask it to do. You need the understanding both to ask sensible questions and to interpret the answers.

15.3 Some definitions

As with just about every field of human endeavour, statistics has its own particular terminology that you need to know and use properly. Confusion is often caused because some words have a very precise and particular meaning when used in a statistical context, which is not always the same as their more general meaning. We will see in the next chapter how a 'significant' result does not necessarily have any significance (see Section 16.2). Some of the most important terms you need to understand are listed in Box 15.1 and explained further below.

In relation to *population* and *sample,* the 'units of study' or 'cases of interest' (to be referred to henceforth as 'cases') that make up your population do not necessarily have to be individual people: you might want to look at households or indeed whole settlements as your units; it depends on the hypotheses you wish to test. In the orchards study, the population is 'community orchards in England' and individual orchards are the cases, whereas in the kitesurfing study the population is 'kitesurfers in the UK' and the cases are individual kitesurfers.

Choosing a sample from a population of interest has been discussed extensively in Chapter 4 and I will not repeat the discussion here. However, it is important that you are clear in your own mind just what the population is from which you are drawing a sample. You do need to be aware, for example, whether a sample group of young people whose environmental opinions you are gathering is intended to be representative of young people around the world, or just in your country, or just students in your country, or just students in your university. Without a clear idea, you would find it difficult to choose a properly *representative* sample

Box 15.1 Key terms: basic statistical terminology

A *population* is the total set of units of study, or cases of interest (see Chapter 4):

- A *parameter* is a numerical summary of the population.

A *sample* is a subset of the population on which the study collects data:

- A *statistic* is a numerical summary of the sample data.

A *variable* is an attribute of the subjects of interest whose value can vary from subject to subject:

  - A *quantitative variable* is one whose values are numerical:
    - a *continuous quantitative variable* can take all values within a given range;
    - a *discrete quantitative variable* is one whose value changes by steps.
  - A *qualitative variable* is one whose values are not numerical.

(see Chapter 4) without which the hypotheses cannot be properly tested, nor population parameters estimated.

In relation to variables, any type of data you record about the cases can be considered in terms of a variable, but it is important to distinguish the different types so that you can use the appropriate graphical and analytical techniques. For example, if your cases were people and you wished to record some personal data about them, then *height, age in years* and *number of children* would be examples of *quantitative variables*, whereas *sex, place of birth* and *level of education* reached would be *qualitative variables*. Furthermore *number of children* is a *discrete* quantitative variable (because you cannot have two and a half children). *Age in years* is also a *discrete* quantitative variable (because it is specified that only full years be counted), whereas *height* is a *continuous* quantitative variable (in fact if we are to be strictly accurate here, although height, like weight and other measurables, is theoretically continuous and should be treated as such for calculation purposes, in practice it is discrete, as measurements are limited by the accuracy of the measuring device used).

15.4 Types of data

You can perhaps see that different *types* of measurement are involved in recording values for the different variables given as examples above. There is a widely used hierarchical classification of types of measurement in statistics, first proposed by S.S. Stevens (1946), and the type of measurement used to collect data for any particular variable determines what can be sensibly done with it. Although there is still some disagreement about the details of this classificatory system, you need to be aware of the different data types (sometimes called *levels of measurement*), so that, at the very least, your analyses are appropriate for your data. You may wonder why the word 'measurement' is used in relation to recording a qualitative variable like 'place of birth' but it is standard usage in this context. The types of measurement are listed in Table 15.3 and discussed further below.

The words nominal, ordinal, interval and ratio refer to different types, or *scales*, of measurement. Data collected using these different scales are thus referred to as *nominal data*,

Table 15.3 Key terms: types of data measurement

Type of 'measurement'		*Quantitative or qualitative*	*Examples of variables that use this type of measurement*
Nominal	Each case is assigned to a labelled category	Qualitative	Place of birth Gender of respondent
Ordinal	Each case is assigned to an ordered category	Qualitative	Level of education Satisfaction rating scale
Interval and ratio	Each case is measured on a numerical scale where intervals or ratios can be compared	Quantitative	Age in years Height

ordinal data, *interval data* or *ratio data* as appropriate and the associated variables as *nominal variables*, *ordinal variables*, *interval variables* and *ratio variables*. You may sometimes see the term *categorical data* used as a collective term for both nominal and ordinal data. However 'categorical' is also sometimes used simply as an alternative to 'nominal', so, to avoid confusion, it is best to stick to 'nominal' and 'ordinal' as appropriate. It is theoretically possible, but unwise, to use different types of measurement in one variable. For example, if you ask people for their age and fail to specify the form in which you would like the answer, you could end up with some people giving you their exact age in years and others, perhaps less willing to reveal their age, or even uncertain as to the exact value, putting themselves in an age group. The former would produce ratio data, the latter ordinal data.

15.4.1 Nominal data

Nominal data are 'measured' by simply putting your subjects into labelled categories. *Gender* and *place of birth* are both instances of nominal data. The possible values of these two variables are labels: just the two values of *male* and *female* for *gender* but more open-ended possibilities for *place of birth*. The only comparison that can be made between two instances of a nominal variable is whether they are equal or not. Are two people in your study both the same gender or not? Do two people in your study come from the same birthplace or not?

15.4.2 Ordinal data

Like nominal data, ordinal data are also qualitative, but with the added sophistication of an order to the categories. *Level of education* is an ordinal variable. You assign a label to each of the subjects with possible values 'no formal education', 'primary school only', 'secondary school' and so on, and there is a clear ordering to the values. Not only can we compare subjects to see if their education level is equal or not (as we could with nominal variables), but now we can also put the subjects in order, from least education to most. *Age in categories* and *Income in categories* (see Question 4 in the kitesurfing study) are also ordinal variables. Thus the comparisons that can be made between two instances of an ordinal variable are whether they are equal, or if not, which one is higher, which one lower.

Provided that they have answered your questions honestly, comparing two people's responses related to the types of ordinal variables mentioned above is not contentious, because the categories are objectively defined. It is generally agreed, for example, that Person A, who went to university, has had more formal education than Person B, who stopped after

> *Box 15.2* Example: questions generating ordinal data from rating scales using words
>
> How would you rate the level of service you received?
> a) Poor b) Fair c) Good d) Very good e) Excellent
>
> *or*
>
> How concerned are you about global warming?
> a) Very concerned b) Slightly concerned c) Not at all concerned

secondary school, and that they have both had more formal education than Person C, who only went to primary school. Rather more problematic however are those variables that seek to measure opinion, or satisfaction levels. If you have ever been asked to fill in a feedback questionnaire, or give your opinion on a contentious topic by means of a questionnaire, you will have encountered questions that are designed to generate ordinal data based on a verbal *rating scale* (see Section 7.3.2.4), such as those in Box 15.2.

When you ask questions like this you are requiring your respondents to put themselves into one of a set of pre-determined ordered categories, and therefore you are collecting data on ordinal variables, because the data are measured on an ordered scale. It is important to emphasize here that neither nominal nor ordinal data are fully numerical. It is perfectly easy (and indeed sometimes necessary when using a statistical package) to code your values numerically such as:

Variable: Gender
  Values: 'Male' = 1, 'Female' = 2
Variable: Service level rating
  Values: 'Poor' = 1, 'Fair' = 2, 'Good' = 3, 'Very good' = 4, 'Excellent' = 5

However, this does not magically turn qualitative variables into quantitative ones. The numbers you have assigned to the values are still just labels, and should not be used for the whole range of statistical techniques described in this and the next chapter.

Numerical rating scales, widely used in social science surveys, are variants on the verbal satisfaction and opinion scales mentioned above, requiring respondents to place themselves directly onto a numerical scale rather than into a verbal category. Numerical rating scales produce ordinal data. Box 15.3 shows an example from the kitesurfing study. It is important

> *Box 15.3* Example: question generating ordinal data from numerical rating scales
>
> 14. Please indicate how strongly you feel about environmental issues. Select the number that corresponds to how you feel; from '1' indicating that you are 'not at all' concerned about environmental issues and '7' indicating that you feel 'very strongly' about environmental issues.
>
> | 1 | 2 | 3 | 4 | 5 | 6 | 7 |
> |---|---|---|---|---|---|---|
> | Not at all | | | Neutral | | | Very strongly |

to note that this still produces ordinal data, even though the categories are labelled in the form of numbers, and that you should therefore only use analyses appropriate for ordinal data. The numbers are not measured in consistent units and it is not possible to say, for example, that the gap between a '1' and a '2' is the same as that between a '4' and a '5', or even less that a score of '6' is worth twice a score of '3'.

The definition of ordinal data is quite wide and exactly how far ordinal data may be treated as numerical is a matter of some disagreement. Data in three or four ordered categories, data in a dozen ordered categories, data in the form of rankings and data derived from numerical rating scales are all classified as ordinal data, although some such data sets clearly contain more numerical information than do others: for example, if there is a large number of categories. In cases such as these, techniques usually reserved for numerical data may be useful. There is therefore a grey area in the extent to which analysis techniques suitable for numerical variables can be used with ordinal variables; to a large extent, it is a matter of common sense, but if in doubt, seek advice from a statistician. However, if you do get it wrong you will be in good company. Numbering ordered categories and then treating them inappropriately as 'real' numbers is one of the most widespread misuses of statistical techniques, perpetrated not only by those who are still learning and might be forgiven, but also by those who really ought to know much better.

15.4.3 Interval and ratio data

I will treat these together, as both are properly quantitative: that is they are measured on scales on which not only can the items be ordered, as with ordinal data, but the intervals between points on the scale can meaningfully be compared. What do I mean by this? If we go back to the 'level of education' example above, it is perfectly sensible to say that Person A, who went to university, has had more formal education than Person B, who stopped after secondary school, and that they both have had more formal education than Person C, who only went to primary school. However, it makes no sense to go on to say that the difference between Person A's and Person B's educational level is the same as the difference between Person B's and Person C's. In contrast, for interval and ratio data such comparisons can be made.

Ratio level measurements are those in which ratios between two different measurements can be calculated. The variables *height, age in years* and *number of children* are all ratio measures. Why? Because it is sensible to ask questions such as 'Is Person A twice as old as Person B?'

The best known example of an interval level measure is temperature. We can say that the interval between 0°C and 10°C is the same as that between 10°C and 20°C. However, it cannot sensibly be said that 20°C is twice as hot as 10°C, because that would assume that the scale only starts at 0°C, which is not the case – the scale continues downwards even though, by convention, the numbers are then assigned labels in the form of negative values. Interval measures have arbitrary zero points and therefore the ratios between different values cannot be calculated, whereas ratio measures have non-arbitrary zero points.

15.4.4 Summary: types of data

The different types of data described above form a hierarchy from the least complex (by convention regarded as the 'lowest') to the most complex (the 'highest') (Box 15.4).

The more complex the data type, the more one can do with the data, and so in theoretical terms it is always preferable to collect your data in the most complex form possible. If necessary

Box 15.4 Data hierarchy

Nominal	**Ordinal**	**Interval**	**Ratio**
Lowest			Highest
Least complex			Most complex

you can transform a ratio level variable into a categorical or ordinal one by grouping the results, but you cannot do it the other way around – you cannot extract a ratio level variable from an ordinal one. For example, if you ask the respondents for their age in exact years, you can always re-code the answers into an ordinal variable whose values are age *categories* if it suits the proposed analysis. In contrast, if the original age data were collected in categories, then you are limited as to how you can use that information. However the choice of data type in asking about a variable such as age must balance these theoretical considerations connected to analysis techniques against practical considerations in terms of likely response errors (see Chapter 7). In many cultures, people are likely to feel more comfortable answering the question 'How old are you?' if they can pick a category rather than give their exact age. Moreover, many people worldwide will not know their exact age and so data collection with this level of precision is not justifiable.

15.5 What are descriptive statistics for?

As the name suggests, the purpose of descriptive statistics is to *describe* the data. The questions that descriptive statistics attempt to answer fall into three broad categories:

1 What patterns do my data make?
2 Do my data points cluster around a central point, and if so what is its value?
3 How spread out are my data values and are there any anomalies?

There are two main reasons for using descriptive statistics:

1 As an exploratory analysis of your data to get some idea of the answers to the questions above before you proceed to testing your hypotheses.
2 To enable your audience to understand just what results you have obtained. Your finished report should include enough descriptive statistics to give a summary.

The most wonderful piece of research loses most of its value if you cannot present the information you have gathered in a way that is meaningful to those who read it.

15.6 Picturing data and seeing patterns: what patterns do my data make?

The use of graphing and pictorial techniques to display data is thought to date back to the late eighteenth century. The website http://www.math.yorku.ca/SCS/Gallery/historical.html (accessed 11 January 2009) contains a fascinating set of historical graphs, together with a discussion of the role of visualization in data presentation. Along with the calculation of statistics, drawing graphs has become very easy when you have access to a spreadsheet or statistical package (although computer-generated graphs can in no way compare to the best of the historical examples mentioned above). If you have ever used a spreadsheet to generate a chart or graph you will have found what seems to be a bewildering array of potential graphs and charts available to you. Most of these, however, are simply decorative elaborations of a few basic types. Each data type has a type of graph most suited to it and you must make sure that you choose a type of graph that suits your data.

For any and all forms of visual presentation, a general rule is that the reader should be able to look at a table (or figure) on its own and understand what it shows. That means that it must have a title that explains exactly what it is about; all columns, rows and axes should be properly labelled (giving the units of measurement, if relevant), and the sample size should be stated. In addition, the main text of a research report should refer to the table and point out key features of the results, but without repeating all the detail. Examples of these details are included in the different parts of this section, with additional guidance as appropriate.

15.6.1 Frequency tables

The most basic way of presenting data visually is as a *frequency table*: a table showing the number of cases in which each possible response was given. Table 15.4 gives an example of a frequency table for nominal data, using the results to question 18 from the orchards study: 'Who owns the land that the orchard is planted on?'

Since there is no logical order to the categories of ownership, you need to decide how best to order them in the table. The best choice is usually the order of frequency, from highest to lowest. A frequency table should include a column for percentage frequency as well as one for frequency counts. The information that 22, the highest frequency, represents 26 per cent of the total is much more useful than the count alone.

The following paragraph shows how the table could be referred to in the text:

> There was considerable variation in the form of land ownership for different community orchards [see Table 15.4]. The most common form was that of a trust or other

Table 15.4 Frequency table for nominal data

Frequencies of different kinds of land ownership for community orchards ($N = 84$)

Type of ownership	*Frequency (no. of orchards)*	*% Freq.* (0 d.p.)
Trust or NFP	22	26
District or Borough Council	14	17
Parish or Town Council	13	16
Private owner	10	12
County Council	8	10
City Council	6	7
Community	4	5
University or school	2	2
Church	1	1
Crown Estates	1	1
Metropolitan Council	1	1
Other	2	2

Table 15.5 Frequency table for ordinal data

Income profile of kitesurfers (N = 99)

Income category (£s per annum)	*Freq. (no. of people)*	*% Freq. (1 d.p.)*	*Cum. Freq. (no. of people)*	*%Cum. Freq. (1 d.p.)*
No income	8	8.1	8	8.1
Under 10,000	9	9.1	17	17.2
10,000-19,999	13	13.1	30	30.3
20,000-29,999	26	26.3	56	56.6
30,000-39,999	19	19.2	75	75.8
40,000-49,999	7	7.1	82	82.8
50,000-59,999	2	2.0	84	84.8
60,000-69,999	4	4.0	88	88.9
70,000-79,999	3	3.0	91	91.9
80,000-89,999	1	1.0	92	92.9
90,000-99,999	0	0.0	92	92.9
100,000 or more	7	7.1	99	100.0

not-for-profit organization, accounting for 26 per cent of all orchards sampled. Other forms accounting for more than 5 per cent of orchards included ownership by different kinds of local or regional councils, and private ownership.

Table 15.5 gives an example of a frequency table for ordinal data, using the results to question 4 from the kitesurfing study: 'If you are working, what is your gross annual income (in pounds)?' (with category responses).

Here the categories, being ordinal, are listed in their logical order, rather than from highest to lowest frequency. Columns for cumulative frequency and percentage cumulative frequency have been added, because, with ordinal data, the cumulative values are meaningful. For example it is easy to read off the table such information as 'More than half the respondents earn less than £30,000 per annum.' Again, the main text should refer to the table and point out key features, without repeating all the detail.

In both examples, note that the columns are properly labelled with their units of measure, and the precision of the percentage calculation indicated ('1 d.p.' means they are calculated to one decimal place). In addition the figures in the body of the table have been typed using a fixed-width, or mono-spaced font. As all the characters in such a font, whether wide like an 'm' or an '8', or narrow like an 'i' or a '1', occupy the same width, its use makes it much easier to align columns of figures properly. (Courier, used in Table 15.5 and originally designed for typewriters, is the best-known of these fonts.) These points may all seem quite minor, but proper attention to such detail makes for a much more readable table.

Tables 15.4 and 15.5 each have 12 rows and this is probably as big as a table ought to be for ease of use. If you wish to present *interval* or *ratio* data in tabular form, then the data need to be re-coded into no more than 12 useful categories.

15.6.2 Graphs

Frequency tables have the advantage of showing actual data values but they are not very visually appealing, nor do they necessarily show data patterns very well. For that you need to draw graphs. The most widely used types of graph are:

1 pie chart
2 bar chart
3 scatter plot
4 line graph

This section looks at what type of graph it is best to use for different types of data.

15.6.2.1 Graphs of nominal data

As nominal data is not numerical, all we can do is count frequencies – how many cases fall in each category – and graph the frequencies with a pie chart or a bar chart.

Consider again the answers to question 18 from the orchards study: 'Who owns the land that the orchard is planted on?' The results have already been shown in a frequency table (Table 15.4). An equally valid alternative is to show them in a graph. However, there are too many response categories in the original question to make for a good graph; the columns and their labels would be too small to see and compare easily. Better to merge some into logical super-categories, and then graph, as in Figures 15.1 and 15.2.

Notice that the bar chart uses frequencies, whereas the pie chart uses *percentage* frequencies. Both forms of graph are valid and both give a clear picture, albeit without the detailed figures. However, the pie chart shows more clearly than the bar chart how the cases are portioned out by a particular variable.

In the pie chart, the percentage frequencies have been written on the pie chart itself (Figure 15.2). When choosing what to include on a graph you need to make sure you include enough information to make your graph useful, but not so much that the basic picture is obscured.

15.6.2.2 Graphs of ordinal data

The same options are available for ordinal data as for nominal data. However, as there is now a logical order to the categories, a bar chart is more useful than a pie chart, as the ordering

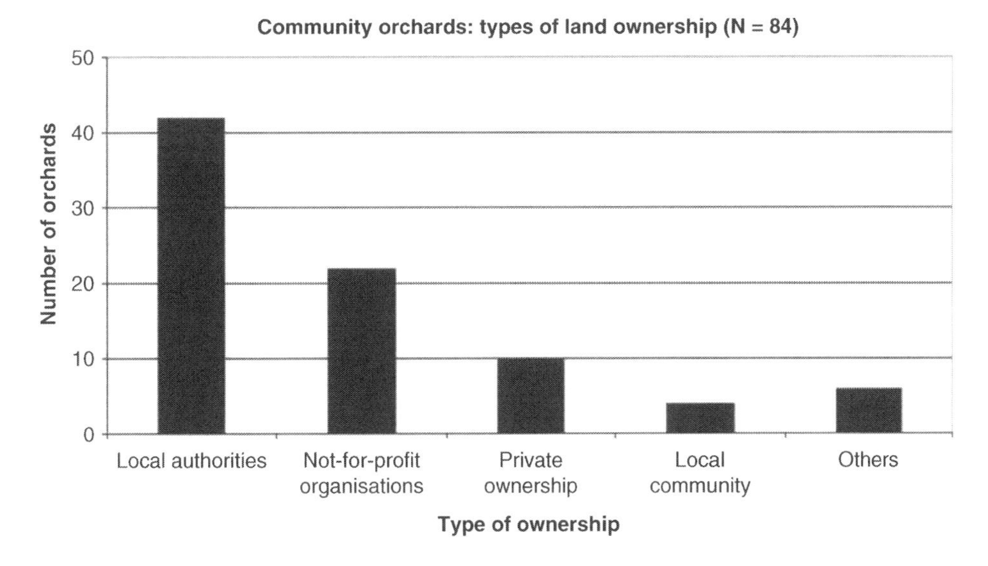

Figure 15.1 Bar chart for nominal data.

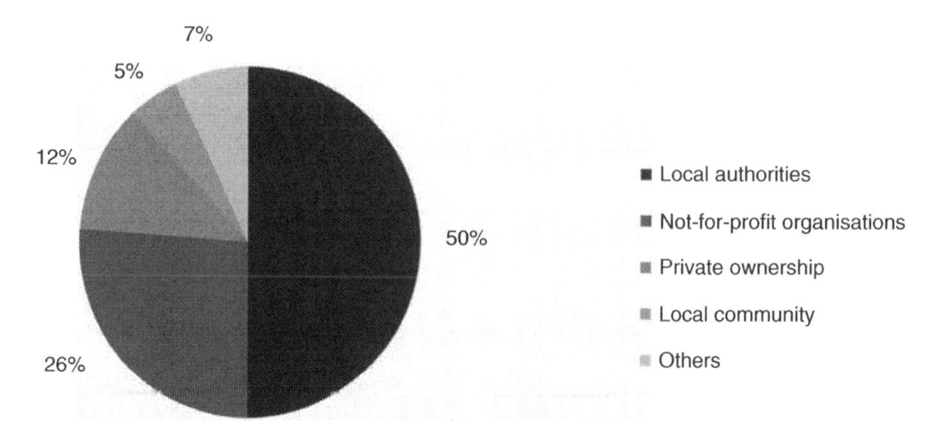

Figure 15.2 Pie chart for nominal data.

is preserved. Consider the answers to question 4 from the kitesurfing study: 'If you are working, what is your gross annual income (in pounds)?' Again, the results have already been shown in a frequency table (Table 15.5) and either a table or a graph is valid. Also again, there are rather too many categories for a good graph, so categories have been collapsed from £10,000 bands into £20,000 bands.

Figures 15.3 and 15.4 show the results, first as a pie chart and then as a bar chart. Both are valid, but the pie chart does not show the distribution pattern as effectively as the bar chart because it cannot show the order of the categories.

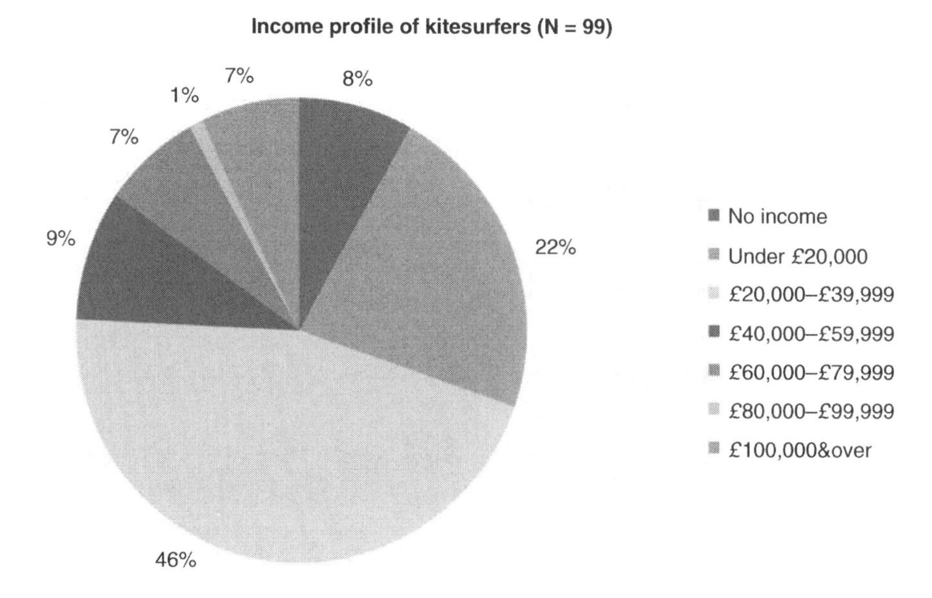

Figure 15.3 Pie chart for ordinal data.

Notice that all the examples above have meaningful titles, and labels and scale values marked on each axis. Unlabelled tables and graphs are not acceptable, whether they are for your own use or for public consumption. As mentioned earlier with reference to tables, the reader should be able to look at a graph on its own and understand what it shows.

The following paragraph shows how the bar chart in Figure 15.4 could be referred to in the text:

> The kitesurfers surveyed had a wide range of incomes, from no income to over £100,000 per annum [see Figure 15.4]. Approximately half of those surveyed came into the modal category '£20,000–£39,000' with the remaining half equally split into those with incomes that were higher than the modal category, and those with incomes that were lower.

15.6.2.3 Graphs of interval and ratio data

In the case of interval and ratio data, graphs can be used either to show the distribution of data on a single variable, or to show how two variables vary together. This section will deal with each of these in turn.

As an example of how to show the distribution of data on a single variable, consider the answers to question 1 from the kitesurfing study: 'What is your age?' They were as follows:

> 43, 27, 37, 46, 33, 41, 30, 37, 27, 25, 33, 25, 28, 23, 32, 26, 34, 40, 35, 17, 35, 49, 25, 28, 29, 35, 23, 45, 31, 35, 24, 38, 34, 32, 19, 39, 21, 43, 40, 46, 35, 31, 38, 27, 45, 35, 31, 34, 39, 20, 22, 16, 45, 30, 30, 33, 25, 31, 39, 21, 13, 32, 32, 31, 46, 15, 31, 37, 13, 42, 27, 27, 35, 36, 31, 28, 45, 22, 24, 43, 49, 29, 29, 27, 26, 15, 24, 41, 25, 30, 36, 49, 34, 44, 26, 32, 22, 27, 14

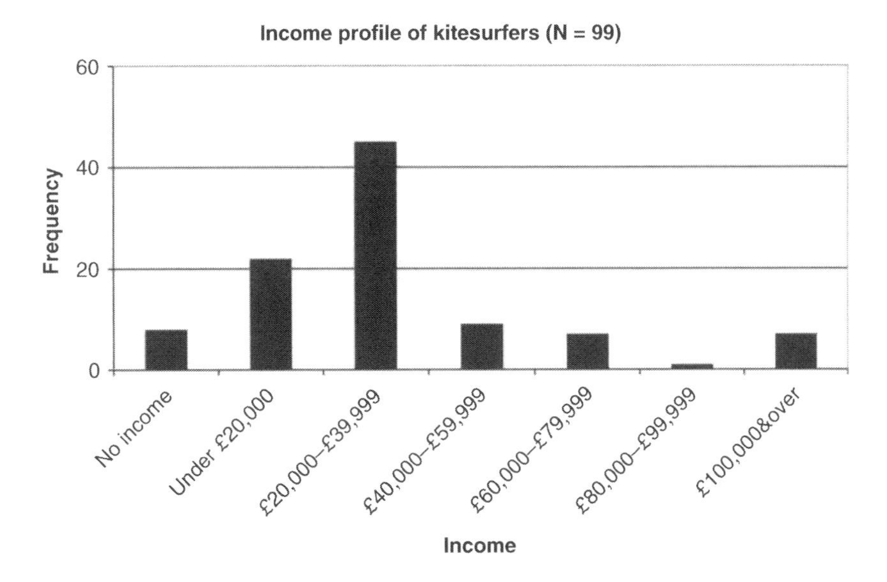

Figure 15.4 Bar chart for ordinal data.

Stem unit: 5 years — Leaf unit: 1 year

Stem	Leaves	Cumulative total
10	3 3 4	3
15	0 0 1 2 4	8
20	0 1 1 2 2 2 3 3 4 4 4	19
25	0 0 0 0 0 1 1 1 2 2 2 2 2 2 2 3 3 3 4 4 4	40
30	0 0 0 0 1 1 1 1 1 1 1 2 2 2 2 2 3 3 3 4 4 4 4	63
35	0 0 0 0 0 0 0 1 1 2 2 2 3 3 4 4 4	80
40	0 0 1 1 2 3 3 3 4	89
45	0 0 0 0 1 1 1 4 4 4	99

Figure 15.5 Stem-and-Leaf diagram.

Note that (1) there are 99 pieces of data, and (2) this is quantitative, ratio level, discrete data. However when the data set is viewed like this as an unsorted list, it is difficult to discern a pattern and see the age profile of the respondents. How can the patterns (if any) be made clearer?

We could first *order* the list:

> 13, 13, 14, 15, 15, 16, 17, 19, 20, 21, 21, 22, 22, 22, 23, 23, 24, 24, 24, 25, 25, 25, 25, 25, 26, 26, 26, 27, 27, 27, 27, 27, 27, 27, 28, 28, 28, 29, 29, 29, 30, 30, 30, 30, 31, 31, 31, 31, 31, 31, 31, 32, 32, 32, 32, 32, 33, 33, 33, 34, 34, 34, 34, 35, 35, 35, 35, 35, 35, 35, 36, 36, 37, 37, 37, 38, 38, 39, 39, 39, 40, 40, 41, 41, 42, 43, 43, 43, 44, 45, 45, 45, 45, 46, 46, 46, 49, 49, 49

It then becomes immediately clear that ages range from 13 to 49, with the 30s as the biggest single group.

One of the easiest paper and pencil methods of representing numerical data like this is the stem-and-leaf diagram, invented by J.W. Tukey, an American statistician who made many valuable contributions to statistics. Figure 15.5 shows a stem-and-leaf diagram of the data above. This diagram was made by:

1. Choosing suitable *stem units:* a convenient number that will split your data set into five to ten categories. I chose five years in this case, as ten years would have given too few categories.
2. Choosing suitable *leaf units:* usually your basic unit of measure. The obvious leaf unit here is one year.
3. Placing the stem units in a column down the left-hand side. As the data range here is 13 to 49, we need stem units ranging from 10 to 45.
4. Placing each individual piece of data in a row to the right of its stem, in order of increasing size. The numbers entered in the rows are the 'leaf' values left over from the 'stem' value of that row. So, for example, the five pieces of data 15, 15, 16, 17, 19 all have the *stem* 15 and *leaves* of 0, 0, 1, 2, 4, respectively.
5. Adding up the number of cases so far for the cumulative total in the right-hand column. So, for example, there are 8 cases with values of less than 20 and 19 cases less than 25.

Without doing any formal calculations at all you now know a lot more about the respondents to this survey. For example:

1 The age groups are not evenly spread throughout the range, but heavily clustered around 30.
2 Almost half the subjects fall within the ten-year range 25 to 34.
3 The mid-point or *median* (see Section 15.7.2) falls within the range 30 to 34.

Stem-and-leaf diagrams are useful for exploring a data set and the shape of its distribution. Although not used as much as they once were, they provide a quick, easy and computer-free way of looking at quantitative data graphically. However, for presentation purposes these results are best treated as ordinal data (using age categories rather than precise age in years) and shown as a frequency table (as in Table 15.5) or a bar chart (Figure 15.6). Note that good presentation often involves losing some of the details of the data in order to present a clearer picture. In Figure 15.6 we have taken *quantitative ratio data* and turned it into *ordinal data* to show the pattern of the data.

We now turn to how graphs can be used to show how two variables vary together. If you want to explore a possible relationship between two numerical variables, a *scatter plot* can be used to show how the variables change together. Figure 15.7 is an example of such a graph. It uses data taken from the World Development Report 2000/2001 (World Bank, 2001). The variable plotted on the horizontal or x axis is gross national product (GNP) at purchasing power parity (PPP) measured in US dollars per capita ('purchasing power parity' means that the actual GNP has been adjusted to take account of the different cost of living in each country and so gives a more realistic comparison of income levels). The variable plotted on the vertical or y axis is carbon dioxide emissions, measured in metric tons per capita (the emissions measured were those produced by the burning of fossil fuels and the manufacture of cement: carbon dioxide is one of the main greenhouse gases). The graph thus shows the relationship, if any, between the comparative wealth of a nation and the amount of a major pollutant emitted.

As might be expected there appears to be a connection between the two variables. There is a general trend: the richer the nation, the greater the emissions. Figure 15.8 shows the same graph with a *trend line* added, which attempts to model the possible relationship. Measuring the strength of a relationship and determining a trend line take us into the realm of *inference*, the subject of the next chapter.

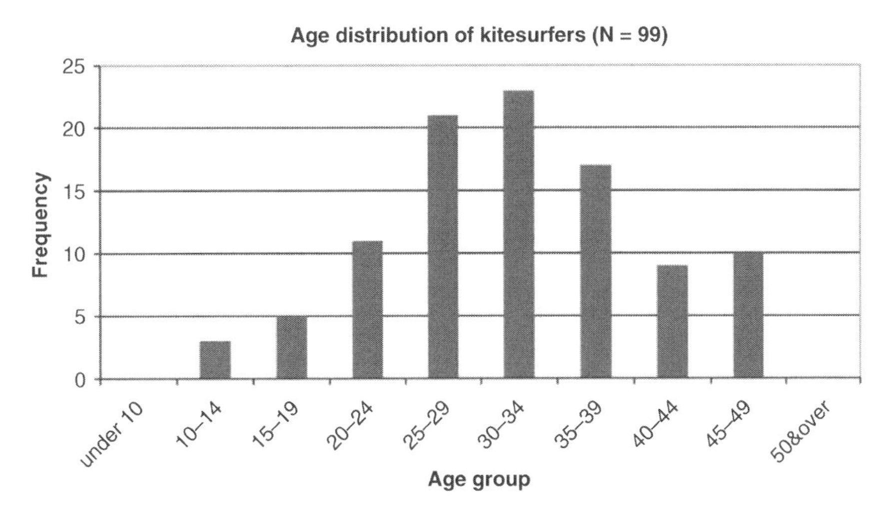

Figure 15.6 Bar chart for ratio data coded into ordered categories.

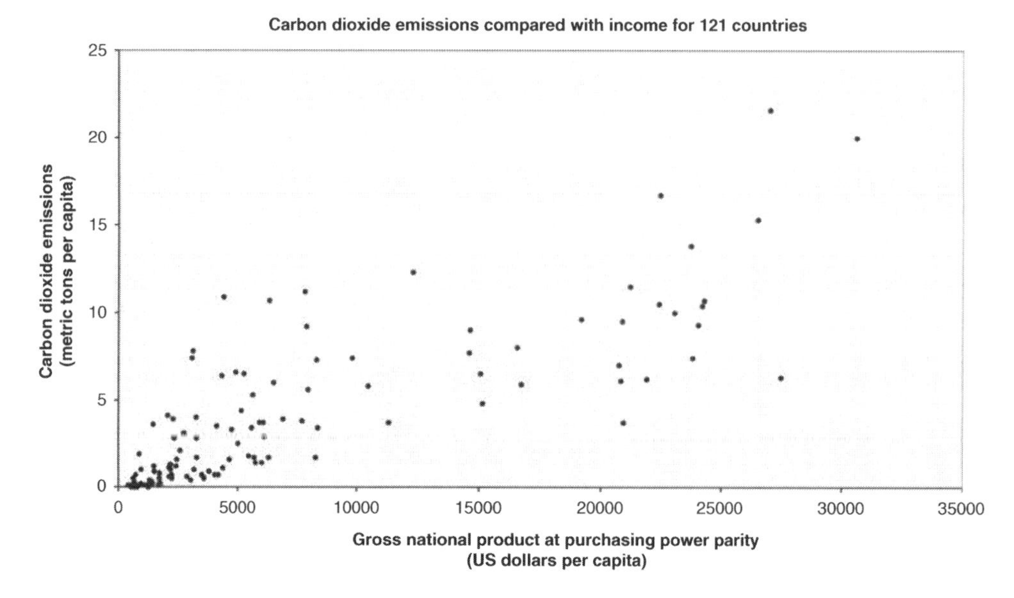

Figure 15.7 Scatter plot showing how two numerical variables vary together.

However, the value of a scatter plot is not limited to showing how two variables may relate to each other: it may also usefully reveal *outliers*, that is, points which lie some distance from the main trend line. Arrowed in Figure 15.8 are two of the richest countries, with almost the same GNP per capita, but very different emission levels.

Note that figure 15.8 is not a *line graph*, even though a trend line has been added to the scatter plot. The points on the graph have not been joined by a single line, because each point

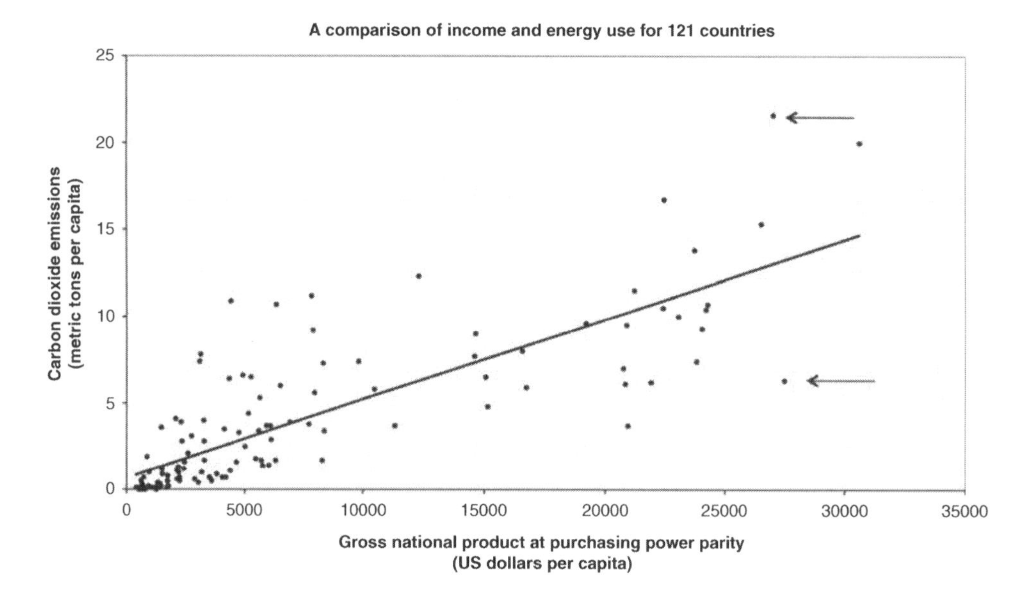

Figure 15.8 Scatter plot with trendline added and some outliers marked.

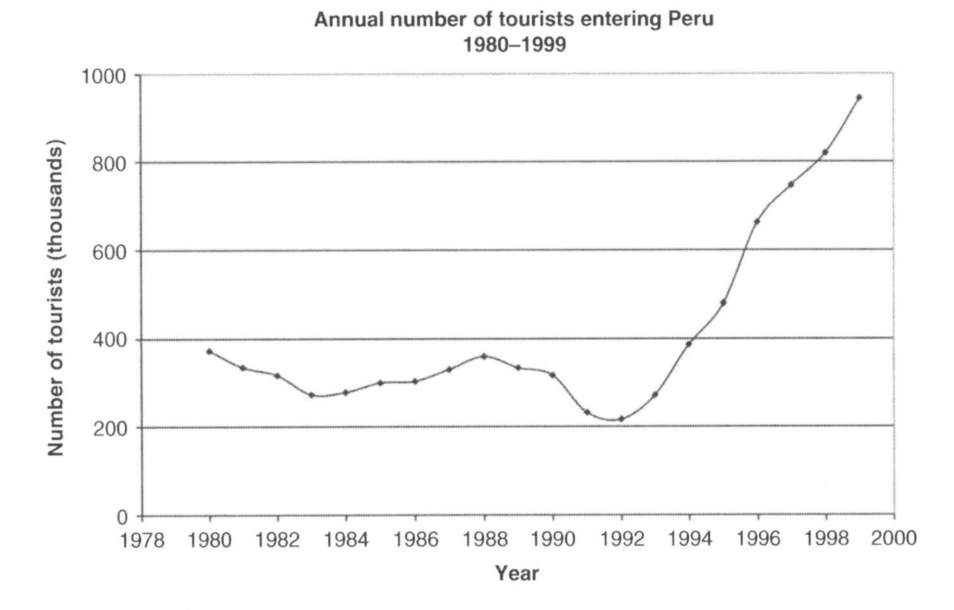

Figure 15.9 A line graph.

represents a different country and there is no rationale to 'joining the dots'. A *line graph* is a particular example of a scatter plot where some logical progression in one of the axes makes joining the points appropriate. This is usually the passage of time, and such line graphs are often called *timelines*. For example, Figure 15.9 plots the number of people entering Peru for tourism purposes annually from 1980 to 1999 (INEI, n.d.). The number of tourists per annum takes the form of discrete, quantitative, ratio level data, while *Year* is an interval level variable. The graph starts with a set of data points, but a connecting line can be drawn because some meaning can be attached to the spaces between the known points. This graph shows the dip in tourist numbers (following a rise in terrorist activity in Peru in the late 1980s and early 1990s) more clearly and effectively than a table of data could ever do.

15.6.2.4 Summary: when to use different kinds of graph

Table 15.6 Summary: when to use different kinds of graph

	Nominal data	*Ordinal data*	*Interval and ratio data*
Pie chart	*Particular use:* To show, for a particular variable, the proportions of cases that hold each value. *Points to note:* Too many categories can make for a confusing graph: recode data into fewer categories where necessary.	*Particular use:* To show, for a particular variable, the proportions of cases that hold each value. *Points to note:* Loses the ordering in the variable, so less useful than a bar chart.	Interval or ratio data can be treated as ordinal data by recoding into a suitable number of ordered categories.

(Continued)

Table 15.6 (*Cont'd*)

	Nominal data	*Ordinal data*	*Interval and ratio data*
Bar chart	*Particular use:* To show the distribution pattern of a variable. *Points to note:* Too many categories can make for a confusing graph: recode data into fewer categories where necessary.	*Particular use:* To show the distribution pattern of a variable, while preserving its ordering.	
Scatter plot	*Not suitable*	*Only suitable if the data is sufficiently 'numerical'.*	*Particular use:* To show how two variables change together.
Line graph		*Particular use*: As for interval and ratio data.	*Particular use:* To show how two variables change together, where there is a logical progression in one of the variables.
Time line			*Particular use:* To show how a variable changes with time.

15.7 Finding the middle: do my data points cluster around a central point, and if so what is its value?

An *average* is a quick way of summarizing a data set in terms of a 'middle' value. Many different averages are defined mathematically, of which three will be of use to you: *mode, median, and mean* (see Box 15.5). As with graphing techniques, which of these you can use depends on the type of data you have.

15.7.1 The mode

The mode is the value that occurs most frequently. Data do not even have to be numerical to have a mode, and so it can be used for all data types, but it is particularly used (and indeed is the only option) for nominal data (see Box 15.6 for some examples).

Box 15.5 Key terms: commonly used averages

- The *mode* is the value that occurs most frequently.
- The *median* is the middle value, if the data are placed in order.
- The *mean* is calculated by adding the data and dividing by the number of data points.

Box 15.6 Examples: the mode

Categorical data

The *modal* community orchard ownership category is *local authorities*. More orchards are owned by local authorities than by any other ownership category.

Ordinal data

The *modal* income category for kitesurfers is £20,000–£29,999. More kitesurfers are in this category than any other.

Ratio data

The age distribution of the kitesurfers is tri-modal: ages 27, 31 and 35 years share top spot.

Note that there can be more than one mode in any particular set of data and indeed a mode may not even exist if all the values are different. Moreover it may change if you merge or re-define your categories.

15.7.2 The median

The median is the middle value if the data are placed in order according to their values. If there is an even number of data points, then the median is halfway between the two middle values. Data do not need to be numerical to calculate the median, but do need to have an order, and so the median can be defined for ordinal, interval and ratio data (see Box 15.7 for examples).

Box 15.7 Examples: the median

Ordinal data

The *median* income category for kitesurfers is £20,000–£29,999.

Ratio data

The *median* age of the kitesurfers is 31 years. In the ordered list in section 15.6, the fiftieth value in the list of 99 is 31.

15.7.3 The mean

The mean (or, more strictly, the arithmetic mean) is the most commonly used type of average, and is usually implied when the word 'average' is used without qualification. It is calculated by adding the data values and dividing the total by the number of data points. It is usually written as $\bar{x}$, where:

$$\bar{x} = \frac{\sum x}{N}.$$

Box 15.8 Example: the mean

Ratio variable

The *mean* age of the kitesurfers is 31.6. This is calculated as follows:

$$\bar{x} = \frac{\sum x}{N} = \frac{43+27+.....+39+21}{99} = \frac{3126}{99} = 31.575757$$

$= 31.6$ (1 d.p.) (i.e. the result has been rounded to 1 decimal place)

The Greek capital letter sigma (Σ) represents summation, x the varying data values and N the total number of data points for that variable. Note that this is not necessarily the same as the sample size because the respondents may not answer all the questions. If, for example 10 of the 99 respondents to the kitesurfing survey had refused to give their age, you would have to calculate a mean age based on 89 data points and report both the answer and the number of pieces of missing data. An arithmetic mean can only sensibly be calculated for interval and ratio data (see Box 15.8 for an example).

Remember that the age in the example in Box 15.8 has only been collected to an accuracy of one year. Therefore it is not sensible to report the arithmetic mean (or any other similar calculation) to several decimal places. A good rule of thumb is to present your calculation to one more decimal place than that to which the original data were collected. You should always indicate, as above, the rounding you have used.

15.7.4 Which type of average to use?

Which average you should use depends on the type of data you have and on how the data are distributed. With nominal data you have no choice but to use a mode. If you have ordinal data you have a choice of using the mode or the median. However, the median is usually the best option, because it is perfectly possible, if your data are oddly distributed, for the modal value to lie at the extreme end of the range of the values. For interval or ratio data you have a choice of mode, median or mean. If the data are quite symmetrical, it does not matter much which type of average you use: the mean, median and mode are about the same. Look back at Figure 15.6, which shows a bar chart of the age distribution of kitesurfers. The data are reasonably symmetrical, with a slight bias towards the older age groups. In this case the modal category (30–34) years, the median age (31 years) and the mean age (31.6) are all quite similar, and all are good representatives of the 'centre' of the data. However if there are extreme values (both high and low) or if the data are distributed unevenly, then the mean, median and mode may be quite different.

An exaggerated example (Box 15.9) shows this more clearly. If we were to quote the average hourly wage as roughly equal at the two companies shown in the example, using the means of £4.00 and £3.90, we would be concealing a deal of difference. In the first instance the mean has been greatly affected by one very large value and the mean of £4.00 represents no one: it is just under 20 per cent of the earnings of the highest paid, yet twice as much as the earnings of everyone else. In the second case, where earnings are more evenly distributed, any of the mean, mode or median could represent the data. The mean still represents no individual's earnings precisely, but it is a fair summary.

Box 15.9 Example: the effect of data distribution on the mean

Consider the hourly rates paid by two companies, each employing ten people:

Slave Drivers plc	£22, £2, £2, £2, £2, £2, £2, £2, £2, £2 mean: £4.00, median: £2.00, mode: £2.00
Workers' Co-op plc	£5, £5, £4, £4, £4, £4, £4, £3, £3, £3 mean: £3.90, median: £4.00, mode: £4.00

This example may seem rather trivial, but it illustrates a serious point about the presentation of summary data. The calculated mean wage for Slave Drivers plc is not wrong – it is calculated perfectly correctly – but it is seriously misleading, and unfortunately, misleading reporting of 'averages' are all too common in 'real' research reporting too. For example, the following statements are both true (as of 2 January 2009: BBC, 2009):

- The average wage for Public Sector workers in the UK is lower than the average wage for Private Sector workers.
- The average wage for Public Sector workers in the UK is higher than the average wage for Private Sector workers.

How can this be? The statements should perhaps read:

- The mean wage for Public Sector workers in the UK is lower than the mean wage for Private Sector workers.
- The median wage for Public Sector workers in the UK is higher than the median wage for Private Sector workers.

You can perhaps guess by looking again at the above example how this anomaly may have arisen. In fact in 2004, the Office for National Statistics in the UK switched to quoting median income, rather than mean income, as their headline figure for average earnings because it better represented the 'centre' of the data. The lesson is that you need to make sure that your summary measures are properly representative of your data, and that you state what kind of average you are using, especially if your work might inform government or local policy directly.

15.8 Measuring the spread: how spread out are my data values and are there any anomalies?

As is apparent from the previous section, looking at the centre of your data does not tell the whole story. You also need to consider how far from the centre the data are spread. Consider an example taken from cricket, where tables of batting averages are routinely compiled and career averages compared as an indication of batting prowess. The batting average for any particular batsman is calculated as total runs scored divided by number of completed innings; thus it represents the *mean* number of runs per innings. The implication is that the higher the average, the greater the batting prowess.

Consider five completed innings for two different batsmen, with the number of runs for each innings and the batting average:

John Steady 40, 45, 50, 55, 60 batting average = 50
Jack Daring 0, 0, 10, 100, 140 batting average = 50

The batting average alone fails to reflect the spread of scores, or the difference in the batsmen's approach that these two sets of scores reflect. We need some measure of spread. There are two useful approaches to measuring spread – the *range* and the *standard deviation* – and these are discussed in the following sections.

15.8.1 Range

The *range* is the distance between the extreme values in a data set. In the example above the ranges are 40 – 60 = 20, and 0 – 140 = 140 runs, respectively. Clearly, for a range to exist the data must at least be ordinal; nominal variables have no order and hence no spread can be defined. However, the range does not have to be strictly numerical. If, as mentioned earlier, you have asked your respondents their level of education, with pre-defined ordered categories to choose from, the range might be from 'Primary school only' to 'Completed doctorate'. The range of the kitesurfers' income – an ordinal variable – is from 'No income' to 'More than £100,000 per annum'.

The problem with using range alone as a measure of spread is that one unusually extreme value can change it completely. The range of the kitesurfers age is 13 – 49 = 36 years. Imagine that, among the kitesurfers, there had been one unusually fit and active 90-year old. The range would now become 13 – 90 = 77 years, even though the others are all 49 years old or less. One alternative is to calculate the *midspread* or *interquartile range,* which addresses this problem by looking at the range of just the middle 50 per cent of the data. The interquartile range is connected to the median, which divides the ordered data into two halves. If we then divide each half into two further halves we have *quartiles*. The lower quartile, (Q_L), is the median of the lower half of the data and the upper quartile, (Q_U), is the median of the upper half of the data. The distance between them is the interquartile range. The lower quartile, Q_L, of the kitesurfers age is 26 and the upper quartile, Q_U, is 37 giving an interquartile range of 12 years. The interquartile range, representing the middle of the data, is unaffected by extreme, unrepresentative values.

15.8.2 Standard deviation

Standard deviation is the most widely used measure of spread or variability. It is defined as

$$s = \sqrt{\frac{\Sigma\left(x_i - \bar{x}\right)^2}{N}}$$

where s is the standard deviation, $\bar{x}$ is the arithmetic mean, x_i are the individual values and N is the total number of values.

A worked example may make this clearer. Consider the cricket example again. For both John Steady and Jack Daring:

N (the number of completed innings) = 5
$\bar{x}$ (the batting average) = 50

Box 15.10 Instructions: how to calculate the standard deviation

Taken step by step:

1 For each value, x_i, calculate the difference from the mean, $(x_i - \bar{x})$.
2 Square each of these, $(x_i - \bar{x})^2$.
3 Add up the squared differences, $\Sigma(x_i - \bar{x})^2$.
4 Find the average (mean) squared value by dividing the total squared value by the number of data points, $\frac{\Sigma(x_i - \bar{x})^2}{N}$.
5 Take the square root of the average squared value $\sqrt{\frac{\Sigma(x_i - \bar{x})^2}{N}}$.

How do their standard deviations compare? Box 15.10 shows how to do the calculation, and Table 15.7 shows the result for this particular data set.

Thus the standard deviations are 7.1 and 58.7 for John Steady and Jack Daring, respectively.

The information:

John Steady: average 50, standard deviation 7.1 (1 d.p.)
Jack Daring: average 50, standard deviation 58.7 (1 d.p.)

is a much more informative summary of their batting scores than just an average. You do not need to know any individual scores to know that Jack Daring's scores vary very greatly, whereas John Steady's tend to be quite similar.

Like the range, the standard deviation is sensitive to extreme values, which can be a disadvantage. However unlike the various ranges it uses all of the data.

Table 15.7 Results of the standard deviation calculation for each batsman

	John Steady			*Jack Daring*	
x_i	$(x_i - \bar{x})$	$(x_i - \bar{x})^2$	x_i	$(x_i - \bar{x})$	$(x_i - \bar{x})^2$
40	–10	100	0	–50	2500
45	–5	25	0	–50	2500
50	0	0	10	–40	1600
55	5	25	100	50	2500
60	10	100	140	90	8100
$\Sigma(x_i - \bar{x})^2$		250			17200
$\frac{\Sigma(x_i - \bar{x})^2}{N}$		50			3440
$\sqrt{\frac{\Sigma(x_i - \bar{x})^2}{N}}$		7.1		58.7	

Box 15.11 Checklist: some principles to guide data collection and analysis

- Gain some statistical understanding *before* you plan your data collection.
- Plan the data collection with your subsequent analysis in mind.
- Use descriptive techniques to:
    - look at data patterns;
    - find the 'centre' and range of the data, if appropriate.
- Use techniques that are appropriate to the type(s) of data you have collected (Table 15.8).
- Do as much preliminary analysis as you wish, but when presenting your finished work include only what is necessary.

15.9 Conclusion

The use of descriptive techniques is a distillation process: your object should be to extract the essential meaning from your data. Box 15.11 summarizes some principles that you should bear in mind when using statistics. Plot as many graphs, calculate as many descriptive statistics as you wish during the exploratory phase, but your finished work should contain only what is necessary to tell a clear story. You do not inform your readers by pouring a torrent of data and output over them: rather you should do so by picking the interesting graph, the appropriate summary that best describes your findings. Remember, 'less is more'.

Finally, Table 15.8 summarizes some of the different techniques that have been described in this chapter and their use for different purposes and different kinds of variable.

Table 15.8 Summary of techniques in descriptive statistics

Type of data	*Suitable graphs*	*Measure(s) of the centre*	*Measure(s) of spread*
Nominal	Bar chart Pie chart	Mode	None
Ordinal	Bar chart Pie chart	Mode Median	Range Interquartile range
Interval and Ratio	Bar chart Pie chart Line graphs Scatter plots (to graph two variables together)	Mode Median Mean	Range Interquartile range Standard deviation

Summary

1 Statistical techniques fall into two broad categories. Descriptive statistics, the subject of this chapter, attempt to summarize data and present them in an easily assimilated form. Inferential statistics, to be dealt with in Chapter 16, seek to draw inferences about the population of which your sample group is a part.
2 There is a variety of calculating tools available to you, from pencil and paper to statistical software packages. The more sophisticated the tool you choose, the more important it is that you understand what you are doing.
3 Statistics has its own particular terminology which you need to know and use properly.
4 Different types of measurement are involved in recording values for the different variables you may wish to measure. Measurement (and data) can be nominal, ordinal, ratio or interval.
5 The purpose of descriptive statistics is to describe the data: in particular to (1) look at possible patterns and (2) find a 'centre' and range (if appropriate).
6 Data can be summarized and presented visually using different kinds of tables and graph, including frequency tables, pie charts, bar charts, scatter plots and line graphs. Each table or graph should be properly labelled so that it can be understood by itself; the accompanying text should refer to it and point out the most important points, without repeating all the detail.
7 Each data type has a type of graph most suited to it and you must make sure that you choose a type of graph that suits your data.
8 Three types of average will be of use to you in describing the 'centre' of your data: mode, median, and mean. As with graphing techniques, which of these you can use depends on the type of data you have.
9 Looking at the centre of your data is not sufficient. You also need to consider how far from the centre your data spreads. Two ways to do so are by calculating the *interquartile range* or the *standard deviation*.
10 Your finished work should contain just enough descriptive output to tell a clear story.

Further reading

The two books listed below both provide in their different ways excellent introductions to statistics and the ways of statistical thinking:

Huff, D. (1991) *How to Lie with Statistics*, London: Penguin. [This book has been continuously in print since first publication over 50 years ago. Concealed within a very entertaining read (laugh-out-loud funny in places) is a serious look at appropriate and inappropriate use of statistical techniques.]

Rowntree, D. (2000) *Statistics Without Tears: An introduction for non-mathematicians*, London: Penguin. [This book does exactly what the title says: provides a clear, straightforward and readable introduction to statistical ideas for non-mathematicians.]

For a list of more standard statistical textbooks, refer to 'Further reading' at the end of Chapter 16.

16 Quantitative analysis

Inferential statistics

C.M. Eagle

Statistics is the art of making informed decisions in the face of uncertainty.
(Anon)

16.1 Introduction

Chapter 15 looked at some of the graphical and numerical techniques that you can use to *describe* your data, for both *exploratory* and *presentational* purposes. However, unless the population of interest is small enough for you to have gathered data from every case, descriptive techniques alone will not answer your research questions. You cannot expect a *sample* to precisely reflect the *population* in every respect however carefully you have chosen it to be as representative as possible. Moreover, if you took repeated samples from the same population, the samples would vary one from another. This is known as *sample variation*. You may find an interesting pattern or effect in your particular sample, but how do you know if this result truly reflects what is happening in the wider population? Is it a 'real' result, or simply sample variation?

If you find the idea of sample variation a little difficult to think of in the abstract, consider the outcome of tossing a coin many times. Over the long run you would expect a fair coin to turn up heads 50 per cent of the time. What conclusion would you draw if you tossed such a coin four times and it turned up heads each time? These four coin tosses can be considered to be a sample from the population of all tosses of that particular coin. Does this sample result mean that this coin always comes up heads? Of course not! – it is one of the possible outcomes you might expect if you looked at repeated samples of four coin tosses. The probability of getting four heads in a row when you toss a fair coin four times is 1 in 16, or 6.25 per cent: low, but nowhere near zero.

It is to answer this vital question – 'Is my interesting sample result a reflection of a real population effect, or just the natural variation expected from a sample?' – that the techniques of inferential statistics exist. There are many different inferential tests, but they each work by calculating a *test statistic* that indicates the probability that a particular pattern in the data is due to chance, to random variation. Without such tests you cannot generalize the results beyond the sample, and so cannot answer your research questions. You may wonder why we need to consider probabilities at all: it is an inevitable consequence of using sample data to draw conclusions about populations, given that no sample is exactly the same as the population it comes from. If the probability is low that a particular pattern is due to chance alone, then it can be concluded, within certain levels of confidence and to certain levels of *statistical significance* (see Section 16.2), that there is a meaningful effect.

This chapter will introduce a range of common tests that are used in inferential statistics. There is a large number of statistical tests available: those looked at in this chapter are the most well-known and commonly used in the context of social science research. There is no space here to describe the theory underlying each test in detail; instead the aim is to give you an understanding of some of the basic principles underlying inferential statistics and clear guidance on how to choose the most appropriate test for your particular research question and data set. For a more detailed account, see a specialized textbook on statistical analysis such as those listed in the further reading at the end of this chapter.

All the tests that are mentioned are available in SPSS and many in the Excel spreadsheet program. However, I would repeat the warning given in Chapter 15. A piece of computer software will simply do as you tell it: it is up to you to *choose the appropriate test* and *interpret the output correctly*.

Which test you should use depends upon four things:

1 The research question.
2 The type of data.
3 The sample size.
4 The distribution of the data.

The first of these comes from the nature of your research: most importantly, what hypothesis you wish to test. The remaining three are important in making a choice between a *parametric* and a *non-parametric* test, which represents the single most important division in inferential statistics (see Section 16.3). In order for you to understand these distinctions, before describing the tests themselves it is necessary to look in more detail first at what is meant by statistical significance (Section 16.2), and second at the importance of different kinds of *probability distribution* (Section 16.3).

In the rest of the chapter, each section and subsection introduces the standard parametric and non-parametric tests for a particular kind of research question and data. Section 16.4 describes some basic techniques for estimating the characteristics of a population based on data from a sample. Section 16.5 describes tests for comparing patterns of variation in a single variable between two or more samples, in order to test whether there is a statistically significant difference between samples. Section 16.6 deals with techniques to test for a significant relationship between two different variables, and Section 16.7 gives a brief overview of some more complex *multivariate* techniques, which are used to analyse the relationships between several variables at once. Finally, Section 16.8 presents a table summarizing which of the parametric and non-parametric tests described in this chapter are appropriate for different kinds of task.

There is no one correct way to present your statistical test results: it depends to some extent on the audience you are writing for. An author of a paper for a statistically aware readership may present results in quite a sparse way, whereas a more explanatory or discursive style may be necessary for a different audience. In Sections 16.4 to 16.7, some of the boxed examples include a sentence reporting the results in appropriate wording for a thesis. Whatever style you use, it should always be clear what particular test you carried out, whether the result of that test is significant or not, and the numbers on which your decision is based.

16.2 Significance levels and hypothesis testing

Sections 1.2.1 and 3.2 described the process in research design of formulating null hypotheses and setting up experiments or comparative observational studies in order to test them.

Regardless of whether or not you set up an explicit null hypothesis in the research design, when you apply an inferential test to a set of data you are in effect testing a null hypothesis: that the interesting patterns in the data are just due to sample variation and do not reflect a real pattern in the population. The calculated test statistic indicates the numerical probability that the actual results you obtained would have occurred *if the null hypothesis were true* – if there were no 'real' pattern. This numerical probability is sometimes referred to as the p-*value* of the test. If the *p*-value is sufficiently low, you can reject the null hypothesis and accept the alternative explanation: that there is a 'real' pattern.

However, we need to first consider the important question 'How low is sufficiently low?' By convention, a value of 5 per cent (or 0.05) is usually taken as the critical value for this probability. This value is known as the *significance level* of the test, usually represented by the symbol α (alpha). If the calculated probability, or *p*-value, is less than the chosen α, then there is a *significant result* (see Box 16.1). Choosing a significance level of 5 per cent is the

Box 16.1 Key terms in inferential statistics: significance, insignificance, power and robustness

Significance

'Significance' in a statistical context simply means that the results from the sample are likely (with a known probability of being wrong) to reflect characteristics of the population from which it comes. Contrary to its everyday usage, it does *not* mean 'interesting' or 'important'.

Insignificance

To report that 'the result was not significant at the 5% level' does *not* mean that you have proved that there is no population effect. You have simply failed to prove an effect with this particular sample. Absence of evidence is not always evidence of absence. It could be that the sample was just too small to show the effect you were looking for. Moreover, an 'insignificant' result is not necessarily uninteresting. If you were investigating attitudes to climate change, firmly convinced that young people were much more concerned about the issue than were older people, yet testing the data showed no significant age effect at all, then provided the sample size was sufficient, this would still be an interesting result. It suggests that, contrary to your expectations, older people are just as concerned as young people.

Power

The power of a test is an indication of its strength: that is its ability to find significant results where they exist. It is related to the type II, or false negative, error rate (β): if β decreases, then the power increases and vice versa (numerically, power equals $1 - \beta$). Calculating the power of a test is not straightforward, as it depends upon the size of the effect you are seeking, which you cannot necessarily know in advance.

Robustness

A 'robust' statistical technique will perform well, and produce reliable results, even when your assumptions about the data do not fit a pre-determined model, or when you have extreme or outlying values. For example, the *median* is a more robust technique for finding the centre of numerical data than is the *mean* because it is far less influenced by extreme values (see Section 15.7).

same as saying 'if there's only a 1 in 20 chance (or less) that my result is just coincidence, I'll assume it's significant'.

Three important points follow from this. First, 'significance' in statistical terms does not mean the same as 'significance' in everyday language (see Box 16.1). Second, in spite of the claims of many quantitative scientists, statistical tests do not *prove* anything beyond doubt; they just calculate probabilities, and accept something as a working hypothesis if the probability of its being a chance effect is lower than 5 per cent (see also Box 1.1: you cannot *prove* a positive). In essence, statistical techniques enable us to quantify the potential errors in our results. Third, you can be wrong in your conclusions in two different ways: either you can conclude that there is an interesting effect when in fact it is 'just coincidence' (known as a *false positive* or a *type I error*) or you can conclude that it is 'just coincidence' when in fact there's something more interesting going on (a *false negative* or *type II error*).

The chance of a type I error is the same as your significance level (α). That means that if you use a 5 per cent level of significance, you can expect a false positive in about one in twenty tests (5 per cent of the time), which is not actually very low. You can reduce this chance by using a lower significance level: a level of 1 per cent (0.01) should only produce a false positive in about 1 in 100 tests (note that the lower the [statistical] significance level, the greater the [practical] significance). Unfortunately, however, as you reduce the significance level and therefore the chance of false positives, you inevitably increase the chance of making a type II error, usually represented by the symbol β (beta) – that is, of missing a genuine effect and so rejecting a 'significant' result.

The *power* of a statistical test is the probability that the test will not produce type II errors: that the test will find the significant results if they are there (see Box 16.1). It is desirable in theoretical terms that a test should be both as powerful as possible (minimizing the chance of false negatives) and as significant as possible (minimizing the chance of false positives). However, it is simply not possible to achieve both these aims and the choice of a suitable significance level is always something of a compromise between significance and power.

The weaker any population effect is, the harder it is to detect, and so the larger the sample size that you need in order to keep both type I and type II errors acceptably low. Unfortunately you cannot know the strength or weakness of any particular effect until you have collected the data and started to analyze them, which is why it is not possible to know in advance what sample size is needed to detect a particular effect (see also Section 4.3.2 on sample size). However having fixed a significance level, it is not considered acceptable to change it to fit what you find. Just as we do not move the finishing line while a race is being run, or change the size of the goal mouth during a football match, so the value attached to 'sufficiently low probability' has to be decided *before* the analysis. If you have planned to test at the 5 per cent significance level and the results come up with a probability of 6 per cent, then you cannot decide, just this once, that you will test at the 10 per cent level and declare the result significant. You must accept the null hypothesis, that you have found nothing statistically significant.

16.3 Probability distributions: deciding between parametric and non-parametric tests

In a statistical context the word *distribution* is often used to describe the way a set of numbers is 'distributed' – the variation in values – particularly those that fall into some sort of pattern. This is important because if you can make certain assumptions about the underlying distribution of the population from which you have drawn the sample, then you can use

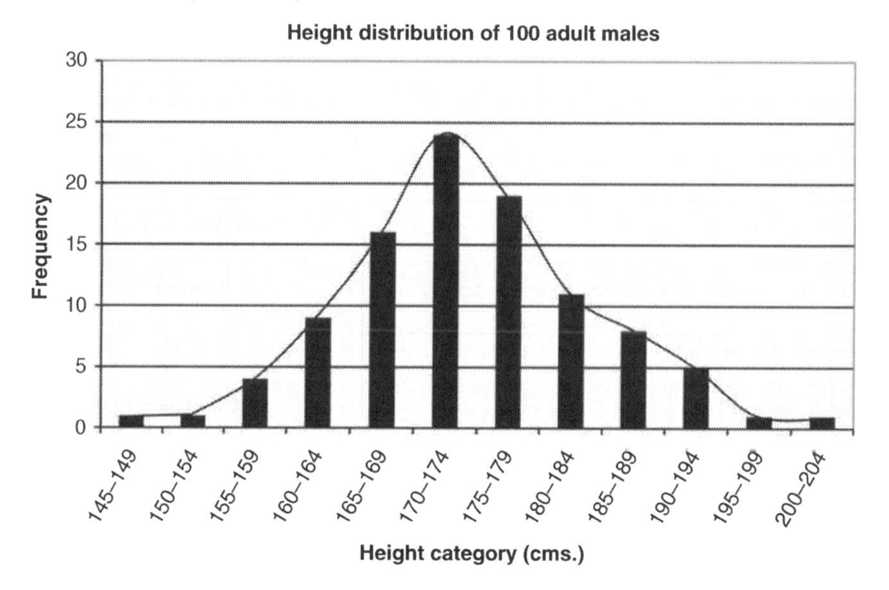

Figure 16.1 Example: a data distribution.

particularly powerful statistical techniques known as *parametric* techniques. If not, you must use *non-parametric* techniques, which do not depend on these assumptions.

We will first look at an example. Figure 16.1 shows a graph of the heights (in centimetres and in categories) of a sample of 100 adult males, with the values on the x axis and the frequency with which they occur on the y axis. The graph shows a *data distribution*. Note that the data are more or less symmetrically distributed about the modal range of 170–4 cm, with decreasing frequencies of values towards the extremes.

This type of pattern, or distribution, is called the *normal distribution* (sometimes referred to as the *bell-shaped curve*) (see Figure 16.2), and is extremely common in the natural world. However the curve in Figure 16.1 does not look *exactly* like the curve in Figure 16.2. This is because it is based on a sample of only 100, and we cannot expect this sample to look exactly like the larger population of which it is a part. If we looked at increasingly larger samples we should expect the asymmetries to be gradually ironed out and the curve to become smoother and more symmetrical.

Given a large group of people or plants or animals, most of the measurable 'natural' things about them will approximate to a normal distribution (for example, shoe size, weight, height or pulse rate). In other words, the values tend to be clustered around an average, but the frequencies decrease in a more or less symmetrical pattern towards the highest and lowest values.

The normal distribution is an example of a *continuous distribution*: it represents continuous, numerical data (see Section 15.3). A very different-shaped distribution arises, for example, when we toss a single (fair) coin and score 1 if it comes up heads, 0 if tails. The probability distribution of the outcome of a single coin toss is a *discrete distribution*, as it can only take one of two possible values, 1 or 0. The probability associated with each of these outcomes is 0.5, or 50 per cent. This probability distribution is shown as a bar chart in Figure 16.3: it bears no resemblance to the bell-shaped curve (the normal distribution).

Both of the distributions described above are examples of *probability distributions*. Probability distributions define *either* the chance of any value of a random variable occurring, if that variable is discrete, *or* the chance of a continuous variable falling between any two values.

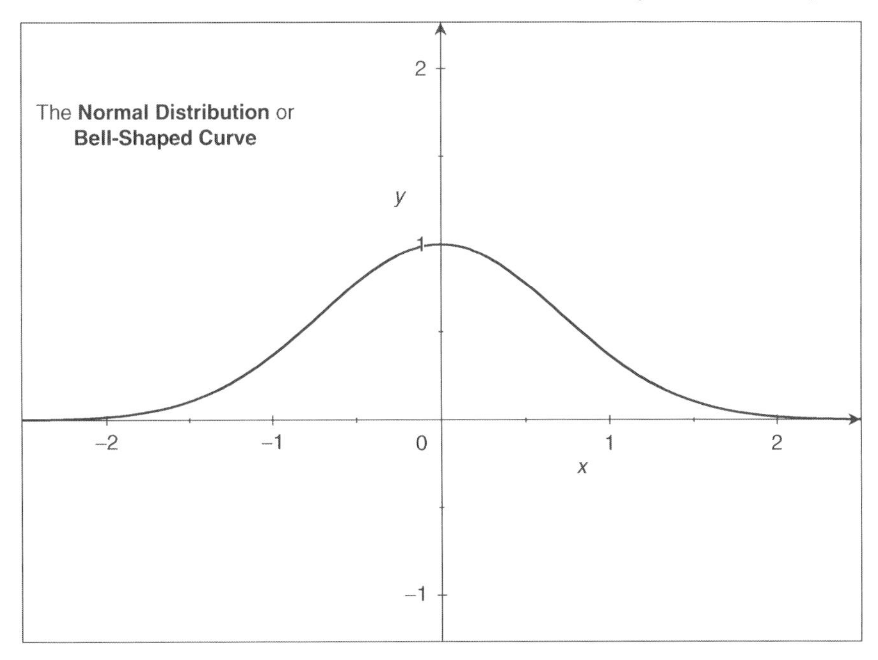

Figure 16.2 The normal distribution.

Associated with probability distributions are *parameters*, numbers that completely define them (see Section 15.3). A normal distribution can be defined by specifying the *mean* (where it is centred: see Section 15.7) and the *standard deviation* (how spread out it is: see Section 15.8). Figure 16.2 in fact shows a particular example of the normal distribution, the *standard normal distribution*, which has a mean of 0 and a standard deviation of 1. The parameters of the coin-tossing distribution are the probabilities, 0.5, associated with each of the two possible outcomes.

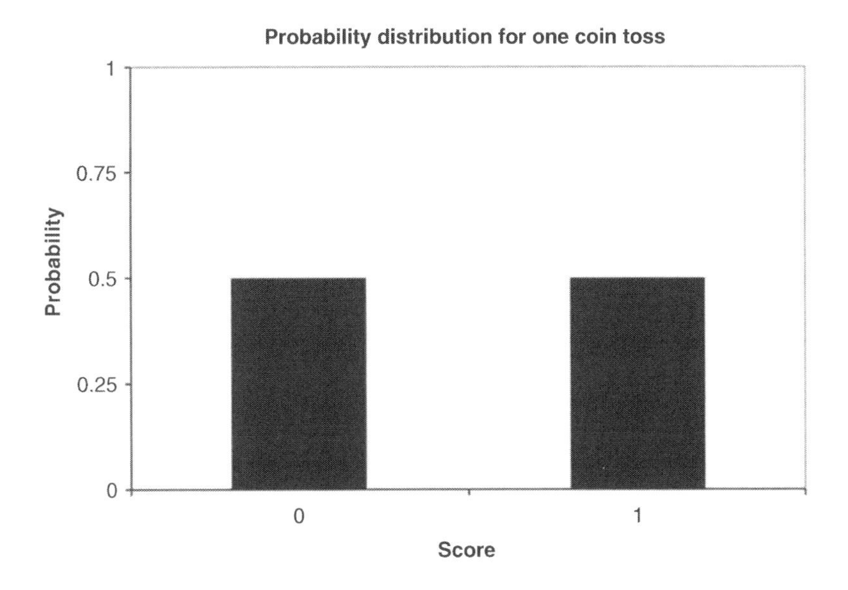

Figure 16.3 Example: a discrete probability distribution.

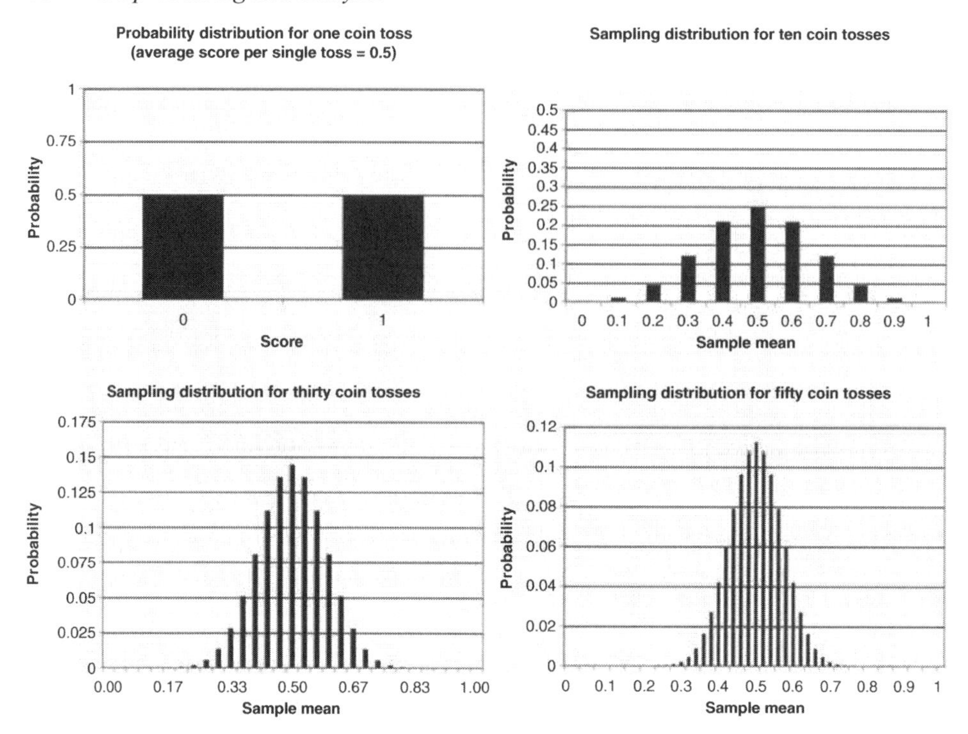

Figure 16.4 Examples: sampling distributions for increasing sample sizes.

However, consider what might be the outcome if we repeatedly tossed ten coins and calculated an average score for each sample of ten. We would not expect to get exactly five heads and five tails (which gives us an average score of 0.5) in each sample. We could have any possible combination from ten heads and no tails all the way to no heads and ten tails, though some of these combinations are much more likely than others. We can calculate the probability of any particular outcome of a sample of ten: a sample mean of 0 (no heads, ten tails), a sample mean of 0.1 (one head, nine tails) all the way to a sample mean of 1 (ten heads, no tails). These probabilities can then be graphed as the *sampling distribution* of the sample mean. A sampling distribution is a second type of probability distribution: rather than graphing the actual data from a single sample, it shows the likely outcomes of repeated samples (of a fixed size) from the same population.

The sampling distributions for 10, 30 and 50 coins are shown in Figure 16.4, along with the original probability distribution. This rather abstract example illustrates a very important theoretical point. The normal distribution and the probability distribution for one coin toss could hardly look more different: one is a continuous, bell-shaped curve, the other a bar graph, with only two possible values. Yet if you look successively at the graphs for samples of size 10, 30 then 50, you can see that the sampling distribution of the mean is getting more and more like the bell-shaped curve. In general, whatever the shape of the original distribution of your variables, if your sample size is big enough, the distribution of sample means approaches a normal distribution. In this context thirty is usually accepted as 'big enough'.

This has practical relevance to you and your work. If the samples are sufficiently large, then you can choose from either parametric or non-parametric techniques. However, if the sample sizes are under 30, then you are restricted to non-parametric techniques, unless you can confidently assume that the variables are normally distributed. The constraint on sample

Box 16.2 Factors to consider when choosing between parametric and non-parametric techniques

Parametric techniques

- Need numerical data.
- Need a sample of 30 or more.
- Or, if your sample is less than 30, it must come from a normal distribution.
- Are more powerful than non-parametrics.
- But less robust.

Non-parametric techniques

- Can be used for ordinal and nominal data, as well as numerical data.
- Make no assumptions about the distribution from which the data comes.
- Are less powerful than parametrics.
- But more robust.

size applies to any possible subgroups you wish to test. For example, if you want to compare men's and women's responses to a question in the form of a rating scale (see Section 7.3.2.4), you cannot make any assumptions about the data distribution and therefore you need at least 30 men *and* 30 women if you want to use parametric techniques (see also Section 4.3.2).

The division between parametric and non-parametric techniques is the single most important division in inferential statistics, and you need to understand the difference. If the assumptions of parametric techniques are met, they will give you more precise results than non-parametrics, and so they are said to be more *powerful*. However if these assumptions are wrong then the results of parametric techniques can be misleading. If in doubt you should use non-parametric techniques: as no prior assumptions need to be made about the data they are more *robust*. Moreover non-parametric techniques can be used for types of data, such as ordinal measures, that are not suitable for parametric techniques.

Box 16.2 summarizes the different factors you need to consider when deciding whether to use parametric or non-parametric techniques.

16.4 Estimating the characteristics of a population

The statistics that *describe* your sample – particularly the measures of central tendency and spread described in Sections 15.7 and 15.8 – can be used to estimate the population *parameters* – the values that characterize the population from which you have drawn your sample. This is done as follows, first in cases where parametric statistics are appropriate and then for non-parametric statistics:

Parametric: The sample mean is the best estimator for the population mean. However, the best estimator for the population standard deviation from a sample of size N is calculated by a similar procedure to that described in Section 15.8. However when inferring the standard deviation of a population from a sample, you divide by $N - 1$, rather than N, when you reach step (4). This is usually referred to as the *sample standard deviation*, to distinguish it from the population standard deviation. The latter describes the spread of an actual set of numbers, rather than an inference. A handheld statistical calculator will have two buttons, marked s_n and s_{n-1} (or perhaps σ_n and σ_{n-1}) that should be used for calculating a population standard deviation and a sample standard deviation respectively, depending on whether you wish to treat your data set as a full population or as a sample.

It is not enough to give just a single estimate for a population mean. You should also quote a *confidence interval* for the point value: a range within which the true population value is likely to lie (a confidence interval, which depends on both sample size and sample standard deviation is a more informative measure than standard deviation alone). The quoted confidence *level* (usually 95 per cent, or 0.95) indicates how likely it is that the confidence *interval* contains the true mean. A 95 per cent confidence level is equivalent to testing at 5 per cent, or 0.05, significance (remember that the significance level is an indication of the probability of error). It is important to realize that the higher the confidence level used, the wider will be the confidence interval. A 99 per cent confidence interval (equivalent to testing at 1 per cent, or 0.01, significance) calculated on the same sample data will be wider than a 95 per cent confidence interval.

For example, the kitesurfing study provides data on some basic characteristics of a sample from the population of kitesurfers. The results for age could be reported as follows: 'The mean age of the kitesurfers is 31.8 years (1 d.p.; $N = 99$), with a 95 per cent confidence interval of 30.1 to 33.5 years (1 d.p).'

As there are 99 respondents in the sample, and age is a numerical variable, a parametric technique is used to describe the population.

Non-parametric: The sample median is the best estimator for the population median.

16.5 Comparing a single variable in two or more samples

16.5.1 Comparing the means of two paired samples: paired t*-test and Wilcoxon signed ranks test*

In some kinds of study, there are two samples that contain corresponding 'pairs' of items. For example, you may take measurements on a single set of respondents (or other units of study) before and after a particular 'treatment' in an experimental study (see Section 3.2.1), or you may apply two different 'treatments' to carefully matched pairs of respondents (or other units) in order to see if the two 'treatments' produce different outcomes. The examples in Box 16.3 may make this clearer. The tests you should use in cases such as these are as follows:

Parametric: A *paired* t*-test*, which looks at the mean difference between the pairs and tests the hypothesis that the true population mean difference is not zero.

Non-parametric: A *Wilcoxon signed ranks* test, which ranks the differences irrespective of sign, then looks for a significant difference between the positive and negative ranks.

16.5.2 Comparing the means of two independent samples: independent t*-test and Mann–Whitney test*

These tests are appropriate when you want to compare two different samples or groups within a sample – perhaps men and women, the inhabitants of two separate villages, or old and young people – and test if the difference in mean scores between them is significant. This is not the same as the situation described in Section 16.5.1 because the members of the

Box 16.3 Examples: comparing the means of two paired samples

1 *'Before' and 'after' scores:*
Imagine that you have asked a group of people to rate the importance of wind power for future energy needs on a scale 1 to 10. They were then shown a short film about the benefits of wind power, and asked for their rating again. The 'treatment' in this case is watching the film. A possible set of some of the 'before' and 'after' scores is shown below. You want to know whether watching the film has made a significant change in people's attitudes, expressed through their ratings.

Table 1	*'Before'*	*'After'*	*Difference (A–B)*
Person 1	5	8	+3
Person 2	6	3	–3
Person 3	4	6	+2
Person 4	8	10	+2

Why choose a paired test? We are interested in the *difference* seeing the film has made ('after' – 'before'), and whether the positive differences significantly outweigh the negative ones overall or vice versa.

Parametric or non-parametric? We can make no assumptions about the distribution of these rating scores and so could only use the appropriate parametric test – a *paired t-test* – if there were at least 30 people in the sample. Otherwise we should use a non-parametric test, the *Wilcoxon Signed Ranks test.*

2 *Carefully matched pairs:*
You might instead want to test two different approaches to informing people about a conservation issue. You identify pairs of people who are matched in terms of age, sex, general outlook on life, and assign one in each pair to watch film A, the other to watch film B (these are the two 'treatments'). After they have watched the relevant film, you test how much they have learnt. An example of the scores is shown below. You want to know which film is the most effective as an educational tool.

Table 2	*Score after watching film A*	*Score after watching film B*	*Difference (A–B)*
Pair 1	7	8	–1
Pair 2	9	6	+3
Pair 3	8	7	+1
Pair 4	5	5	0

Why choose a paired test? We are interested in the *difference in effect* the two films have on pairs of people who are otherwise well-matched, and whether film A or film B has been the most effective overall, as measured in the post-film test. Do the positive differences significantly outweigh the negative ones, or vice versa, or is there no significant difference at all?

Parametric or non-parametric? As for the 'before' and 'after' example, we can make no assumptions about the distribution of these rating scores and so could only use a (parametric) *paired* t-*test* if there were at least thirty people in the sample. Otherwise we should use a non-parametric test, the *Wilcoxon signed ranks test.*

two groups are not paired together in any way. You do not need to have the same number in each group. The appropriate tests are as follows:

Parametric: An *independent* t-*test*, which looks at the difference of means between the two groups and tests the hypothesis that the true population mean difference between these two groups is not zero.

Non-parametric: A *Mann–Whitney* test, which ranks the data overall, then looks for a significant difference in ranks between the two groups, in order to test the hypothesis that the medians of the two groups are different.

16.5.3 Comparing the means of more than two independent samples: analysis of variance (ANOVA) and the Kruskal–Wallis test

If you want to compare three or more groups in a sample, it is not considered good practice to do repeated pairwise tests on every possible combination. If, for example, you want to compare three groups, A, B, and C you should not simply use the tests in Section 16.5.2 to separately compare A and B, B and C and A and C. To do so would mean you were failing to look at the data as a whole and the more tests you do, the more you increase the possibility of false positives (type I errors). Instead you need to use one of the tests designed for the purpose, which are as follows:

Parametric: An *analysis of variance* (ANOVA) tests the hypothesis that the group means are not all equal. If this test is significant, you then need to use a post hoc test ('post hoc' means 'after the event'), such as the *Tukey test,* to find out which pairs of means are significantly different. An analysis of variance on two groups is mathematically equivalent to the independent t-test and will produce the same result.

Non-parametric: A *Kruskal–Wallis analysis of variance* test works in a very similar way, except that actual scores are replaced by ranks and the test looks for a significant difference in rank sums between the groups, in order to test the hypothesis that the medians of the groups are different. A Kruskal–Wallis analysis of variance on two groups is mathematically equivalent to the Mann–Whitney test and will produce the same result.

Box 16.4 Example: comparing the means of two independent samples

Imagine that you have asked a sample of teenagers, 21 boys and 25 girls, to take an 'environmental awareness' test. The overall results gave a mean of 62.3 per cent, median 68 per cent, standard deviation 18.2. However split by gender, the results were:

Boys: mean 57.7 per cent, median 63 per cent, standard deviation 14.7 ($n = 21$)
Girls: mean 62.8 per cent, median 67 per cent, standard deviation 13.2 ($n = 25$)

The sample result appears to show a difference between the sexes. You want to test its significance and see if the difference we have observed in the sample means reflects a real difference in the population from which the sample is drawn.

Why choose an independent test? The aim is to compare two independent groups, the boys and the girls.

Parametric or non-parametric? There are fewer than thirty in each group, so we must either have good reason to assume the scores are normally distributed, in which case we can use a parametric *independent* t*-test,* or else the non-parametric test, the *Mann–Whitney* U *test*, should be used.

In fact, neither test produces a significant result on the above data. The p-value of the t-test is 22 per cent, meaning that the probability of obtaining such a difference in means by chance is 22 per cent or less, which is not nearly low enough to satisfy the significance criterion discussed above. The p-value of the Mann–Whitney, the less powerful test, is even higher at 27 per cent.

Examples of how to report the results
(Note: t and U are the test statistics calculated).

> An independent t-test showed that the mean difference of 5.1 per cent between the girls' and boys' score was not significant at the 5 per cent level ($t = -1.25$, $p = 0.22$).

Or:

> There was no significant difference between the scores of girls and of boys (independent t-test, $t = -1.25$, $p = 0.22$).

> A Mann–Whitney test showed that the difference in rankings between the boys and the girls was not significant at the 5 per cent level (U = 212.5, $p = 0.27$).

Or:

> There was no significant difference between the scores of girls and of boys (Mann–Whitney test, U = 212.5, $p = 0.27$).

16.6 Looking at the relationship between two variables

16.6.1 Correlation

In general usage, if two things are correlated, then they are related in some way. The statistical usage is more precisely defined: *correlation* is a measure of the *strength* and *direction* of the *linear* relationship between two variables.

Box 16.5 Example: comparing the means of more than two independent samples

Imagine you have calculated a numerical index 'knowledge about elephants' (see Section 7.2) for a set of respondents who come from four different villages. You wish to test whether knowledge about elephants, as expressed in your index, is the same across the four villages. Your exploratory descriptive analysis has shown that, taken village by village, the mean index score for elephant knowledge is different in each village. The results, expressed as the mean, median and standard deviation, are as follows:

Village A: 12 respondents, mean score 9.3, median 9.5, standard deviation 2.10
Village B: 10 respondents, mean score 10.4, median 10.5, standard deviation 2.27
Village C: 11 respondents, mean score15.0, median 15.0, standard deviation 2.97
Village D: 11 respondents, mean score 18.5, median 18.0, standard deviation 1.97

Is the score sufficiently different, village by village, to be statistically significant? As discussed above, you should *not* test your villages against each other in a pairwise fashion, but use either the ANOVA or the Kruskal–Wallis to test for significance.

Parametric or non-parametric?
There are fewer than 30 in each group, so we must either have good reason to assume the scores are normally distributed, in which case we can use an *analysis of variance* (test statistic: F), or use the non-parametric equivalent, the *Kruskal–Wallis analysis of variance*.

Either test produces a significant result on the above data, with p-values of less than 0.001 in both cases. That is, the probability of such a result happening by chance if there is no real difference between the villages is less than one in a thousand, well within the accepted significance level.

In order to find out which pairs of villages are significantly different a post hoc test should be used. One of the most common parametric post hoc tests to follow a significant ANOVA result is the *Tukey* test. When used on the above example it shows that all the possible pairwise differences are significant *except* the difference between Villages A and B.

How to proceed with post hoc testing using the Tukey test
You only need post hoc testing if the analysis of variance gives a significant result. If so:

1. Calculate the difference between each pair of means, ignoring signs.
2. List the differences in order, from highest to lowest.
3. Apply a post hoc Tukey test to the largest mean difference.
4. Going in order down the list, test each mean difference in turn.
5. Stop when you have a non-significant result.
6. The remaining lesser mean differences will also be insignificant.

There are other post hoc tests used, including non-parametric procedures suitable for use after a significant Kruskal–Wallis test. For further details see Field (2009: 372–5 and 565–8).

Example of how to report the results

An analysis of variance showed that the mean scores were significantly different between villages ($F_{3,40} = 36.5$, $p < 0.001$). A post hoc Tukey test showed a significant difference between all villages at the 5 per cent level except in the case of Villages A and B.

If you are interested in how two variables change together, you should first draw a *scatter plot* (see Section 15.6.2.3). Only if that indicates a possible linear relationship, proceed further, and calculate a *correlation coefficient*. The tests that you should use are as follows:

Parametric: a *Pearson product-moment correlation coefficient* is calculated by measuring the way the variables change together in order to test the hypothesis that there is a correlation in the population.

Non-parametric: a *Spearman's rank correlation coefficient* is calculated in just the same way as the Pearson correlation (testing the same hypothesis), except that actual scores are replaced by ranks.

Both types of correlation coefficient are always numbers between −1 and +1. A positive value indicates that the two variables increase together, with +1 being a perfect positive relationship. A negative value indicates that one variable increases while the other decreases, with −1 being a perfect negative relationship. A value of 0 (zero) indicates no correlation at all. Figure 16.5 shows some examples.

It is particularly important to test the statistical significance of any correlation you find. If you generate (say) two sets of ten random numbers and calculate the correlation coefficient between them you will find that it is very unlikely to be zero. After I wrote this last sentence, I did just this five times and found a correlation coefficient as high as 0.43 between my sets of random numbers. Although this is seemingly large, it is not statistically significant. For a sample of 10 pairs, a correlation coefficient needs to be greater than or equal to 0.632 (or less than or equal to −0.632) to be considered significant. The smaller the sample you have, the larger the (absolute) value of the correlation coefficient needs to be, in order to be significant. Conversely, with a larger sample, it is possible to have a small or *weak* correlation coefficient that is nonetheless significant. A sample of 100 pairs need only have a correlation coefficient of 0.2 (or −0.2) to be significant. This is a particular example of 'significant' not being significant: a small correlation coefficient such as this may well be statistically significant, in that it represents a real effect, but it is often too small to be of any importance or practical significance. If you use statistical software, such as SPSS, to calculate a correlation coefficient, the output will include not only the value of the correlation coefficient, but also whether it is statistically significant for your particular sample size.

It is also important never to forget that correlation does *not* imply causation. If you measured reading ability and shoe size for a sample of primary school children, you would almost certainly find a significant correlation. However, this would not imply that an increase in reading ability brings on a spurt of foot growth, nor vice versa. It is clear that these variables will both be significantly correlated with a third variable, age. Sometimes it is simply not possible to identify any causative third variable: the relationship might just be coincidence, or at least the causation difficult to find. Box 16.6 gives a 'real' example of a significant correlation.

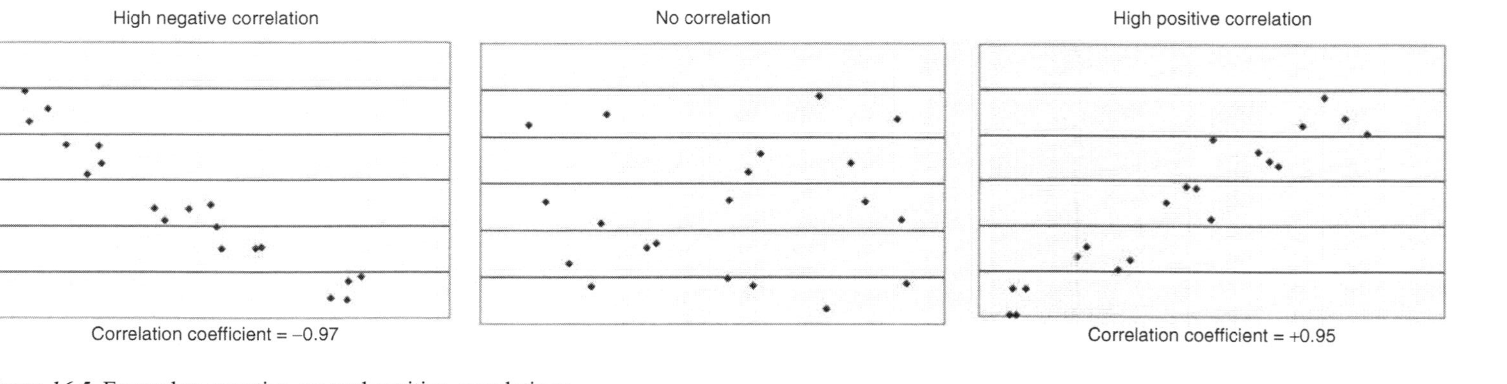

Figure 16.5 Examples: negative, no and positive correlations.

Box 16.6 Example: correlation – carbon dioxide emissions compared with income

The scatter plot in Figure 15.7 (and shown again in Figure 16.6) showed a possible relationship between carbon dioxide emissions and per capita income for 121 countries.

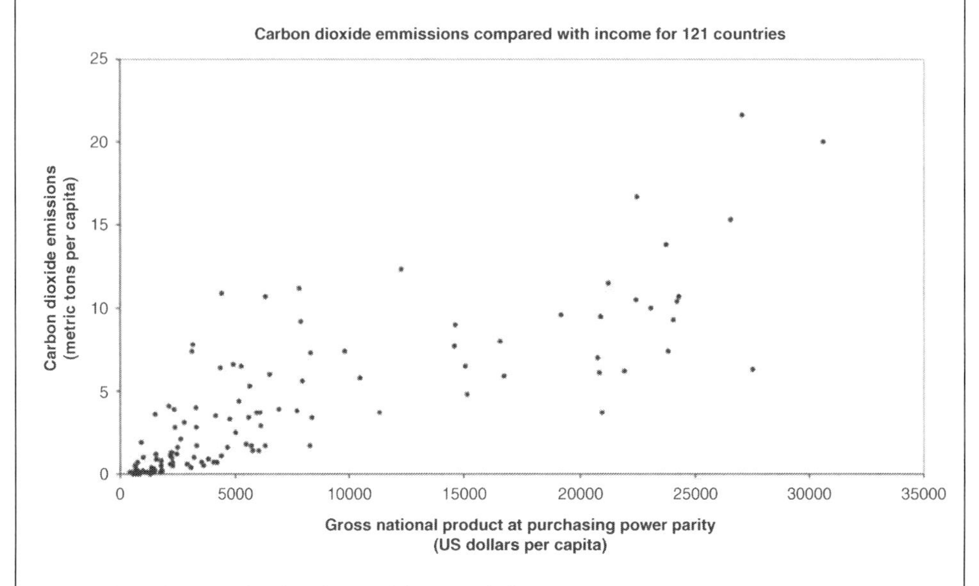

Figure 16.6 Scatter plot showing positive correlation.

The scatter plot shows an apparent linear trend for countries with higher levels of income to have higher levels of emissions. Is this a significant correlation?

Parametric or non-parametric? As there are 121 pairs of data, we can use the *parametric correlation coefficient*, the *Pearson*, without needing to assume a normal data distribution.

The correlation coefficient for these data is $r = 0.815$, which has an associated probability of < 0.0001 (that is, the probability of this result arising by chance is less than 0.01 per cent).

Example of how to report the results

> Carbon dioxide emissions were found to be positively correlated with GNP at PPP (Gross National Product at Purchasing Power Parity) (Pearson's correlation, $N = 121$, $r = 0.815$, $p < 0.0001$).

Note that although common sense suggests a causal relationship between income levels and carbon dioxide emissions, it is only possible to test for a correlation, and therefore it would be incorrect to report the results of the test in terms of causality.

16.6.2 Linear regression

If you have two variables that are correlated as described in Section 16.6.1, you may want to quantify the relationship further and add a straight line equation, a 'line of best fit', to the scatter plot. Many of you may be familiar with this from schooldays: drawing the best line through a set of points that are roughly in a straight line by wiggling a ruler around it until it looked about right. This is a simple form of *regression analysis,* which enables the prediction of a value of one variable, given a value for the other. A more formal, mathematical linear regression quantifies the way that one variable changes with the other. Usually (though not always) it aims to explain the behaviour of a *dependent* variable (which is affected) in terms of an *independent* or *explanatory* variable (which causes the effect) (see Box 7.2). In a graph, it is usual to assign the independent variable to the x-axis and the dependent variable to the y-axis. The tests used are as follows:

> *Parametric:* a *Least Squares Regression* calculates a line of best fit by minimizing the squared distance from the observed data points to the line (see Box 16.7 for an example). The line is tested for significance by testing whether its slope (usually referred to as B) is significantly far from 0 (zero). The test statistic is the F statistic referred to in Box 16.5.
>
> *Non-parametric: Logistic Regression* can be thought of as a form of *non-parametric regression*, although it is more often seen as the multivariate equivalent of the chi-square test (see Section 16.7.4). It is a method of predicting an outcome from a combination of numerical and categorical variables.

16.6.3 Cross-tabulations and the chi-square test

All the tests we have looked at so far need numerical data at some level (though some only require ordinal data). However, much of the information you are likely to gather will only be

> *Box 16.7* Example: linear regression by the least squares method
>
> Smith et al. (2003) looked at the relationship between biodiversity and corruption. The graphs in Figure 16.7 show the percentage change in the population of (1) African elephants, and (2) black rhinoceroses between 1987 and 1994 plotted against a 'governance' score (which assigns a maximum score of ten to the least-corrupt countries). Both graphs show an increase in animal populations where corruption is lower and a decrease where corruption is higher. A (significant) regression line has been fitted to both graphs.
>
> In each case the *explanatory variable* is governance score, which helps *explain* the dependent variable, the change in animal populations.

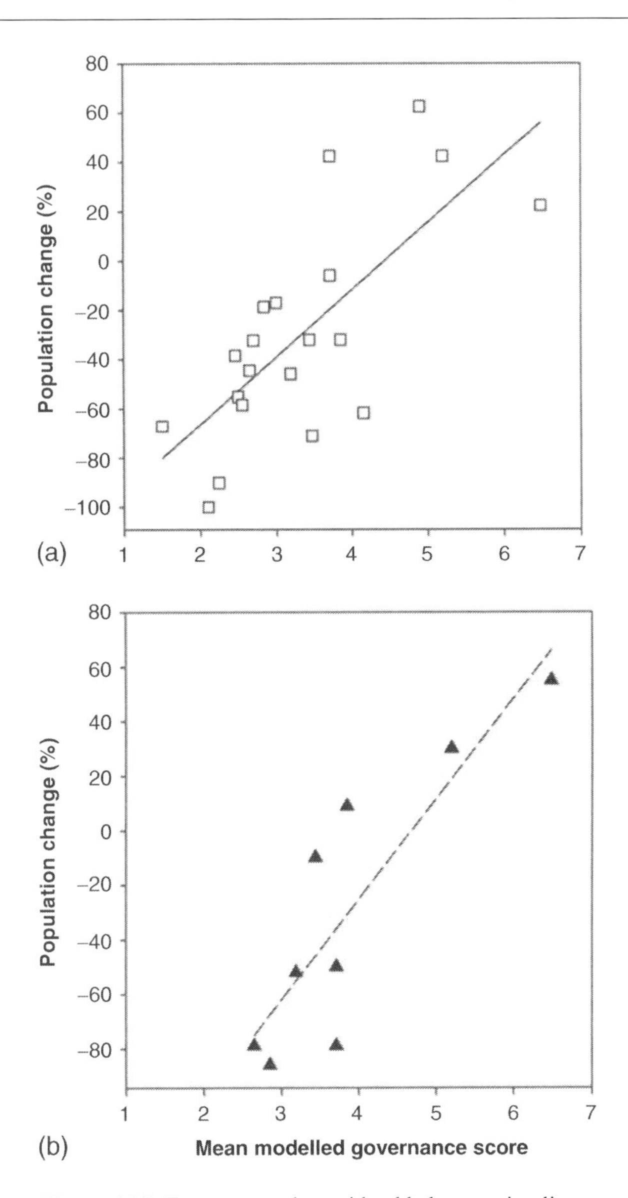

Figure 16.7 Two scatter plots with added regression lines.

Example of how to report the results

Changes in numbers of African elephants and black rhinoceroses between 1987 and 1994 are both significantly related to governance score (see figures a and b; $F_{1,20}$ =20.38, B = 27.37, $p < 0.001$; $F_{1,20}$ =20.33, B = 36.93, $p = 0.003$).

'measured' at the nominal level. For example, the following are all examples of questions that do not produce numerical answers:

> 'What sex are you?'
> 'What is your marital status?'
> 'Have you ever taken part in a wildlife safari?'
> 'What was the main reason for your visit to the Amazon?'

Cross-tabulations and the *chi-square test* are techniques for looking at the potential patterns in frequency counts of nominal variables. An example should make this clearer. Imagine that you have surveyed a sample of 100 people – 40 men and 60 women – and asked them a number of questions, including whether or not they are vegetarians. The results show that 30 of them are vegetarians, 70 are not. How can you test whether or not there a significant difference in rates of vegetarianism between men and women?

Each of the 100 people in the sample fits into one of four categories:

1 Male vegetarian
2 Female vegetarian
3 Male non-vegetarian
4 Female non-vegetarian

The first thing you might do to illustrate this possible association is to draw a *cross-tabulation* showing the number of cases in each of the above four categories (known as the *observed values*). Some possible data are shown in Table 16.1. In this case, gender is the independent variable and vegetarianism the dependent variable; it is usual to place each category of the independent variable down a column and each category of the dependent variable across a row.

The *column totals* and the *row totals* give the overall frequency counts for each value of each variable (male and female for gender, 'yes' and 'no' for vegetarian), and the column totals and the row totals should each add up to the *grand total*, the total number in the sample.

There seems to be a pattern in the cross-tabulation: more women are vegetarian than men. However, since the sample sizes are different for men and for women, this can be seen more clearly by looking at the *proportions* rather than the absolute frequencies. Table 16.2 shows the proportions as the percentage split down each column (note that the percentages are

Table 16.1 A cross-tabulation: observed values

Gender differences in vegetarianism
Observed values

Vegetarian?	Male	Female	
Yes	5	25	30 ← Row totals
No	35	35	70 ← Row totals
	40	60	100 ← Grand total
	↑ Column totals	↑ Column totals	

Table 16.2 A cross-tabulation: observed values with column percentages

Gender differences in vegetarianism
Observed values
Column percentages

Vegetarian?		Male	Female	
	Yes	**5** *12.5%*	**25** *41.7%*	**30** *30%*
	No	**35** *87.5%*	**35** *58.3%*	**70** *70%*
		40 *100%*	**60** *100%*	**100** *100%*

calculated for each value of the *independent* variable, so that they add up to 100 per cent for each column, but not for each row). If there were no difference between men and women in terms of the proportion who were vegetarian, the percentage yes/no split would be the same for both men and women, and the same as the overall split, 30/70 per cent.

Looking at Table 16.2, it seems as though there is a 'real' difference between men and women: only 12.5 per cent of the men in the sample are vegetarian compared to 41.7 per cent of the women. But this is a sample from a larger population. Is what you observe just sample variation, or an indication of a 'real' difference in the population? This can be tested using a chi-square test.

Assuming you have already assembled the table in the previous section, the next step in order to carry out a chi-square test is to construct a table of the values that could be expected if there were no difference in rates of vegetarianism between males and females. In this case, given that 40 men and 60 women were interviewed and a total of 70 respondents were vegetarians, what values could be expected in the different cells of the table? The *expected value* for each cell can be calculated very simply, as the row total multiplied by the column total divided by the grand total. Thus for the first cell in the table the value is $(30 \times 40)/100 = 12$. Table 16.3 shows the expected values and column percentages for each cell.

Table 16.3 A cross-tabulation: expected values with column percentages

Gender differences in vegetarianism
Expected values
Column percentages

Vegetarian?		Male	Female	
	Yes	**12** *30%*	**18** *30%*	**30** *30%*
	No	**28** *70%*	**42** *70%*	**70** *70%*
		40 *100%*	**60** *100%*	**100** *100%*

Of course, even if there is no real difference in the population it is very unlikely that your sample would produce exactly the expected results shown in Table 16.3. How far from the 'expected' results do the 'observed' results have to be before you can safely say that there is a 'real' population effect? A *chi-square* test statistic (sometimes written χ^2) is calculated to measure this probability – the probability of obtaining the observed results if there is no population effect.

For Table 16.1, chi-square = 9.7, and the associated probability (p-value) is 0.2 per cent (0.002). This is well below the accepted significance level of 5 per cent, and so we can accept the hypothesis that there is a relationship between gender and rates of vegetarianism, and moreover that women are more likely to be vegetarians than men. The chi square test does not test for the directionality of the relationship – in this case, in which gender vegetarianism is more common – but if it produces a significant result, it is usual to report the directionality as well, based upon examination of the cross-tabulation.

Given the nature of much of the data collected in social survey research, cross-tabulations, together with chi-square tests, are some of the most valuable techniques available. In addition to comparing two nominal variables, as in the example above, they also offer one of the few ways to combine a nominal variable and a numerical variable in the same analysis. This is done by transforming the numerical variable into ordered categories (see Section 15.4). Box 16.8 summarizes three conditions that must be met in order to use a χ^2 test.

16.7 Multivariate techniques: looking at the relationship between more than two variables

So far we have looked at tests for *univariate* (one variable) or *bivariate* (two variables) data sets. However the majority of data sets collected by researchers are *multivariate* (more than two variables) and it is sometimes useful to consider the variables simultaneously, rather than try to isolate and deal with them separately. For example, this is often the case in analyzing questionnaire results, where a single variable – say, attitudes to the environment – may be influenced by one or more of several other variables on which data are collected, such as age, gender, level of education, membership of conservation organizations and so on.

A word of warning: multivariate tests are complex and easy to mis-apply. The tests discussed in the previous sections are all 'doable' by hand, or at least with the help of a calculator

Box 16.8 When to use a chi-square test

There are three rules to follow:

1. Categories must be mutually exclusive so that each case (here, each person) appears in one cell only. For example, in the case above you could not put any person into both the 'yes' and the 'no' category on the grounds that they were occasional vegetarians. If you had people in that category, better to expand your definition of 'vegetarian' to 'yes', 'no' and 'sometimes'.
2. The 'expected' number of items in each cell must be at least five. There are other tests, such as *Fisher's exact test*, to use when some expected values are very low.
3. The tests must be carried out on the *actual numbers* of items in each cell, not on derived proportions or percentages.

(although much quicker with computer aid), whereas multivariate analyses depend upon computing power and would be almost impossible to do otherwise. They thus become something of a 'black box' into which data disappear and from which results emerge. Even more than with the simpler tests, you need a clear understanding of what you are doing and why. This section aims simply to explain what they can offer, so that you know when they might be appropriate for your own analysis. It does not give detailed instructions. Before using these tests you should consult a more specialist text such as those listed at the end of the chapter and if possible, consult a statistician.

16.7.1 Multiple regression

Linear regression, described in Section 16.6.2, models a possible linear relationship between a single independent (or explanatory) variable, x, and a dependent (or response) variable, y. *Multiple linear regression* is simply an extension of linear regression that can deal with several explanatory variables at the same time. In order to understand what it does it is useful first to explain how linear regression works.

Linear regression calculates the equation of the line of best fit for a given set of data, which can be expressed as $y = a + bx$. The parameter a is the point where the line cuts through the y axis and the parameter b is the slope of the line. Testing the regression for significance is equivalent to testing whether the slope of the line (b) is significantly different from 0 (zero): that is, whether there is a slope at all.

In multiple linear regression the equation is expanded as follows:

$y = a + b_1x_1 + b_2x_2 + b_3x_3 \ldots\ldots + b_nx_n$, for n explanatory variables

The warning at the end of the last section applies particularly for multiple regressions. You can seriously weight the outcome of a multiple regression by including explanatory variables that are highly correlated with each other. The usual way round this problem is to do a *stepwise regression* in which you put in or take out explanatory variables according to some pre-determined plan, until you have explained the maximum possible amount of the variation in the data using a minimum number of variables.

The regression examples in Box 16.7 on biodiversity and corruption in fact started out as multiple regressions, which also tested for links with three other explanatory variables: gross domestic product, population density, and human development index scores were also included. A stepwise regression showed that in each case, the governance score alone was sufficient to explain the changes in elephant and rhinoceros numbers.

16.7.2 Principal components analysis

Principal components analysis (PCA) is an alternative approach to dealing with several correlated variables. It is a technique that transforms a number of highly correlated variables into a set of uncorrelated (abstract) variables, each of which is a linear combination of the original variables. The new variables are derived in decreasing order of importance, so that the first principal component (PC) accounts for as much of the variation in the original data as possible, the second accounts for as much of the remaining variation as possible while also being uncorrelated with the first, and so on. It is usually possible to account for most of the variation in the data with only a few principal components and hence simplify later analysis.

The first PC is usually an index of the overall 'size' of the data: a kind of weighted average of all the variables that have gone into the mix. The second and subsequent PCs show up contrasts. The results of PCA can be used for data exploration, to reveal patterns in the data and if required, for further analyses, such as multiple regression and cluster analysis.

16.7.3 Cluster analysis

This is a classification technique used to *cluster* cases on measures of either similarity or difference. The output is often presented as a dendrogram (or tree diagram). The data used can be collected directly, or obtained by measuring 'distances' between cases using a previous principal components analysis.

16.7.4 Logistic regression

Logistic regression can be thought of as the multivariate equivalent of the chi-square test, or as a form of *non-parametric regression*. The chi-square test only compares two variables at a time, but you may want to look *as a whole* at how one nominal variable relates to several others. For example, you may want to look at how vegetarianism varies with a range of factors, such as gender, income, age, smoking habits and so on. *Logistic regression* (also known as the *logit model*) is a method of predicting an outcome from a combination of numerical and categorical variables.

16.8 Conclusion

Table 16.4 lists the different kinds of task for which inferential statistics is commonly used, and indicates the appropriate parametric and non-parametric test for each task. The techniques that together make up 'statistics' constitute a powerful set of tools. Like all tools they can be misused or well used: and like all tools, you must learn their proper use before you can use them to the fullest.

Table 16.4 Summary of common tasks in inferential statistics and the appropriate tests to use

What do you want to do?	*Parametric test*	*Non-parametric equivalent*
Compare the means of paired samples	Paired *t*-test	Wilcoxon signed rank test
Compare the means of two independent groups	Independent *t*-test	Mann–Whitney test
Compare the means of more than two independent groups	Analysis of variance (ANOVA)	Kruskal–Wallis analysis of variance
Examine how two numerical variables change together	Pearson product-moment correlation	Spearman rank correlation
Examine the relationship between two nominal or ordinal variables		Chi-square test
Model the linear relationship between a dependent variable and one explanatory variable	Linear regression	
Model the linear relationship between a dependent variable and several explanatory variables	Multiple linear regression	Logistic regression

Summary

1 The techniques of inferential statistics exist to answer the question 'Is my interesting sample result a reflection of a real population effect, or just the natural variation expected from a sample?'
2 Statistical tests do not prove anything beyond doubt: rather they enable us to quantify the potential errors in our results.
3 The assumptions that you can make about the distribution of your data determine whether you can use the more powerful parametric tests, or the more robust non-parametric tests.
4 The choice of statistical test is also determined by the hypothesis you wish to test, which comes from the nature of your research, and your research question.
5 There is no one correct way to present your statistical test results because it depends to some extent on the audience you are writing for. However whatever style of presentation you use, it should always be clear what particular test you carried out, whether the result of that test is significant or not, and the numbers on which the decision about significance is based.
6 The tests described in this chapter are a selection of the most widely used ones. Consult a specialist statistics textbook for further information about these and other statistical tests.
7 Statistical techniques are tools that can be misused or well used: you must learn to use them properly.

Further reading

There is a variety of good textbooks written to explain basic statistical techniques to social science undergraduates, and many manuals on using the statistical software, SPSS. The books listed below have all proved especially useful in particular ways.

Agresti, A. and Finlay, B. (2008) *Statistical Methods for the Social Sciences*, 4th edn, Upper Saddle River, NJ: Pearson Education. [This book is particularly clearly written.]

Brace, N., Kemp, R., and Snelgar, R. (2009) *SPSS for Psychologists*, 4th edn, Basingstoke: Palgrave Macmillan. [This book provides very good explanations of statistical techniques, as well as SPSS instruction. It also has detailed examples of how to write reports and present results.]

Everitt, B.S. and Dunn, G. (2000) *Applied Multivariate Analysis*, 2nd edn, London: Hodder Arnold. [Most general statistical textbooks do not cover multivariate techniques in detail, and you may need to consult a specialist book, of which this is one of the best.]

Field, A. (2009) *Discovering Statistics Using SPSS*, 3rd edn, London: SAGE Publications. [Another book that provides very good explanations of statistical techniques, along with SPSS instruction.]

Fowler, J., Cohen, L., and Jarvis, P. (1998) *Practical Statistics for Field Biology*, 2nd edn, Chichester: John Wiley & Sons. [Although this describes itself as a book for field biologists (and the examples are drawn from that area), the clear descriptions of basic techniques make it very useful for a wider audience.]

Pallant, J. (2007) *SPSS Survival Manual: A Step by Step Guide to Data Analysis Using SPSS for Windows,* 3rd edn, Milton Keynes: Open University Press. [Another book that provides very good explanations of statistical techniques, along with SPSS instruction.]

Section V

Writing up, dissemination and follow-up

17 Writing up the report

> The time to begin writing an article is when you have finished it to your satisfaction. By that time you begin to clearly and logically perceive what it is you really want to say.
>
> (Mark Twain (in Paine 2006: 380))

17.1 Introduction

The write-up of a research report requires very different skills from the fieldwork, and many novice researchers find this the most challenging part of a research project. At the end of data collection you may have piles of documents, photographs and perhaps computer files or recordings. Your mind is full of ideas and impressions of what you have found. Turning the material into a coherent report requires organization and clear thinking. There are several different kinds of report that may be required – for example a thesis, a consultancy report, or a report to 'users' or participants. This chapter will focus on the thesis, with additional guidelines about writing a non-academic consultancy report.

Three important differences between a thesis and a consultancy report relate to the audience you are writing for, the way in which they are likely to use the report, and the kind of information they are likely to be looking for. The principal audience for a thesis are the examiners, who should know something about the broad subject area and should read the thesis from cover to cover (except perhaps the appendices). In addition to the reported information they will be looking for evidence that you know the literature, understand the principles of research design and the methods you have used, and are able to analyze the data and interpret the results appropriately. Do not be tempted to try to cover up any methodological weaknesses; your examiners probably read many theses each year and they will spot them without your help. Provided that you demonstrate that you understand the problems, you can still get good marks.

The audience for consultancy reports varies enormously in terms of knowledge of the subject (and of the research process), and rather than read the report from cover to cover, busy managers are likely to flick through it and either read sections that catch their eye or use it as a reference work to look up particular topics or items of information as they need them. Therefore a consultancy report needs to be written in a style that is accessible to non-specialists, and structured and presented on the page in a way that makes it easy to find sections on a particular topic.

A basic principle in writing is that you should make the reader's task of following and understanding what you have to say as easy as possible without oversimplifying the content.

That means structuring the report clearly and writing in an appropriate style. Structuring will be discussed in the next section. The writing style should be factual and formal, with properly structured sentences (do not write in note form). It should also be appropriate to the reader's level of technical knowledge. Even thesis examiners may not be specialists in the subject of your thesis, and therefore you should define any but the most general of technical terms; obviously this is even more important when writing for a non-academic audience. One specific stylistic point that deserves mention is the 'voice' with which you write. In many disciplines there is a tradition of using the passive voice ('the site managers were interviewed') or using the third person to refer to yourself ('the researcher interviewed the site managers') rather than using the first person ('I/We interviewed the site managers'). However, using the passive voice sometimes makes the text unnecessarily cumbersome, and in most disciplines it is acceptable to use the first person occasionally if it will help to make the meaning clearer. Unlike this book, a formal report should *never* use the second person ('you').

What differentiates research writing from most other forms of writing is that throughout, each statement or assertion should be backed up by some form of 'evidence' so that the readers can judge for themselves whether the statement is justified. The evidence usually consists of a reference either to the literature or to your own data; examples are given in the following sections. In general, unsubstantiated statements (statements made without any supporting evidence) should be avoided. This does not mean that you cannot express your views, but you should make it obvious when you are doing so (for an example, see Box 14.13) and keep such statements to a minimum. Similarly, you should avoid emotive language or judgemental statements; even if you do have strong moral views, you should present the findings as factually as you can. If they support your views it will come across much more convincingly in this way. Good consultancy reports should present evidence for each statement just as thoroughly as academic reports do, although the way that this is done is rather different (see Section 17.4).

Box 17.1 Practical tips: writing a research report

- Use a formal, concise style.
- Be consistent in the 'voice'. Probably the best option is to use the passive voice and occasionally the first person where the sentence structure would otherwise be awkward.
- Avoid or explain technical language.
- Construct a logical 'argument' and present it in the form of a continuous storyline or narrative that presents each step in the argument in turn.
- Use 'signposting' to help the reader follow the structure and the narrative line.
- Back up statements with 'evidence' – either from the literature or from your data.
- Avoid emotive language and unsupported statements based on your own views. Your job is to report the findings of the research as factually as possible.
- Back up your work regularly and save old drafts.

One final, vitally important point: back up your work regularly. This is obvious, but it is astounding how many people do not do it. Every year I meet students who are panicking towards the end of their thesis write-up because they had a problem with their computer and had not backed up their files for some time. I back up the document I am working on every time I get up from the computer, either by copying it to a flash drive or by emailing it to myself. Everything is also backed up on an external hard drive about once a week. I also rename the file each time I make major changes to the structure or delete sections. Right now I am working on a Word file called 'chapter 17 draft 03'; drafts 01 and 02 are still saved in the same folder.

The rest of this chapter describes the writing-up process step by step. Section 17.2 discusses the tasks that should be done on return from the field, and Section 17.3 focuses on the process of writing the first draft. Section 17.4 outlines exactly how to cite the literature and format the reference list. Section 17.5 describes the process of revising the first full draft and Section 17.6 details the final tasks you need to do – write the abstract, proof-read everything and finalize the layout and presentation. Finally Section 17.7 draws some general conclusions.

17.2 Initial tasks: what to do when you get back from the field

There are four things that you need to do when you return from the field:

- Organize all your papers and other materials systematically.
- Write an outline of the structure of the report.
- Develop a storyline.
- Make a timetable for the whole of the writing-up period.

The following sections discuss each of these in turn.

17.2.1 Organizing your papers

If your field site is at a distance from your institutional home and all the fieldwork is carried out in a single trip, then when you get back from the field you are likely to have a big pile of documents – not only those with the data, such as notebooks and data sheets, but also published and unpublished reports, maps and so on that you have collected along the way. You may also have photographs and recordings. The first task is to organize all the different types of material. Catalogue any background documents you have picked up, so that as you write each section of the report you know what documents you have that are relevant. Put each data set in a separate file – each set of completed questionnaires or interview notes or workshop flipcharts – and note down how many there are in each set. If you have not already done so, number all your field notebooks and their pages, and index their contents. File your papers systematically so that you can find them easily as you need them.

17.2.2 Planning the structure of the report

Next, you need to write an outline of the broad structure of the report (for example, by writing a draft of the contents page). The standard structure for a thesis is set out in Table 17.1, together with some common variations. It consists of a short abstract; separate chapters for the introduction, methods, results and discussion; a reference list, and if appropriate,

Table 17.1 Standard structure for a thesis

Section	*Description*	*Common variations*
Title page	Includes thesis title, your name, the year, the name of the degree, department and institution	
Acknowledgements	Usually to funders, collaborators, advisors and any other individuals or institutions that have provided substantial assistance, as well as friends or family who have been particularly supportive	
Contents and lists of tables and figures	After the contents, there is usually a separate, numbered list for (1) the tables and (2) the figures	There may also be a list of acronyms (abbreviations) – for example WWF or CBC (community-based conservation)
Abstract	Concise summary of the thesis	
Chapter 1: Introduction	Introduces the broad topic, gives background and context, and states the specific aims, objectives and/or research questions	The description of the study site may form a separate chapter if it is extensive
Chapter 2: Methods	Outlines the overall methodology and describes individual methods in detail	
Chapter 3: Results	Presents the results on each subtopic systematically	May be divided into several chapters, each addressing one subtopic
Chapter 4: Discussion	Interprets and comments upon the results with reference to the aims and objectives and to existing literature	May include brief conclusion/ recommendations, or these may form a separate, final chapter
References	A list of all references cited in the thesis	
Appendices	May include additional background materials and extra details of methods (such as copies of questionnaires), statistical analyses, or results (such as raw numerical data on key points or extracts from qualitative interviews)	

appendices containing more detailed material. Within each chapter there should be numbered or titled subsections. There is usually a word limit (not counting appendices) of something like 10,000 words for an undergraduate thesis, 15,000 to 20,000 words for a Masters thesis and between 80,000 and 120,000 words for a PhD thesis, and as a basic rule, the longer the thesis, the more chapters there should be. Thus in Masters or PhD theses there is often an additional chapter describing the study site, and the results may be divided into several chapters, each dealing with a different subtopic. However, both the word limit and the specific requirements for the structure vary, so check the regulations for your own degree programme. It is also very helpful to look at some distinction-level theses by previous students on the same programme; it should give you a clear idea of what is required.

The nature of consultancies in social aspects of conservation varies hugely, and therefore the nature of the report also varies. In terms of content it is likely to contain less theory and more practical points than a thesis, and the results may contain sections of quite dense descriptive material that would be omitted in a more theoretical report. Sometimes the structure is specified in the terms of reference but often this is not the case; indeed, the people commissioning the work may not have a clear idea themselves of what kind of report they expect at the end. The best way to get a feel for what kind of report is appropriate is to examine similar reports done by other consultants in the past – if possible for the same people. If you are really unsure, at the beginning of the write-up period you can send the sponsors a draft outline of the report structure you propose to use and ask for their comments.

However, some general principles apply to any research report. It should begin by introducing the study, giving enough background to put it in context and stating the aims and objectives. It should then describe what was done, what was found out and what the findings mean (in practical or theoretical terms). The structure of a consultancy report should allow the reader to use it as a reference document, and therefore apart from a brief introduction and conclusion, the text may be divided into short, issues-based sections rather than by the chapter headings that are used in a thesis. The text is often broken up with illustrations and photographs, and boxes may be used to separate out brief examples or technical details. Since the report may not be read in its entirety, there is often a substantial, stand-alone executive summary rather than a shorter abstract (see Section 17.6).

17.2.3 Developing a storyline

A good report should tell a story, moving from broad description to specific detail to analysis and keeping a clear focus on the central topic of the research throughout. Each section should follow on smoothly from the previous section, adding another piece of information to the storyline. The broad outline of the story is as described above (introduction and context, then what you did, what you found out and what it means), but the detailed content and structure (except for the methods section) should be framed according to the main issues reported in the results. If the study has a very specific research design then the main issues are known in advance of data collection and the storyline should be obvious. However, in broader or exploratory forms of research – especially inductive research – they cannot be defined precisely until you return from the field, or sometimes until you have at least partially analyzed the data.

A common problem is that once you start writing up the results, you get immersed in the detail and cannot see what is important to include or what you should be writing about next. A related common problem is that if the research has gone well every detail seems important; you cannot bear to leave anything out. However, a research report almost never makes use of *all* the data that has been collected; if it did, it would be too crowded and piecemeal. Developing a storyline before you start is invaluable in deciding what to include and what to leave out, how to structure the results, and how much detail or analysis is appropriate.

Box 17.2 suggests some techniques that may help in developing a storyline. If you are really stuck, move on to another task for a few days such as finishing off the data processing or writing up the methods or the study site description; you may find it easier to see the big picture when you come back to it. The detail of the storyline will probably continue to evolve as you write, so refer back to it regularly as you write each section. If you find you have strayed from the plan, then you need to decide whether to revise the storyline to include the new material or cut out what you have been writing and get back on track. Always keep

Box 17.2 Practical tips: five methods to produce a storyline

1 Review your original research design, edit it to reflect any changes that have been made, and then plan out the different sections to reflect the different aims, objectives or research questions.

Or:

2 In qualitative research, code your data using hierarchical codes and then structure the text by including a section on each top-level code, with subsections as necessary for each of the subcodes.

Or:

3 Write a summary of what you think you have found out, from memory. Identify the different sections in the summary and make sure that the order creates a smooth storyline without sudden changes in direction. Then structure the report accordingly.

Or:

5 Try to summarize what you have done and what you think you have found out to a friend or colleague. Someone else may more easily be able to discern a common thread, because their head, unlike yours, is not cluttered with all the detail.

Or:

6 Draw a 'mind map': Write the central topic of the research in the middle of a piece of paper and draw a line around it so that it is in a box. Then brainstorm on all the different issues that have come up, writing each one down at a distance from the central box, and draw a line around each one. Join the different boxes together with lines to show how the different topics relate to one another. Then for each issue, brainstorm for relevant sub-issues and again, join them together as appropriate. You will end up with a diagram showing the different topics and subtopics in clusters, and showing how they fit together. Then prioritize – decide which ones are central and which are marginal – and decide on the order in which you will write about them.

a copy of any material you decide to cut out, just in case later on you find that you need it after all.

A simple way to make the storyline easy for the reader to follow is to use *signposting* (see Box 17.3). Signposting should not be formulaic or over-prominent or it can become very tedious, but when used appropriately it can greatly increase the clarity of the writing.

17.2.4 Writing a timetable

Once your papers are in order and you have decided on a structure for the report, you should make a timetable outlining all the different tasks you need to do. Apart from writing, the tasks may include data processing and analysis; supplementary reading; editing drafts; putting together a reference list or bibliography, and presentational aspects – finalizing the

Box 17.3 Key term: 'signposting'

'Signposting' consists of telling the reader what you are going to write about and in what order, so that they can follow the storyline. Signposting should be used at strategic points throughout the report. It is often used at the end of the first section of the report as a whole (for example, see the end of the first chapter of this book) and also near the beginning of each chapter in order to tell the reader what the chapter will contain (see the last paragraph of Section 17.1). It can also be used at the beginning of a lengthy section within a chapter, especially if there is a sudden shift in the narrative.

layout, proofreading, inserting photos and other illustrations, assembling the list of contents, tables and figures, and printing and binding. The proportion of time you should allow for the different tasks varies enormously from project to project. Inputting data from questionnaires into spreadsheets for analysis can take a *lot* of time, and should have been started during the fieldwork if possible. Similarly, copying notes from interviews or focus groups into electronic format or especially transcribing recordings is *extremely* time-consuming (but not always necessary – see Chapter 14). Coding and indexing qualitative data is relatively quick; I have coded a 100-page notebook in less than 2 days. However, these are all tasks that take a lot of concentration, and rather than try to work on them all day long it can be more productive to intersperse them with other tasks such as writing the methods chapter (which is usually straightforward) or doing additional reading. Once the basic processing is completed, further data analysis may either be done in advance of the write-up, or it may be done for each subtopic and dataset in turn before writing the corresponding sections in the report.

Table 17.2 shows a possible timetable for a 10-week writing up period. This is not meant as a blueprint; the process is too variable for that. However, note that in this example only

Table 17.2 Basic timetable for writing up

Week	*Tasks*
1	Planning and organising; basic data processing.
2–3	Basic data processing Write the methods chapter and study site description. Do any supplementary reading that is necessary.
4–5	Data analysis and writing-up of results. Supplementary reading if necessary
6–7	Finish the results. Write the introduction, discussion and conclusion.
8	If possible take a couple of days off, then review and edit the full draft and write the abstract.
9	Send the whole draft to your supervisor or line manager if required. While waiting for their comments assemble the contents and reference list, check figures and tables are properly presented, insert photographs or other illustrations, write the acknowledgements, and finalize the appendices. When you receive your supervisor's comments do any final editing and proofread.
10	Tidy up the layout; finalize the abstract, contents and lists of tables and figures; print and bind the report.

about five weeks out of ten are dedicated to writing. Note also that the timetable has a suggested break of a few days after the first complete draft is produced. If you can afford the time to do this, it allows you to come back to the report fresh and review the draft with a clear head, uncluttered by detail; it may then be easier to see the overall picture and decide which sections can be cut, what needs to be added, or how you need to adjust the wording to maintain the storyline. You may be expected to send the full draft to your supervisor or line manager to review before it is finalized, in which case agree the date with them well in advance or there may be a substantial delay in receiving their comments. While you are waiting for their comments you can work on your reference list, figures and illustrations, and appendices.

17.3 Writing the first draft

Once you have defined the broad structure of the report and a draft storyline, you are ready to start writing. Probably the most common practice is to start with the methods and results, then write the introduction and discussion, then the conclusion and recommendations, and at the very end the abstract or executive summary. The following sections describe each chapter of a thesis in turn. A consultancy report may not include the same headings or divide the material up in exactly the same way but it should include similar kinds of information, and therefore most of what follows still applies.

17.3.1 Introduction

The introduction should begin by stating the nature of the research topic and explaining the rationale for the study (the reason for doing it). There should then be a more detailed background section, which may include a literature review (especially in an academic thesis) and a description of the study site. The literature review should be based on the literature search that was done prior to data collection (see Section 2.4), topped up with additional literature that is relevant as a result of what the results have revealed. It should summarize current knowledge and understanding of the subject, setting out the key conceptual and theoretical issues, discussing and critiquing them along the way, and illustrating different points with examples from particularly relevant case studies (see Box 17.6 for an example). Each point should be supported with reference to the literature. Throughout, it should also include occasional comments that show how the material you are presenting is relevant to your own particular study.

As far as possible, the text should be worked into a smooth, well-structured argument that starts broadly and narrows down in focus, so that by the end of the chapter it leads seamlessly into a statement of the specific aims, objectives and research questions of the research project. If the aims, objectives and research questions have changed as a result of what happened during data collection, it is acceptable to edit them for the write-up to reflect what you ended up doing, even if this is quite different from what you originally planned to do. At the end of the chapter there may be a closing statement signposting how the rest of the report is structured.

17.3.2 Methods

The methods chapter should begin within an overview of the methodology, explaining the theoretical approach, research strategy and research design if necessary, and stating what different methods were used and why (for a simple example, see box 17.4). There should

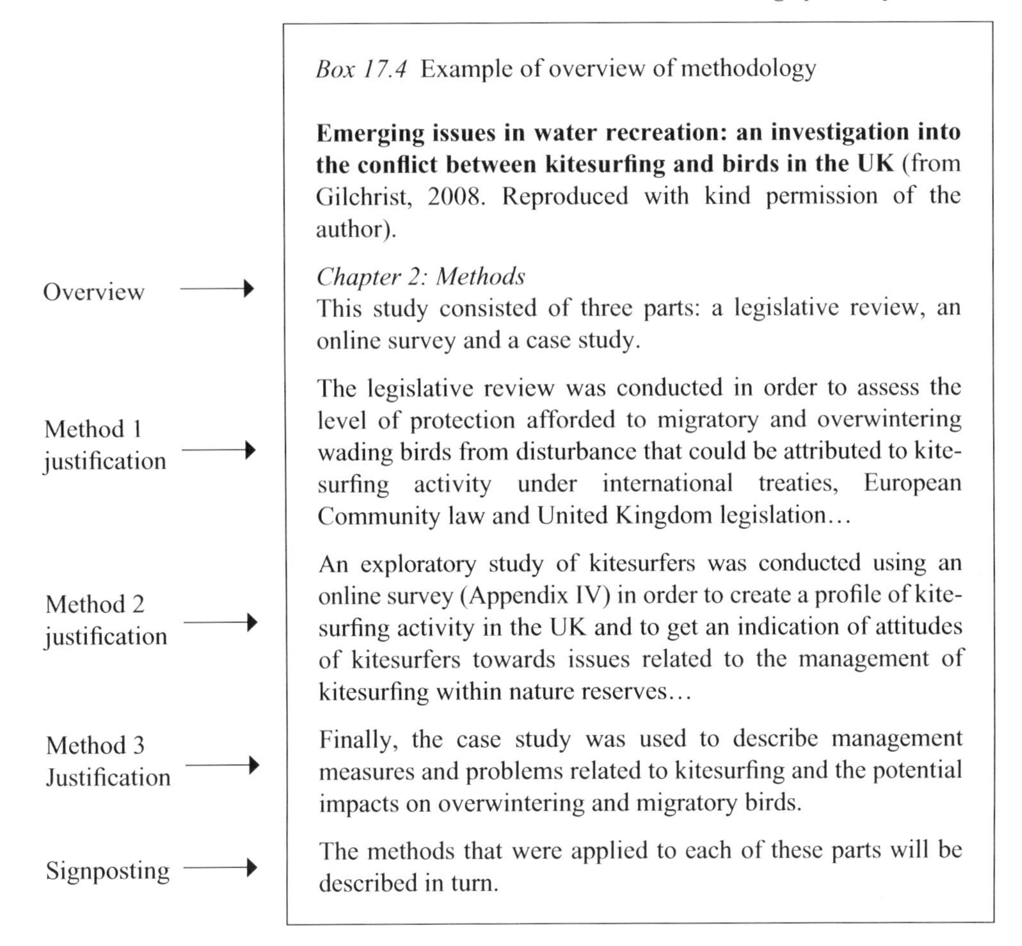

Box 17.4 Example of overview of methodology

Emerging issues in water recreation: an investigation into the conflict between kitesurfing and birds in the UK (from Gilchrist, 2008. Reproduced with kind permission of the author).

Chapter 2: Methods
This study consisted of three parts: a legislative review, an online survey and a case study.

The legislative review was conducted in order to assess the level of protection afforded to migratory and overwintering wading birds from disturbance that could be attributed to kitesurfing activity under international treaties, European Community law and United Kingdom legislation…

An exploratory study of kitesurfers was conducted using an online survey (Appendix IV) in order to create a profile of kitesurfing activity in the UK and to get an indication of attitudes of kitesurfers towards issues related to the management of kitesurfing within nature reserves…

Finally, the case study was used to describe management measures and problems related to kitesurfing and the potential impacts on overwintering and migratory birds.

The methods that were applied to each of these parts will be described in turn.

then be a section on each method that was used, describing what was done in enough detail to enable someone else to repeat the research in the same way (including sampling strategies and, if necessary, analysis techniques). Lengthy details such as full-length questionnaires and interview guides can be placed in appendices. You should also mention any major practical problems you encountered either here or at the beginning of the results chapter, and discuss any resulting methodological weaknesses. The text should be factual and concise, and references should be limited to those that provide information informing the research design.

17.3.3 Results

The results section should begin with a summary explaining what different datasets were collected (stating the final sample sizes), and should then consist of a series of subsections addressing different subtopics, starting with broad description and progressing to specifics, in order to present a clear line of argument throughout. The different subsections are not necessarily organized according to the different methods used; they should be organized according to the line of argument. Therefore in multiple methods studies, where two or more methods provide information on the same thing, the findings may be combined or at least presented in a single section that tells the reader everything that has been discovered about that subtopic. For example, if there are data from sections of both qualitative interviews and questionnaires about people's attitudes to elephants, then there should be a section on people's attitudes to elephants that makes use of both data sets, either mixed together or in subsections. Data on other issues from the same sets of interviews or questionnaires should be reported in other sections. Only data from your own work should be reported in this chapter; if other studies provide relevant information then they should be mentioned in the discussion chapter.

Within each section, the results should be worked into a continuous narrative. This does not mean that it must be written in beautiful literary prose, but simply that at this level, too, there should be a clear line of argument, expressed in concise, complete sentences that follow on logically from one to the next. Where a series of graphs or tables is presented they should be linked together by text that leads the reader logically through the findings, introducing and commenting upon each one in turn and drawing attention to some of the most important features. The results of statistical tests should be reported in the formats outlined in Chapters 15 and 16; negative results should be reported as well as positive ones if they are important to the storyline. Each 'result' should be presented in only one form: there is no need to show the same data in both a table and a graph, and the text should *not* duplicate details that can be seen in the table or graph itself. The reader should be able to understand each table or figure alone, without referring to the main text; it should have a caption saying what it is about, an indication of sample size if relevant, and in the case of graphs, properly labelled axes and a key to colours or shading (see examples in Chapter 15). The writing style in reporting qualitative results is usually more discursive than for quantitative results; it takes the form of more detailed textual description interspersed with supporting quotes, accounts of particular events, diagrams or illustrations. Section 14.3.1 gives some examples and Box 17.5 gives instructions for the correct formatting and positioning of quotes.

Box 17.5 Instructions: how to use quotes

- Quotes should be used to illustrate particular points made in the preceding narrative. They should be used sparingly; it is rarely necessary to use two in a row.
- The exact formatting for quotes varies slightly from publisher to publisher. If there are no specific instructions, a common approach is as follows:
    - Quotes longer than about a line should be put in a separate paragraph, indented and in italics and quotation marks (see example in Box 14.12).
    - Shorter quotes can be included in the main text and put in quotation marks (see example in Box 14.13).
- In reporting your data, the attribution (who the quote is from) does not usually give the *name* of a person, but gives their *role* – man aged 40–50, representative of an NGO, tourist, and so on. This is partly to preserve anonymity. Alternatively no attribution is given (see example in Box 14.13: the quote is introduced with 'a common response… was…').
- Quotes from the literature should be formatted in the same way as quotes from respondents (see Box 17.6 for an example), with the exception that when quoting from the literature, the precise source must *always* be given in the form of an in-text citation (see Box 17.7).

17.3.4 Discussion

Many people find the discussion the hardest section to get right. Its overall aim is to comment upon and interpret the results in order to show how they have addressed the research aims, objectives and questions. The results should also be compared with the published academic literature (at least in a thesis) and, in applied research, the implications for management or policy should be discussed explicitly.

The neatest way to structure the discussion is by addressing each of your original objectives or research questions in turn, and discussing what the results tell you about them. However, you should *not* repeat detailed material that you have presented earlier, and neither should you introduce new data; rather, you should summarize material from the results chapter only as much as is necessary to provide a context for your comments and interpretations. As you discuss each subtopic, you should also comment on how your findings relate to the published literature. Do they fit in with current thinking or contradict it? If the latter, why might this be? Do they raise new questions that need investigation? Much of the time you can do this by referring back to the material in the introductory chapter, but new literature should be introduced that is particularly relevant in the light of your results.

Finally, the discussion is the section of the report where it is most acceptable to include your own hunches and impressions. As long as they are reported appropriately and you make it clear that this is all that they are, they are useful in pointing to issues that should be considered in management or could be followed up in future research.

17.3.5 Conclusion and recommendations

The conclusion may form the final section of the discussion chapter, or it may be a standalone chapter. It is usually brief, and serves to bring the report to a close. It should summarize the overall significance of the research, preferably with direct reference to the original aims and objectives. It may also point to policy or management implications. Especially in a consultancy report there may then be a bullet-pointed list of concise recommendations, including both practical recommendations aimed at policymakers or managers of the study site and also recommendations for further research.

17.4 Citing the literature: in-text citations, notes, reference lists and plagiarism

This section describes the correct format for citing references and also discusses the issue of plagiarism – making use of other people's work without adequately acknowledging the source. The most common system for referring to ('citing') sources of information in academic writing is the Harvard referencing system, which involves inserting the author's surname and the date of publication in the body of the text and giving full publication details in a list of references at the end of the document. The reference list should include *all* of the sources you have cited in the body of the report and none that you have not cited. Consultancy reports (and, for some reason, academic lawyers) often use footnotes or endnotes to refer to the literature rather than in-text citations.

Box 17.6 reproduces the introductory paragraph from a paper describing my own research on community conservation in Amazonian Peru. Note the way that references to the literature are used to back up each major statement. Without the references, the text would appear to simply express my own opinion and therefore it would carry little weight. The principle is

Box 17.6 Example of an introduction: extract from 'Unpicking "Community" in Community Conservation: Implications of Changing Settlement Patterns and Individual Mobility for the Tamshiyacu Tahuayo Communal Reserve, Peru'

Perhaps the core premise of community conservation is that people who have permanent, exclusive rights to land and resources are more likely to manage resources sustainably on the long term (McCay & Acheson 1987; Acheson 1989; Oglethorpe 1990; Ostrom 1990; Lynch & Alcorn 1994; Hanna et al. 1996). Such an approach is based on a clear definition of those who hold permanent or long-term resource rights to a specific area. However, mobility and migration – the subject of this book – represent a basic challenge for this approach. Rural communities are not fixed, bounded entities; they move in location, change in composition as people move in or move away, and do not necessarily have a clearly defined membership. Whilst these issues are increasingly recognized in the academic literature (Uphoff 1998; Agrawal & Gibson 1999; Kumar 2005), they are often still overlooked in the implementation of community natural resource management projects, which have typically treated the 'community' as 'a distinct social group in one geographical location, sharing common cultural characteristics, in harmony and consensus: images that actually may be quite misguiding reflections of reality' (Kumar 2005). At a time when collaborative approaches to conservation are subject to an increasingly strong critique (Oates 1999; Terborgh 1999; Terborgh & Peres 2002), there is an urgent need to move beyond this rather simplistic approach to 'community'. The first step is to build up a body of case studies that unpick the concept of 'community' in community conservation, in order to inform the development of a more realistic framework for community conservation projects. This chapter attempts to provide such a case study, with a particular focus on changing settlement patterns and individual mobility of local residents, based on communities neighbouring the Tamshiyacu Tahuayo Communal Reserve in Amazonian Peru.

(Source: Newing, 2009)

that if the reader doubts what you are saying or wants to find out more, they can look up the references in order to check further.

The use of references should not be carried to extremes. It is not necessary (and indeed can be distracting) to cite every single reference you have come across that is relevant to what you are saying; if you make a broad statement, you should just select a few key references that adequately back it up. Not every statement needs a reference; you do not need them for common knowledge or for statements that follow logically from the previous material. In the example in the box, the second sentence relates directly to the first one and is covered by the same references and the third and fourth sentences lead into the fifth sentence and the references cited there. In a longer literature review there may be a whole page or even a couple of pages without extra references, especially if you are commenting upon and developing the material you have already presented, or if you are summarizing a particular paper from the literature in some detail (in which case this must be absolutely clear: for example state 'the following section is summarized from Jones [1999]').

Box 17.7 gives guidelines and examples for correct formatting of in-text citations and Box 17.8 gives guidelines and examples for formatting the full publication details in the

Box 17.7 Instructions: formatting in-text citations

- Where the citation is integrated into the sentence structure, enclose the date in brackets.

  Example: According to Kumar (2005), the term 'community' has often been used unrealistically in the natural resource management literature

- Where the citation is not part of the sentence structure, enclose the author's surname and date in brackets together.

  Example: The term 'community' has often been used unrealistically in the natural resource management literature (Kumar 2005).

- If there are two authors, separate their names with an ampersand ('&') if it is in brackets, or with 'and' if it is not.

  Examples: (Caira & Ellis 2005)
  According to Caira and Ellis (2005), …

- For citations with more than two authors use *et al.* in the in-text citation (but write out all authors in the normal way in the reference list).

  Example: (Velieu et al. 1996).

- List multiple citations in brackets chronologically from oldest to newest and separate them with a semicolon. Papers published in the same year should be listed alphabetically by author.

  Example: (McCay & Acheson 1987; Acheson 1989; Oglethorpe 1990; Ostrom 1990, Lynch & Alcorn 1994; Hanna et al. 1996).

- If there are multiple papers by the same author simply list the years and separate them by commas.

  Example: (Crone 2001, 2002).

- If you refer to two or more papers by the same author in the same year, refer to them as a, b, c… and use the same labelling in the reference list. This applies to citations in different parts of the report as well as cases where you cite the two papers together.

  Examples: (Crone 2001a, b)
  (Crone 2001a)… (Crone 2001 b)

- If (in the whole report) you refer to two authors with the same surname but different initials, insert their initials as well.

  Example: (M. Jones 2008).

- Papers not yet published but in press can be referred to as follows:

  Example: (Newing, in press).

- If you are referring to material on a specific page or pages in a book or long article, include the page number(s) so that the reader can find the relevant section easily. This is not usually regarded as necessary for a short article (which is why no page number is given in the above reference to Kumar, 2005).

  Example: (Kirkpatrick & Kirkvliet 2006: 132–134)

- Unpublished data or things people say to you can also be cited, although this should be kept to a minimum. No year or entry in the reference list is necessary.

  Examples: (W. Hickson, unpublished data).
  (T. Fellowes, personal communication).

- If you are tempted to cite a paper that is referred to elsewhere but that you have not read, consider whether you should get hold of a copy and read it. However, if it is unavailable or only marginal to your argument you can cite it indirectly.

  Examples: (Kumar 2005, cited in Newing, 2009).
  'According to Kumar (2005, in Newing, 2009)…'

Box 17.8 Formatting of full publication details in the reference list

The references should be put in alphabetical order by the surname, and if necessary then the initials, of the first author. Where more than one article is included by the same author, they should be in order of date.

Book or report: Include the author(s), date, title, publisher and place of publication. Where there are multiple authors, the correct layout of the surnames and initials is shown in example 3.

Example 1: Harvey, G. 1998. Writing with sources: a guide for students. Hackett Publishing Company, Indianapolis.

Example 2: Wells, M.P. 1997. Economic perspectives on nature tourism, conservation and development. Environmental Economics Series Paper 55. World Bank, Washington.

Edited book: As above but insert 'editors'.

Example 3: Western, D., and P. Wright. Editors. 1994. Natural connections: perspectives in community-based conservation. Island Press, Washington.

Chapter in edited book: Note the different ordering of surnames and initials for the editors.

Example 4: Seymour, F.J. 1994. Are successful community-based conservation projects designed or discovered? Pages 472–498 in D. Western and P. Wright, editors. Natural connections: perspectives in community-based conservation. Island Press, Washington.

Journal article: The volume of the journal is in Bold, followed by a colon, space and page numbers. If there are several issues per volume and each issue starts with fresh page numbering, include the issue number in brackets (see example 7).

Example 5: Salafsky, N., and E. Wollenberg. 2000. Linking livelihoods and conservation: a conceptual framework and scale for assessing the integration of human needs and biodiversity. World Development 28: 1421–1438.

Web-based journal: This does not refer to online copies of articles from journals that are also produced in hard copy but only to journals that are exclusively web-based. Use the same format as for other journals but include the web site. Unlike other web-based materials, no access date is needed.

Example 6: Shanley, P. 2006. Science for the poor: How one woman challenged researchers, ranchers, and loggers in Amazonia. Ecology and Society 11(2): 28. http://www.ecologyandsociety.org/vol11/iss2/art28/

Web sources other than journals: Include the author (if given; otherwise see example 8), year, and title in the same format as for a book. In place of the publisher add the name of the organization responsible for the website and their physical location. Also include the full website address and the date you accessed the site.

Example 7: Lassen, B., Newing, H., and Borrini-Feyerabend, G. 2007. Strengthening Community Conserved Areas. Report on international workshop, Kastamonu, Turkey, 1-5 October 2007. IUCN – CEESP, Gland, Switzerland. Available from http://www.iucn.org/themes/ceesp/CCA/Kastamonu/Report_CCA_workshop_Kastamonu.pdf (accessed January 2008)

Institutions as authors: for both paper sources and websites, if there is no indication of the individual authors, put the institution as author. The way you refer to the institution should match how it is cited in the text (in this example, FPP (2009) or Forest Peoples Programme (2009).)

Example 8: FPP (Forest Peoples Programme). 2009. Forest Peoples Programme: facing the challenge. FPP, Moreton-in the Marsh.

Or:

Forest Peoples Programme (FPP). 2009. Forest Peoples Programme: facing the challenge. FPP, Moreton-in the Marsh.

reference list. Different publishers have very specific guidelines for the exact punctuation, layout and spacing that should be used for in-text citations and reference lists. The examples in this section follow the guidelines for the journal Conservation Biology (http://conbio.net/publications/consbio/instructions/Style.cfm). It is not essential that you follow the precise formatting used here, but it is absolutely essential that within your report you are consistent, right down to the level of the punctuation. If you intend to persevere with academic writing it is well worth investing in a specialist referencing program such as EndNote, which allows you to export lists of references in different formats.

Box 17.9 Example of the use of footnotes

[1]Central Africa is defined in this report as including Cameroon, the Republic of Congo, Equatorial Guinea, Gabon, the Democratic Republic of Congo (DRC), and the Central African Republic (CAR); [2]Jackson (2005) gives a comprehensive account of the current situation of indigenous peoples in Central Africa; [3]The zoning process is based on guidelines established by the International Tropical Timber Organisation (ITTO 1990; Foahom 2001); [4]Minnemeyer (2002): 9; [5]Giles-Vernick 2002 : 176,190; [6]No information is available on Equatorial Guinea.

(Source: adapted from Newing, 2007)

Box 17.9 gives an extract from some footnotes in a consultancy report. The notes include not only the references but also comments or additional technical details that would interrupt the main storyline. Often the full bibliographic details are included in the notes themselves, but it makes things much easier for the reader to use the format in the box and to include a full list of references at the end of the article as well.

Other than providing evidence for your statements, referencing also serves to acknowledge the use you are making of other people's work so that you do not take credit for statements, ideas or concepts that are not your own. Plagiarism – the use of other people's work without fully acknowledging the source – is regarded as serious misconduct in research, and is a growing problem in all forms of writing, partly because the ease with which you can cut and paste from electronic documents makes it all too easy to plagiarize accidentally. If you do cut and paste sections of text from electronic sources into your own files as you do your reading, always mark the text (for example by enclosing it in quotation marks) so that when you come back to it later you will know that it is a direct quote. In order to use it in your own text you should either summarize the material in your own words (and still cite the source) or if you decide to quote short extracts directly, put them in quotation marks and cite the source (the example in Box 17.6 includes a direct quote from Kumar [2005]). In this case, the rules for formatting quotes are the same as those in Box 17.5, with the exception that the precise source must *always* be cited, using the formats for in-text citations in Box 17.7.

Students often find it difficult to grasp what counts as plagiarism. Plagiarism includes not only unacknowledged use of someone else's *words,* but unacknowledged use of *any* material from other people's work – which includes information, ideas, concepts and so on – even if you rephrase what they have said. Even if you cite the source, you may commit plagiarism if your writing misleads the reader into thinking that more of the work is yours than is actually the case.

Box 17.10 contains several possible ways of referring to sections of text from the example in Box 17.6; before you read on, see if you can judge which of the examples would constitute plagiarism and which would be correct citation of the work. The bottom line in all these examples is whether it is obvious to the reader what is your own work and what is taken from elsewhere. This applies not only to the words you use, but also to ideas and information. For a full guide on referencing and how to avoid plagiarism, see Harvey (1998).

Box 17.10 Making use of someone else's work: plagiarism or correct citation?

Original text:
Perhaps the core premise of community conservation is that people who have permanent, exclusive rights to land and resources are more likely to manage resources sustainably on the long term (McCay & Acheson 1987; Acheson 1989; Oglethorpe 1990; Ostrom 1990; Lynch & Alcorn 1994; Hanna et al. 1996). Such an approach is based on a clear definition of those who hold permanent or long-term resource rights to a specific area.

Citations: Correct usage or plagiarism?

1. The core premise of community conservation is that people who have permanent, exclusive rights to land and resources are more likely to manage resources sustainably on the long term.
2. According to Newing (2009), the core premise of community conservation is that people who have permanent, exclusive rights to land and resources are more likely to manage resources sustainably on the long term.
3. One of the principles of current approaches to community conservation is that local people are most likely to manage natural resources sustainably if they constitute a clearly defined population and have long-term, exclusive rights to those resources (Newing, 2009).
4. One of the principles of current approaches to community conservation is that people are most likely to manage natural resources sustainably if they have long-term, exclusive rights to those resources (McCay & Acheson 1987; Acheson 1989; Oglethorpe 1990; Ostrom 1990; Lynch & Alcorn 1994; Hanna et al. 1996).
5. According to Newing (2009):

   'Perhaps the core premise of community conservation is that people who have permanent, exclusive rights to land and resources are more likely to manage resources sustainably on the long term. Such an approach is based on a clear definition of those who hold permanent or long-term resource rights to a specific area.'

Original text:
...they are often still overlooked in the implementation of community natural resource management projects, which have typically treated the 'community' as 'a distinct social group in one geographical location, sharing common cultural characteristics, in harmony and consensus: images that actually may be quite misguiding reflections of reality' (Kumar 2005).

Citations: Correct usage or plagiarism?

6. The term 'community' has often been used unrealistically in the natural resource management literature to mean 'a distinct social group in one geographical location, sharing common cultural characteristics, in harmony and consensus: images that actually may be quite misguiding reflections of reality' (Kumar 2005, cited in Newing, 2009).

7. The term 'community' has often been used unrealistically in the natural resource management literature to mean 'a distinct social group in one geographical location, sharing common cultural characteristics, in harmony and consensus: images that actually may be quite misguiding reflections of reality' (Kumar 2005).

(for the correct answers, see box 17.11).

Box 17.11 Answers to examples in box 17.10: correct usage or plagiarism?

Only examples 3, 5 and 6 are correct usage. Examples 1, 2 and 4 constitute plagiarism, and example 7 could be either plagiarism or correct usage, depending on what it is based on.

Example 1: Plagiarism. Direct quote with no mention of the source.

Example 2: Plagiarism. Gives the source but the text is a direct quote and there is nothing to indicate this. If you were reading this as part of a longer article you would know that the underlying idea came from Newing (2009), but you would probably assume that the words were the author's.

Example 3: correct citation. It is clear that the idea being expressed comes from Newing (2009), but the words are the author's own.

Example 4: Plagiarism. The idea is expressed in the author's own words but the list of cited references is identical to the list in Newing (2009). The reader would assume that the author had found these references and brought them together into a list by him or her self, but it is extremely unlikely that they would come up with exactly the same list as Newing (the same is not true for a single reference – see example 7).

Example 5: correct citation. There is no ambiguity here; it is absolutely obvious that the text is a direct quote from Newing (2009).

Example 6: correct citation. This passage contains a direct quote from Kumar (2005) that is reproduced in Newing (2009), but both the original source and the source from which the author took the quote are acknowledged.

Example 7: this one is a trick question. If the author took this quote from Newing (2009) without looking at the original source, then this is plagiarism – the format in example 6 should be used. On the other hand, if the author has also read Kumar (2005) – whether independently of Newing's article or as a result of seeing Newing's reference to it – then this is a correct citation. In this case the text is based on Kumar, not Newing, and the extract is clearly marked as a direct quote.

17.5 Revising the full draft

Once you have produced a full draft of the report, you need to read through it and make sure that everything is clear and targeted, that the different sections fit together into a coherent whole, and that you have addressed the stated aims and objectives explicitly. As you read through, also check the detail of the writing. Look out for long, confusing or incomplete sentences, unsubstantiated statements (statements with no supporting evidence) and sudden skips in the narrative or long sections of descriptive material that are only marginally relevant. You may be able to correct the latter two problems by changing the order of the sections, inserting signposting or otherwise modifying the wording, or you may decide to cut some sections altogether. Alternatively you may notice more significant faults and decide to redraft whole sections that are not clear. Table 17.3 lists some common weaknesses in the first draft of each main section and, where it is not obvious, indicates how these may be corrected.

17.6 Final stages: the abstract, proofreading and presentation

The final stages of writing up involve writing a short summary, proofreading, and finalizing the presentation of the report. The summary takes the form of an abstract in a thesis, and often, a longer executive summary in a consultancy report.

Table 17.3 Checklist: some common weaknesses in the first draft of the write-up

Introduction	• No clear statement of the research problem • Launches straight into a background section without first telling the reader why it is relevant. • Massive, unfocused literature review in which you have tried to include *everything* you have read. Cut it down to include only information that is directly relevant to the storyline. • Extended description of the study site, including much detail you will never refer to again. Cut it down to a short overall description and further details that are relevant to what you will say later on. • Too long; as a general rule, the introduction should form no more than one-quarter to one-third of the whole report. Cut it down.
Methods	• Launches straight into descriptions of the individual methods with no explanation of the research design or overall methodology. • Insufficient information on scope and sampling, or insufficient detail of each method.
Results	• Lack of an introductory overview or signposting. • Division by method even where this separates out material on a single topic. Restructure, bringing together results according to the topic rather than the method. In quantitative research: • Lack of information on final sample sizes and response rates. • Lengthy sections on background variables such as demographics of the respondents. Cut down to a short summary of the demographic profile. • Formulaic run through every possible statistical test. Statistical testing should be targeted to address the specific research questions and hypotheses. In qualitative research: • Too much descriptive detail that does not contribute to your aims and objectives. In an academic report, cut it down. In a consultancy report it may be required, but make sure it is clearly presented and divided into relatively short sections for ease of reference. • Presents a mixture of your own results and substantial information from published sources. Move the latter to the introduction or discussion, as appropriate.

(Continued)

Table 17.3 (*Cont'd*)

Discussion	• Fails to address the original aims, objectives and research questions directly. This is a major fault. If necessary restate them at the beginning of the chapter and then address each one in turn. • Repeats details of the introduction or the results. Delete, or reduce to brief summaries with cross-references where this is necessary to put what you want to say in context. • Introduces fresh data. Delete or move to the results chapter. • Fails to place your research in the context of the existing literature. This may be acceptable in a consultancy report, but not in a thesis or other academic writing. • Conclusions and recommendations summarize principles of good practice rather than being informed by your research findings. In a thesis or other academic report this is unacceptable – they should follow logically from your own data – whereas a consultancy report may include a mixture of general principles and more specific recommendations informed by your findings.
References	• Inconsistencies in formatting. • Some references incomplete – for example, lacking page numbers, publishers' names or place of publication. • Reference list includes articles that have not been cited in the report, or excludes articles that have been cited. See Box 17.13.

Box 17.12 gives instructions for writing an abstract. It should be very short and concise; every sentence should contribute a specific piece of information about your study. Probably the commonest mistake is to include too much background material; the abstract should be about *your* research and should only give background in as much detail is necessary to set your research in context.

An executive summary is considerably longer than an abstract; typically, its length in page numbers is between five and ten percent of the length of the full report. The executive summary is the only part of the report that will be read by a high proportion of the people to whom it is distributed, and therefore it should stand on its own and provide substantial information. It is usually divided into very short, numbered sections (often, paragraphs).

Box 17.12 Instructions: how to write an abstract

Check whether there is a specified word limit for the abstract (for example, in the thesis regulations or consultancy terms of reference). Word limits vary, but are usually between 200 and 800 words.

- Start with a few sentences stating the general aim and rationale of the study.
- If necessary, state the specific objectives.
- The majority of the abstract should describe the methodology and summarize the results.
- End with a brief concluding statement about the wider implications of what you have found or suggesting further action.

Box 17.13 Instructions: how to check the reference list against the in-text citations

- Search the main text electronically for in-text citations. All in-text citations should include brackets, so this can be done by searching for an opening bracket '(' .
- For each in-text citation, check whether it is included in the reference list. If so, tick it off on the list; if not, circle it in the text.
- When you have searched the whole document, go back through, find each reference you have circled, look up the full bibliographic details, add it to the list, and tick it off.
- Finally, delete any references on the list that have not been ticked.

If it gives the page or section numbers in the report that give more detail on each issue, the reader is more likely to turn to the relevant sections to find out more.

Proofreading consists of reading through the whole text with careful attention to detail in order to pick up typing and grammatical errors. Word processing programmes have built-in spell-check and grammar check facilities, so make full use of these. As well as checking the main text, make sure that figures, tables and appendices are correctly presented and that they are numbered sequentially and referred to correctly in the text. Review the reference list, making sure that all the references cited are included in the correct format and that the list includes all references that are cited in the text and none that are not cited (see Box 17.13).

The presentation should be consistent and formal and the text should be well spaced on the page. Thesis regulations often specify details such as fonts, justification, and line spacing. If not, a formal font such as Times New Roman (size 12), full justification and one-and-a-half times or double line spacing is appropriate. Styles for headings and subheadings should also be consistent and margins should be at least 2.5 cm wide. Each chapter should start on a fresh page. This is also the time to insert photographs and write your acknowledgements. Once the layout is finalized you can write out the contents, giving page numbers for each section.

Students are often surprised at how long these final tasks can take. For a 40-page report, ideally you should allow at least a week to ten days for thorough proofreading and attention to presentation and layout. If you need to print the report, do not leave it until the last possible day and risk finding that some of the figures do not print properly or that the printer is broken or – common for thesis deadlines – that there is a long queue for the printer as all your peers try to print their theses at the last minute as well.

17.7 Conclusion

For many people, report-writing is the hardest part of the research process. Making sense of a mass of data and structuring the results into a coherent report is a challenging task. The best research reports have a single, coherent storyline running from the introduction through to the conclusion. The worst reports have no clear storyline, are difficult to follow, and may include a mass of descriptive information that leaves the reader wondering why any of it is relevant (and probably feeling extremely bored). Whatever the audience you are writing for, good structure, clear, concise writing, and good presentation makes a major difference to how your report will be received.

Summary

1 There are several different kinds of report – a thesis, a non-academic consultancy report, or a report to 'users' or participants.
2 A basic principle in writing is that you should make the reader's task as easy as possible without oversimplifying the content.
3 The writing style should be formal and tailored to the audience you are writing for.
4 Each statement or assertion should be backed up by 'evidence' consisting of references either to the literature or to your own data.
5 Before you begin writing, you should organize the materials you have brought back from the field, plan the structure and storyline of the report, and write a timetable.
6 The standard structure for a thesis consists of separate chapters for the introduction, methods, results and discussion. Non-academic reports vary in their structure, and often have an executive summary rather than an abstract.
7 Commonly, the methods and results are written up first, followed by the introduction and discussion.
8 The whole report should have a single, coherent storyline running from the introduction through to the conclusion.
9 Correct formatting should be used for citing the literature and plagiarism must be avoided.
10 The first full draft should be revised to make sure that everything is clear and targeted, that the different sections fit together into a coherent whole and that you have addressed the stated aims and objectives explicitly.
11 Finally the whole report should be proofread, an abstract or summary should be written, and the presentation and layout should be finalized.

Further reading

Becker, H.S. and Richards, P. (2007) *Writing for Social Scientists: How to Start and Finish Your Thesis, Book, or Article*, Chicago, IL: University of Chicago Press. [A discursive, accessible text on the process of writing.]

Harvey, G. (1998) *Writing with Sources: A Guide for Students*, Indianapolis, IN: Hackett Publishing Company. [A useful booklet on how to refer to different sources – not only how to cite references, but also more broadly how to work the information into your own narrative and avoid plagiarism.]

Wallace, M. and Wray, A. (2006) *Critical Reading and Writing for Postgraduates*, London: SAGE Publications. [A very clearly written and informative book that gives excellent guidance on critical reading and writing.]

18 Final tasks

Dissemination and follow-up

18.1 Introduction

Once you have completed your research report, there are various additional tasks to be considered before moving on. First, if you have just finished an academic thesis then you may face an examination by oral presentation (see Box 18.1). Second, you may need to report back to funders or collaborators and participants (see Section 18.2). In addition, the research may merit disseminating to a wider audience (see Section 18.3). Conservation is an advocacy discipline and the purpose of research should be not only to further academic understanding but also to inform current management and policy (for example, see Nelson and Vucetich, 2009).

Finally, you should spend a few days organizing and archiving all your papers – the data, the references and your own notes. They should be filed systematically so that if you need to refer to them in the future you will be able to do so. Otherwise much valuable information will be lost, and if you follow a career in conservation research you will find yourself going back over the same ground needlessly. An early research project may form the basis for further research projects for some years to come, so you need a filing system that will last for years. Section 18.4 discusses the research cycle – the process by which successive research projects build from one to the next.

18.2 Reporting back

In addition to a thesis or consultancy report, you may need to prepare additional materials to feed back your results to collaborators, funders and participants. In some cases this will be a formal requirement, but it is as regarded as good practice even where this is not the case. It is one way to repay them for the time and hospitality they have given you, and may also provide them with information that can inform future choices concerning the subject of your research. Moreover, if you do not do provide any feedback, local people may become unwilling to cooperate in future research (whether it is carried out by yourself or by others). Many local communities across the world are extremely cynical about visiting researchers because in the past they have taken up a lot of time, made a lot of demands and promises, and then failed to provide any feedback or keep in touch. Each time this happens it makes it harder for researchers to work there in the future.

In the case of a consultancy report, the written report is of course the principal form of feedback that is required by the funding organization. However, the funder may also ask you to come and give a presentation on your findings. In the case of a thesis, some of the funding for fieldwork may have been provided by small societies or local businesses, or the whole

Box 18.1 The oral examination

Oral examinations are usually held with students on taught programmes only when their marks just miss a higher overall grade for their degree; the purpose is to assess whether the student merits being awarded the higher grade (they cannot result in your grade being lowered). The examination may be on one particular piece of work (often, but not always, the thesis), or it may be more wide-ranging.

For postgraduate research students, the oral presentation is a standard part of the examination process and may last for anything from one to four hours. In the UK this is known as the *viva voce* examination or 'viva'. It takes the form of a face-to-face interview in which two or more senior academics question you on your written thesis in order to test your understanding. There are no written guidelines about exactly how a viva should be conducted; it is up to the examiners to agree to an approach. Very often they will start with a broad, open question ('Tell me about your research' or 'tell me what you think are the most interesting points in your research' or 'tell me what you think is the significance of your research'). This gives you an opportunity to frame the discussion, so prepare some points that you would like to talk about.

At some point they will move on to the fine detail. They may quiz you on theoretical issues and test your knowledge of the literature; pick up on methodological and analytical details, especially weaknesses; or ask for further explanation or comments on the results and interpretation. Throughout, you should pay careful attention to exactly what they ask, answer them directly and honestly, and be open about the weaknesses as well as the strengths of your work. They already know what you have done (from the written thesis); what they are testing in the viva is your knowledge and understanding rather than the piece of research itself.

The most important preparation you can do for your viva is to make sure that the details of your work are fresh in your mind and think how you would defend it in terms of its strengths, weaknesses and possible interpretations. The viva should not be seen just as an ordeal, however; it is probably the only time in your life when you will be able to discuss a substantial piece of your own work with two or more experts in the field who have dedicated several days to evaluating it in fine detail, so think about what you would like to ask them too. You will be expected to engage in debate rather than simply to give a series of short answers.

project may have been funded by larger trusts and foundations, conservation organizations or, in the case of PhD research, research councils. Academic funders may require a copy of the thesis itself or may have their own format for reporting; they are likely to be more concerned to receive confirmation that the funding has been used as intended than to receive a detailed account of the results. Smaller or more informal funding organizations may require a non-technical summary or a verbal presentation. Even if this is not a formal requirement, feedback of this kind is often very welcome. There may also be institutional requirements to provide a copy of your report or thesis to government departments who granted research permits, or to site managers.

In addition to formal requirements of this kind, it is good practice to report back to collaborators and participants in the research. Workshops are often organized to report back to large numbers of participants. Local experts or key informants who have acted as advisors, and conservation organizations that have provided support, may appreciate a copy of your report or thesis, but for many, a less theoretical or technical form or feedback is more appropriate. Non-specialists may not be familiar with technical terms, or may simply be less interested in theory for its own sake than hard data and its implications for management. Busy conservation executives may not have time to read through a whole report. Research students who work in close collaboration with non-academic partners sometimes write a consultancy-style report in addition to their academic thesis to inform managers or collaborating conservation organizations.

In participatory action research, reporting back to the participants is absolutely essential and may be an integral part of the research process – their comments on initial findings may even be treated as a source of additional data and triangulation. If you live near the study site or will visit it again, then feedback to participating communities is straightforward, and often takes the form of workshops, meetings, verbal presentations or slideshows. In literate societies, posters, leaflets and information booklets can also be used to disseminate results on the longer term (but you will need a budget for their production). The hardest situation is when you live far from the study site; in this case, if possible, you should consider providing a short, written report summarizing what you have done and outlining some preliminary findings, or else arrange a feedback meeting near the end of your fieldwork. The alternative is to try to persuade a local collaborator to agree to act as an intermediary and pass on feedback from you later on, based on materials you send out once the data are fully analyzed.

18.3 Dissemination: communicating the results to a wider audience

Forms of dissemination vary from verbal presentations at meetings or workshops, posters, leaflets, media coverage and articles in popular magazines to formal academic journal papers, seminars and conference presentations. This section describes three of the most common forms of dissemination – verbal presentations, posters and journal papers.

18.3.1 Verbal presentations

Verbal presentations come in many forms and serve many purposes. They include academic seminars and conference presentations, presentations to sponsors or policymakers, popular talks to local societies and schoolchildren, and feedback to local communities in community meetings or participatory workshops. Whatever the occasion and whoever the audience, verbal presentations should be designed rather differently from written accounts of research. People simply cannot take in as much detail in a verbal presentation as in a written account, because they cannot go back over sections that are unclear or take a break when their concentration wavers. It is also harder to follow a complex structure in a verbal presentation. A common piece of advice for presentations is that you should focus on no more than two or three key points that you want to get across. Therefore if your research covers several distinct subtopics, it may be best in any one talk to focus in on only one or two of them rather than attempt to crowd them all in and risk confusing your audience. As with a written report, you need a clear narrative line that provides a brief background and context, a statement of the topic, and then a description of what you did, a summary of what you found out, and some statements on why it is significant. You also need signposting: within the first few

Box 18.2 Practical tips: making a verbal presentation

- Focus on two or three points that you want to get across and build the presentation to create a storyline around them.
- Fit the level of formality and technical detail to the audience.
- As with written reports, 'signposting' makes it much easier for people to follow what you are saying. Within the first few minutes you should tell people exactly what you are going to talk about and at the end you should summarize what you have said.
- Engage with the audience. Do not read from a prepared script, eyes down, talking at the floor. The more you can speak from memory or better still, spontaneously, the better it will come across.
- Speak slowly and clearly and if necessary, ask the people at the back of the room whether they can hear you.
- If the audience is small, invite people to stop you and ask questions if they do not understand what you are saying.
- For talks of less than 15 minutes, practise giving the talk to yourself over and over again until it is within the acceptable time limit.
- Use stories and examples to break up the theoretical material and catch the audience's attention.
- For longer talks, illustrations such as stories and examples also offer a means to adjust the length of your talk as you go along.
- Be ready for questions after the talk. Acknowledge each question – repeat it back if necessary – and answer it as directly as possible.

minutes you should tell people exactly what you are going to talk about (which may include one or more of your research objectives, but does not always do so), and at the end you should wrap up by summarizing what you have said. How much detail you include along the way depends partly on the amount of time you have. Conference presentations may be as short as ten minutes, in which case you can do little more than talk around your abstract and show a few key results. Academic seminars or talks to local societies generally last from half an hour to an hour.

In longer talks, the theoretical or conceptual material should be broken up with ample examples and illustrations. Telling stories – narrating particular events – is a very good way to do this. Many speakers make a habit of starting their presentation with a story that illustrates the research problem and sparks the audience's interest. Alternatively, you could tell a story or show a set of photographs of a particular example to make a break in the middle of the talk. The more informal the talk and the more non-specialist the audience, the more you should use these techniques, but even stuffy academics like a good story and an eye-catching photograph, so do not feel that you should not use them in academic settings. Just make sure the substance is there as well.

Even many experienced speakers tend to prepare far too much material for a presentation and go over time. For short conference presentations the only way to get the timing right is to practise giving the talk to yourself over and over again until it is within the acceptable time limit. For longer talks with more flexible time limits, illustrations such as stories and photographs offer a means to adjust the length of your talk as you go along. If you are short of time you can cut some of them out, or alternatively if you have a few extra examples in your mind, you can introduce them if you are in danger of finishing embarrassingly early.

In terms of actually delivering the presentation, try to develop a style that engages with the audience. Do not read a prepared text, eyes down, mumbling into your chest (obvious, but people do). The more you can speak from memory or better still, spontaneously (and still keep to the point), the better it will come across. Speak slowly and clearly and if necessary, ask the people at the back of the room whether they can hear you. If the audience is small, invite people to stop you and ask questions if they do not understand what you're saying. And in all cases, also be ready for questions after the end of the talk. Acknowledge each question – repeat it back if necessary – and answer it as directly as possible. It is all too common for speakers to ramble on in response to questions and not really address the point the questioner raised, either because they were not listening properly or because they are too keen to make their own points regardless of what questions are asked.

PowerPoint presentations are the norm for most kinds of verbal presentation nowadays. They can be excellent when well-designed but are widely despised because of a tendency to use them to flash up endless bullet-pointed lists, or else to fill the slides with dense text that nobody in the audience has time to read and digest. Keep text to a minimum and take advantage of PowerPoint's capacity to present visual materials – graphs, diagrams, photographs or even video clips. Design the presentation for the audience rather than using it as your lecture notes. Box 18.3 gives some extra tips on the use of PowerPoint presentations.

Box 18.3 Practical tips: PowerPoint presentations

Designing the presentation

- Keep text to a minimum and use plenty of photographs and illustrations. If necessary, prepare additional notes for yourself that are not included on the presentation.
- Use a clear font that is easy to read from a distance, such as Arial (font size 24 and above).
- Use bullet points for signposting and for lists of items, but do not use them on every slide.
- Be wary of putting text on top of the pictures – it is often unreadable.
- Use animations sparingly – overuse can be distracting.
- Keep an eye on the overall file size. High resolution photographs, video clips and animation all increase it dramatically, which can slow it down to an unusable level or even cause the computer to crash. Many conferences specify a maximum file size.
- When you have a full draft, run through the presentation on your computer a few times using the slide show setting (under the 'view' menu) and as you do so, give the presentation to yourself. Edit any parts that do not quite fit into the narrative line, make notes to yourself of points you keep forgetting, and check how much time it takes you to 'do' the presentation.
- When you are happy with it, copy the final file onto a USB stick or a CD (or both), ready to take with you.

Preparing to make the presentation

- Conferences often require you to send the file well in advance so that they can load it up. Check the final date of submission.
- Check equipment details with your hosts. Is the computer you will be using a PC or a Mac? What operating system? What version of PowerPoint? If there may be an incompatibility issue, take a laptop as well.
- Also take a printout of a 'handout' (go to 'File' – 'print' – and select 'handout' in the drop-down box titled 'print what'). You can use it as a reminder of what is coming next as you make the presentation, and of course you can also make copies to give to people.
- Arrive early to make sure all the equipment works (especially if you are taking your own laptop) and that your file functions properly.
- Be prepared for the worst scenario. If the technology fails (as it all too often does), will you still be able to give the talk, based on your 'handout'?

18.3.2 Posters

Posters are a common way to communicate the basic outline of your research to a broad audience. Most academic conferences invite contributions in the form of posters as well as spoken presentations, and dedicate a special area for poster exhibition. Posters are also an excellent means of communication to other audiences – you can place them in a public area of your institution's office, in village halls or zoos or environmental education centres, or send them back to communities and organizations that took part in the research.

The key to good poster design lies in getting the right balance between detail (usually in the form of text) and presentation (including the general layout and the use of illustrations). As a general rule, the less specialist the audience, the less detail you should try to include. Many educational posters aimed at the general public have only a few lines of text; it is the artwork that gives them impact. On the other hand, a flashy poster containing only a few lines of text would be inappropriate for an academic conference. However, even with academics, if you include too much detail and produce a poster that looks as though it will be hard work to read, then very few people will read it. At an academic conference there may be hundreds of posters, all displayed side by side in an exhibition hall or marquee. The conference participants stroll along glancing at each poster as they go past until they come to one that interests them. Therefore you only have perhaps 10 or 15 seconds to catch their attention. Once they have stopped, people are most likely to read a short summary (so you should include one in a prominent position – usually immediately under the title) and to look at each illustration in turn, so convey as much of the information as possible in illustrations and figures. Box 18.4 includes some additional tips for designing posters.

18.3.3 Journal papers

Journal papers are the ultimate output for academic research, giving the highest prestige value and reaching the widest academic audience. If your research has gone well, then you should consider sending an article to an academic journal. If you are a student, ask your supervisor's advice about whether your work is of a sufficient standard to merit publication.

Box 18.4 Practical tips: designing academic conference posters

- You do not need specialist software to design a poster; you can use PowerPoint. To specify the poster size, go to 'page setup', select 'custom' in the drop-down box titled 'slide sizes for…', and then insert the required dimensions.
- Conferences usually specify the appropriate size for posters; if not, standard sizes are A2 (42.0 × 59.4 cm) or A1 (59.4 × 84.1 cm).
- The poster must be readable from a distance of two to three metres. Use a font size of about 50 for the title and font sizes of 24 and above for the main text. Acknowledgements/references can be smaller – down to font size 14.
- The title should communicate exactly what the poster is about so that it catches the attention of those for whom it is relevant.
- Put your name immediately below the title and your institution and contact details either there or at the bottom of the poster, in smaller font.
- Include a brief summary, prominently placed, that makes sense on its own. Many people will only read the title and the summary, then look at the pictures, graphs, diagrams and other illustrations.
- Include lots of photos, graphs and other illustrations and not too much text.
- Photos should be high resolution. Graphs and other illustrations should be clear and well-labelled so that they make sense on their own.
- Break the text up into short sections using separate text boxes so that it does not look too dense.
- Check that all text is readable – especially if it on top of pictures. Use a font colour that stands out from the background colour.
- If the work was carried out in collaboration with partner institutions, send them a draft of the poster and ask for comments.
- Include your contact details so that people who are really interested can get in touch later.
- Acknowledge funders and collaborators on the poster and consider including their logos (but ask them first).

Undergraduate theses do occasionally result in publications, although it is rare. A good Masters thesis should support at least one journal article, though not necessarily in the highest ranking journals. On the other hand a PhD thesis should contain sufficient material for several journal articles including at least one in a leading international journal. If you are not an academic and you wish to publish in an academic journal, then if possible discuss the possibilities with an academic whose judgement you trust.

The article should consist of a brief introduction to the relevant literature, a sound research design, and (for quantitative studies) appropriate sample sizes to support your interpretations. The standards are highest for the top-ranking international journals, but so is the competition – many of them have very low rates of acceptance. Therefore you should make a realistic judgement about how high to aim. There are also many practitioners' journals that require less theory and reference to the literature and probably less rigorous methodological standards, but focus more on immediate implications for practice and policy. Box 18.5 lists some broad-scope international academic journals that publish articles on social aspects of conservation and also two practitioners' journals. There are also many more

Box 18.5 International journals covering social or interdisciplinary aspects of conservation

Journals specialising in conservation

- *Biodiversity Conservation*
- *Biological Conservation*
- *Conservation Biology*
- *Conservation and Society*
- *Ecological Applications*
- *Ecology and Society*
- *Environmental Conservation*
- *Journal for Nature Conservation*
- *Society and Natural Resources*

Other academic journals that frequently publish articles on social aspects of conservation

- *Environment, Development and Sustainability*
- *Global Environmental Change*
- *Human Ecology*
- *Human Organisation*
- *World Development*
- *Development in Practice*
- *Community Development Journal*
- *Development and Change*
- *The Geographical Journal*

Conservation practitioners' journals

- *Oryx*
- *Conservation in Practice*

specialist journals reflecting specific subject areas, disciplinary approaches or geographical regions.

A good way to make a shortlist of the journals that might be appropriate is to look through the reference list at the end of your report and see which journals you have cited most frequently. Then look at each journal's website, check its scope in terms of subject areas and disciplinary approaches, and also look at the description of the type of articles they accept and the word limits for each type of article. Obviously your article will be most widely disseminated if it is published in a highly prestigious international journal, but be realistic, both in terms of the quality of your research and also in terms of the broadness of interest. Unless your research addresses a topic that is of interest internationally, you have a better chance of getting published if you target a journal with a narrower audience and a more specific topical scope. If you are unsure, send an abstract to the journal editor and ask whether it fits in with the journal's remit.

Once you have chosen a journal, look more closely at the style guidelines. Different journals require different structures and formatting and each journal has very specific guidelines.

Checking these details before you begin to write can save you a lot of time later on, and may even make the difference between having your article accepted or rejected.

18.4 The research cycle: developing further research proposals

If, as is often the case, your research raised more questions than it produced answers, then why not look for funding for another research project? If you failed to get a statistically significant result but you think this may be due to a small sample size, plan a new study with a larger sample size. If a new factor emerged that seemed to influence the results but was not studied systematically, then seek funding for a new research project focusing on that factor. If you have just completed a consultant's report, it may be possible to persuade the sponsors to fund a follow-up project to look at some aspects in more detail. If you have just completed an undergraduate thesis, perhaps you should apply for a Masters degree and develop the research further in the Masters thesis; similarly a Masters thesis can act as the starting-point for a PhD study. If you have completed a PhD, then you can find out what postdoctoral grants are available for your own research.

Designing another research project on a related topic is simpler than changing to something completely new, because you should already be familiar with the literature and methodology and you may already be in contact with other researchers working on the same topic in other places. There are also great advantages to returning to the same research site again and again, in that your local knowledge and contact networks should make everything run much more smoothly than at a new site. Most research careers are built in this way – each researcher follows a specific research interest over several years or even over the whole of their career. Some people continue to carry out research at the same site throughout their careers, which should allow a highly detailed understanding to be built up; on the other hand, moving to a different site from time to time can give you a fresh perspective and lead to some interesting comparisons between different sites. Many people change the subject of their research radically at some point in their career as their interests change or as new issues arise. However, in all cases, the practice of research forms a cycle of which the actual data collection is only one part. You develop an idea for a project, write a proposal and apply for funding (which may take a year or more to come through), carry out the research, write it up, and while writing, think of further directions and write funding proposals for research to fill in the next piece in the puzzle.

18.5 Conclusion

More than in most disciplines, the applied nature of the metadiscipline of conservation and the urgency of the need to stem the biodiversity crisis mean that dissemination of results to different audiences should be regarded as an important responsibility by any researcher who wishes to contribute to the 'mission' of conservation. Too many research projects produce nothing but a hefty report that is only read by a handful of people – if that. If your research has produced some interesting results, and especially if the results have implications for policy and practice, then it deserves to be disseminated as widely as possible – to policymakers, perhaps to the general public, and of course to other academics.

Summary

1 Once you have completed your research report, there are various additional tasks to be considered.

2 If you have just finished an academic thesis, then you may face an examination by oral presentation.
3 Reports to funders and often to government offices or site managers may be a formal requirement, but even if not, it is good practice to provide them.
4 It is also good practice to report back to other collaborators and participants.
5 Conservation is an advocacy discipline and if the research has produced some interesting results, it may merit disseminating to a wider audience.
6 Other than the research report or thesis, forms of reporting back include verbal presentations, workshops, leaflets, information booklets and posters.
7 Dissemination to the public may also include media coverage and articles in popular magazines.
8 The principal forms of dissemination to a wider academic audience are seminars, conference papers and posters and academic journal papers.
9 If you wish to continue in conservation research, it may be possible to secure funding for a follow-up project on the same topic.

Appendix 1

Online survey: kitesurfing and the environment

(Gilchrist, 2008, reproduced with kind permission of the author)

[Online survey template] Open between 2 September and 15 October 2008

Welcome

Thank you for volunteering to take part in this study, your time is most appreciated. By completing this questionnaire:

a) You will be helping to create a profile of kitesurfers and kitesurfing activity in the UK;
b) You will be contributing to what is known about the effects of kitesurfing on the environment;
c) You will be helping to identify the most effective means through which to disseminate information to kitesurfers about environmental issues; and
d) You will be helping to identify the most effective means to manage kitesurfing for both people and wildlife.

This questionnaire takes approximately 20 minutes to complete.

All responses are anonymous and confidential.

The project has been approved by British Kitesurfing Association (BKSA).

Note that once you have clicked on the CONTINUE button at the bottom of each page you can not return to review or amend that page.

General Information:

1. What is your age?————

2. What is your gender? (Please tick)
   a) Male
   b) Female

3. Which of the following best applies to you (tick one only):
   a) Working full-time
   b) Working part-time
   c) Registered unemployed
   d) Not registered unemployed but seeking work
   e) At home/not seeking work
   f) Retired
   g) Full-time education
   h) Part-time education
   i) Other (specify)

4. If you are working, what is your gross annual income (in pounds)? (Please tick one)
   a) Less than 10,000
   b) 10,000–19,999
   c) 20,000–29,999
   d) 30,000–39,999
   e) 40,000–49,999
   f) 50,000–59,999
   g) 60,000–69,999
   h) 70,000–79,999
   i) 80,000–89,999
   j) 90,000–99,999
   k) 100,000 or more.

5. For how many years have you been kitesurfing?————

6. What do you enjoy most about kitesurfing?

   ____

   ____

   ____

7. Do you do any beach sports other than kitesurfing? (Please tick)
   a) Yes
   b) No

7a. Which other beach sports do you practice?

   ____

   ____

   ____

8. Please list the main factors you consider when looking for a site to kitesurf:
   a) ______________________________
   b) ______________________________
   c) ______________________________

9. Is guidance from your local council or site management authority a factor in your choice of kitesurfing sites? (Please tick)
   a) Yes
   b) No

9a. Please describe how guidance from your local council or site management authority is a factor in your choice of kitesurfing sites:
   ____
   ____
   ____

10. Do you receive any information pertaining to environmental site regulations (Please tick)?
   a) Yes
   b) No

10a. Please indicate from which sources you receive this information (Please tick; you may select more than one answer):
   a) The local council
   b) Natural England
   c) Site management authority
   d) Local non-governmental organization (NGO)
   e) BKSA
   f) Local kitesurfing club
   g) Online kitesurfing fora
   h) None.

10b. In what forms do you receive this information? (Please tick; you may select more than one option):
   a) Newsletters
   b) Website
   c) Handouts
   d) Signs
   e) Informal conversations
   f) Meetings

11. In your opinion, what is the most effective method for disseminating information regarding environmental site regulations to kitesurfers?
   ____
   ____

12. Other than the BKSA, are you a member of any other kitesurfing clubs or associations? (Please tick):
   a) Yes
   b) No

12a. Please specify the other kitesurfing club(s) or association(s) that you are a member of:

____

____

13. Are you a member of an environmental organization? (Please tick)
    a) Yes
    b) No

13a. Which environmental groups are you a member of?

____

____

____

____

14. Please indicate how strongly you feel about environmental issues. Circle the number that corresponds to how you feel; from '1' indicating that you are 'not at all' concerned about environmental issues and '7' indicating that you feel 'very strongly' about environmental issues.

1 2 — 3 4 — 5 6 7

Not at all — Neutral — Very strongly

Kitesurfing activity and management

15. Please indicate in which months you go kitesurfing by selecting YES or NO and, of the months that you have indicated, please rank them in order of activity. Place 1 in the box next to the month that you kitesurf most often, 2 next to the month that you kitesurf the second most often, and so on. Please do not place the same number in more than one box.

Table A.1.1

	Month	*Do you kitesurf in....*		*Ranking*
		Yes	*No*	
a.	January			
b.	February			
c.	March			
d.	April			
e.	May			
f.	June			
g.	July			
h.	August			
i.	September			
j.	October			
k.	November			
l.	December			

16. Why do you kitesurf more in the months that you have indicated?

____

____

17. What is the name of the site where you usually kitesurf?

____

17a. How far do you have to travel to get there (in miles)?————

17b. How many other sites do you use for kitesurfing? (Please write a number from 0 to 12):

17c. Of these, how far away is the furthest one (in miles)?————

17d. How do you usually get there? (Please tick one only):
a) Bus
b) Train
c) Car
d) Other (please specify): ————

18. In order of importance, starting with the most important, what are the three most typical ways by which you acquire initial information when you are looking for new sites to kitesurf?
a) ____________________
b) ____________________
c) ____________________

19. Are you aware of any possible negative environmental impacts of kitesurfing? (Please tick):
a) Yes
b) No

19a. Please list the most important possible negative environmental impacts that you are aware of, starting with the most important:
1 ____________________
2 ____________________
3 ____________________

19b. How did you become aware of these possible negative impacts?

____

____

20. Are you aware of any conflicts between kitesurfers and other user groups? (Please tick):
a) Yes
b) No

20a. Please list three user groups with which kitesurfers are in conflict:
1 ________________________________
2 ________________________________
3 ________________________________

20b. Please describe the conflicts that exist between kitesurfers and the other user groups that you have listed previously.

____

____

21. At what state of the tide do you usually kitesurf, and why?

____

____

22. Do you think that kitesurfing disturbs birds? (Please tick):
a) Yes
b) No

22a. In what ways do you think kitesurfing disturbs birds?

____

____

22b. Please describe any observations that you have made where birds have been disturbed by kitesurfing activity.

____

____

____

22c. Do you have any suggestions as to what could be done to minimize disturbance to birds from kitesurfing?

____

____

____

23. Do you ever consider bird disturbance when choosing a site to kitesurf? (Please tick):
a) Yes
b) No

23a. Can you tell me about how and why you consider bird disturbance when choosing a site to kitesurf?

____

____

____

____

24. Relative to the following recreational activities, how much of an impact do you think kitesurfing has on birds?

Table A.1.2

	Activity	*Kitesurfing has LESS of an impact (1)*	*Kitesurfing has the SAME level of impact (2)*	*Kitesurfing has MORE of an impact (3)*	*Don't know (4)*
a.	Horse riding				
b.	Dog walking				
c.	Motor boats				
d.	Jet skis				
e.	Windsurfing				
f.	Sailing				
g.	Shore angling				
h.	Bait digging				
i.	Swimming				
j.	Bird watching				
k.	Four wheelers				
l.	Dirt bikes				
m.	Land boarding & kite buggying				

25. Please indicate how much you AGREE or DISAGREE with the following statements regarding the management of kitesurfing in protected areas:

Table A.1.3

		Strongly disagree (1)	*Disagree (2)*	*Neutral (3)*	*Agree (4)*	*Strongly agree (5)*
a.	All kitesurfers should be members of their local club.					
b.	Danger to other water users by kitesurfers is overestimated.					

(Continued)

c.	The 'Guides for Water Users' are followed by most kitesurfers.					
d.	Failure to follow local kitesurfing regulations should be a legal offence.					
e.	Kitesurfers should be allowed anywhere they can kitesurf.					
f.	Kitesurfers should work together to self-regulate the sport.					
g.	Kitesurfing should be allowed in nature reserves.					
h.	Kitesurfers should follow voluntary codes of conduct.					
i.	Kitesurfing should be kept to designated sites (zoning).					
j.	Kitesurfing zones in environmentally sensitive areas should be demarcated using sign buoys and/or signs.					
k.	Permit systems should be implemented at all kitesurfing locations.					
l.	Kitesurfing in environmentally sensitive sites should be limited to specific times of the year.					

26. Did you read and/or participate in the online discussion that was posted on the www.kiteboarder.co.uk Forum titled 'Does kitesurfing have an impact on birds?'? (Please tick):
    a) Yes
    b) No

27. If you would like to add anything else relating to the topics of this questionnaire you may do so in the box below:

    ____
    ____
    ____

28. Would you mind if I contacted you should I have any further questions? (Please tick)
    a) Yes
    b) No

28a. Contact Information:

____

____

____

____

Appendix 2

Questionnaire: community orchards

(Johnson, 2008, reproduced with kind permission of the author)

Questionnaire:

Thank you for taking the time to fill out this questionnaire. It should only take about 10 minutes of your time.

The purpose of this questionnaire is to provide a deeper understanding of traditional and community orchards in the England in the context of conservation. The information that you provide will be used for a Master's degree dissertation project and your anonymity is assured if requested.

Part 1: General information

1: What is the name of the orchard?

____

2: What is the name of the village that is nearest to the orchard?

____

3: How large is the orchard?
(please enter a number in acres or hectares and select the appropriate units)

O ———(number)
O Acres
O Hectares
O Don't know

4: What year was the orchard planted in? ———

If unknown, approximately when? ———

5: Is the orchard designated or perceived as a community orchard?

O Yes
O No
O Don't know

5A: If yes, is the orchard:

O Designated as a community orchard
O Perceived but not designated as a community orchard
(Please tick one option)

5B: Who initiated the process of designating the orchard as a community area? (Tick one):
- O 1–2 private owners of the orchard
- O A city council
- O A parish council
- O A village hall
- O A non-profit organisation (e.g. charitable trust)
- O The local community or a local community group
- O A private company
- O The national government
- O Other (please specify) ——————————————
- O Don't know

5C: Which of the following were considered reasons for designating the orchard as a community space? (Tick all that apply)
- O Fruit growing
- O Habitat for wildlife
- O Green space
- O Park for community gatherings
- O Cultural heritage
- O Recreation
- O Other (please explain) ——————————————
- O Don't know

6: Which types of fruit are planted in the orchard? (Tick all that apply)
- O Apples
- O Pears
- O Plums
- O Cherries
- O Quince
- O Other ————
- O Don't know

7: Does the orchard share a border with an allotment?
- O Yes
- O No
- O Don't Know

Part 2: Community involvement

8: Is the orchard open to the public at all times?
- O Yes
- O No
- O Don't know

8A: If no, how often is it open to the public? ————

9: Have there ever been community events in the orchard?
- O Yes
- O No
- O Don't know

If no or don't know please skip to question 10

9A: Are community events held in the orchard every year?
O Yes
O No
O Don't know

9B: How often are events held in the orchard? (Tick one only)
O Once every two years
O Once or twice a year
O 3–6 times a year
O 7–12 times a year
O 13 or more times each year
O There are no events held in the orchard

9C: If yes, what sorts of community events are held in the orchard? (Tick all that apply)
O Wassailing
O Blossom festivals
O Seasonal festivals
O Fruit picking events
O Tree planting
O Educational activities
O Other (please list) ____

____

9D: On average, how many community members participate in orchard events? (Tick one only)
O 1–30
O 31–50
O 51–100
O More than 100
O More than 200

9E: How many from outside of the community? (Tick one only)
O 1–30
O 31–50
O 51–100
O 101–200
O More than 200
O Don't know

9F: If you are having any events in the coming months would it be permissible for a researcher involved in this study to attend?
O Yes
O No

If yes – please provide the date and time, website where information would be available on the events or contact details

10: Is the orchard used as a classroom for local school children?
O Yes
O No
O Don't know

11: What sorts of activities do community members do individually at the orchard? (Tick all that apply)
O Rambling
O Cycling
O Dog walking
O Picnicking
O Fruit picking
O Wildlife watching
O Animal grazing
O Skill sharing
O Other ———
O Don't know

12: What other benefits are provided by the orchard (if any)?

13. Has the ecological condition of the orchard changed since it was designated as a community orchard?
O Yes
O No
O Don't know

If yes, please describe the changes:

Part 3: Conservation value:

14: Does the orchard have any conservation designations?
O Yes
O No
O Don't know

If no/don't know, continue to question 15

14B: If yes, which designations does the orchard carry?
(Tick as many as apply)

O National Nature Reserve
O Local Nature Reserve
O Site of Special Scientific Interest (SSSI)
O National Park
O Area of Outstanding Natural Beauty
O Special Area of Conservation
O City Conservation Area
O County Conservation Area
O Other (please specify) ———

15: Are any measures taken to create wildlife habitats?

O Yes
O No
O Don't know

15A: If yes, what measures? ____

____

____

____

16: Has the orchard ever been surveyed or monitored for wildlife?

O Yes
O No
O Don't know

16A: Is there on-going monitoring of plant or animal wildlife in the orchard?

O Yes
O No
O Don't know

16B: If yes, are there endangered species present in your orchard?

O Yes
O No
O Don't know

If yes, what species? ____________________

____

____

17: Does the orchard have a:

O Biodiversity Habitat Plan?
O Biodiversity Action Plan?
O Other?

(please specify: ____)

O None of the above?

PART 4: Management: This section aims to understand the different levels of maintenance and other duties that exist in community orchards.

18: Who owns the land that the orchard is planted on?
(Tick one only)

- O The National Trust
- O A wildlife trust
- O A county council
- O A borough, city or town council
- O A parish council
- O A village hall
- O A non-profit organisation (e.g. a charitable trust)
- O 1–2 individuals (privately owned)
- O The local community or a local community group
- O A private company
- O The national government
- O Other (please explain) ____________________
- O Prefer not to answer

19: Who makes the primary decisions regarding the orchard's maintenance, care and future?
(Tick all those that apply)

- O The National Trust
- O A wildlife trust
- O A county council
- O A borough, city or town council
- O A parish council
- O A village hall
- O A non-profit organisation (e.g. charitable trust-please specify)
- O 1–2 individuals
- O The local community or a local community group
- O A school
- O A private company
- O Don't know
- O Other (please specify) ____________________

19A: How was this person or group appointed to carry out this duty?

____

____

20: Who carries out the administrative duties for the orchard? (Tick all that apply)

- O The National Trust
- O A wildlife trust
- O A county council
- O A borough, city or town council
- O A parish council
- O A non-profit organisation (e.g. charitable)

- O 1–2 individuals
- O The local community or a local community group
- O A school
- O A private company
- O Don't know
- O Prefer not to answer
- O Other (please specify) ——————————————

21: What types of maintenance are carried out in the orchard? For each, who conducts the activity (i.e. Community volunteers, paid contractors, land owners, etc.)? (Tick all that apply)

- O Mowing
- O Pruning
- O Management of grazing stock animals
- O Spraying of pesticides
- O Spraying of herbicides
- O Bee keeping
- O Fruit Picking
- O Clearing of old fruit, debris
- O Irrigation or watering
- O Other ————
- O Other ————
- O None

21A: Who carries out these activities? (Tick all that apply)

- O Community volunteers
- O The land owners
- O Paid contractors
- O Council employees
- O Individuals conducting community service
- O Other ——————————————

22: What are the objectives of the maintenance performed in the orchard? (Tick all that apply)

- O Fruit production
- O Aesthetic purposes
- O To maintain for community use
- O To maintain as a wildlife habitat
- O Other (please specify:) ——————————————

23: Are there profits made by the orchard?

- O Yes
- O No
- O Don't know

If you don't know, please give me the contact details of the person I should contact to find out, if you know it: ——————————————

——————————————

23B: If yes, what are the profits used for? ___

24: Does the orchard receive funding (not commercial profit) for management, maintenance and/or upkeep?
- O Yes
- O No
- O Don't know

24A: If yes, what type of funding is provided for the orchard?
(Tick all that apply)
- O Private funding
- O Local community fund raising
- O DEFRA grant
- O Heritage Lottery Fund
- O Living Spaces fund
- O Other ___

Part 5: Information on the responding individual and closing questions

25: You are:
- O Male
- O Female

26: What is your position or primary involvement with the orchard?

___

27: In your own words, what is your vision for the future of the orchard? (please use the back of the page if you need additional space)

28: How long have you been involved with the orchard? (Tick one only)
- O Less than 1 year
- O 1–3 years
- O 4–6 years
- O 7–10 years
- O 11–15 years
- O 16–30 years
- O More than 30 years

29: Where do you obtain information on the best techniques for going about your tasks related to the orchard?
(Tick all that apply)
- O From family, friends or acquaintances who have practical experience with orchards

- O From Common Ground's publications and handbooks
- O From other publications (please list)

____

- O From university training
- O From organisations such as countryside projects or trusts
- O Other (please list)

____

30: Is it permissible for a researcher involved in this study to contact you individually to ask further questions?

- O Yes
- O No

31: If any of the information you provide is cited specifically in reports or publications, do you wish for the orchard's name to be included?

- O Yes
- O No

32: If you would like to make any further comments please use the space below. You can also send any additional documents with further information to the researcher at [email adress]

Thank you very much for your time. If you have any questions regarding this questionnaire please feel free to contact [email address] or at the mailing address below.

Best regards,

Kira

[address]

References

Adams, W. (2004) *Against Extinction: The Story of Conservation*, London: Earthscan.

Agresti, A. and Finlay, B. (2008) *Statistical Methods for the Social Sciences*, 4th edn, Upper Saddle River, NJ: Pearson Education.

Amazon Conservation Team. (2010) *Participatory Ethnographic Mapping: Mapping Indigenous Lands*. Online at www.amazonteam.org Accessed on 13 May 2010.

Appleton, S. and Booth, D. (2005) 'Strong fences make good neighbours: survey and participatory appraisal methods in poverty assessment and poverty reduction strategy monitoring', pp. 119–135 in J. Holland and Campbell, J. (eds), *Methods in Development Research: Combining Qualitative and Quantitative Approaches*, Swansea: Practical Action Publishing.

Balmford, A. and Cowling, R. (2006) 'Fusion or failure? The future of conservation biology', *Conservation Biology* 20: 692–95.

Banks, M. (2007) *Using Visual Data in Qualitative Research*, London: SAGE Publications.

Barnes, J. (1979) *Who Should Know What? Social Science, Privacy and Ethics*, Cambridge: Cambridge University Press.

Barry, A., Born, G. and Weszkalnys, G. (2008) 'Logics of interdisciplinarity', *Economy and Society* 37: 20–49.

BBC (2009) 'More or Less', (broadcast on Radio 4 2 January 2009), available http: <http://news.bbc.co.uk/1/hi/programmes/more_or_less/7798152.stm> (accessed 29 November 2009).

Becker, H.S. and Richards, P. (2007) *Writing for Social Scientists: How to Start and Finish Your Thesis, Book, or Article*, Chicago, IL: University of Chicago Press

Bell, J. (2005) *Doing Your Research Project: A Guide for First Time Researchers in Education, Health and Social Science*, Maidenhead: Open University Press.

Berlin, B. and Kay, P. (1969) *Basic Color Terms: Their Universality and Evolution*, Berkeley, CA: University of California Press.

Bernard, R. (1994) *Research Methods in Anthropology*, 2nd edn, Thousand Oaks, CA: SAGE Publications.

—— (2006) *Research Methods in Anthropology*, 4th edn, Walnut Creek, CA: Altamira Press.

Bernstein, J.H., Ellen, R.F. and Antaran, B.B. (1997) 'The use of plot surveys for the study of ethnobotanical knowledge: A Brunei Dusun example', *Journal of Ethnobiology* 17(1): 69–96.

Bierce, A. (2007 [1966]) *The Collected Works of Ambrose Bierce, Volume VIII. Bibliolife.*

Blaikie, N. (1993) *Approaches to Social Enquiry*, Cambridge: Polity Press.

—— (2000) *Designing Social Research*, Cambridge: Polity Press.

Borgatti, S.P. (1996) *ANTHROPAC 4*, Natick, MA: Analytic Technologies.

Borgerhoff Mulder, M. (2007) 'Interdisciplinary collaboration: painting a brighter picture and identifying the real problem', *Conservation Biology* 21: 903–4.

Borgerhoff Mulder, M. and Logsdon, W. (1996) *I've Been Gone Far Too Long: Field Trip Fiascos and Expedition Disasters*, Oakland, CA: RDR Books.

Borrini-Feyerabend, G., Farvar, M.T., Nguinguiri, J. and Ndangang, V. (2000) *Co-management of Natural Resources: Organising, Negotiating, and Learning-by-doing*, Heidelberg: GTZ and IUCN.

Borrini-Feyerabend, G., Pimbert, M., Farvar, M.T., Kothari, A. and Renard, A. (2004) *Sharing Power: Learning-by-doing in Co-management of Natural Resources Throughout the World*, Tehran: IIED and IUCN/CEESP/CMWG/Cenesta.

Boster, J.S. (1986) 'Exchange of varieties and information between Aguaruna manioc cultivators', *American Anthropologist* 88(2): 428–36.

——(1985) 'Reqiuem for the omniscient informant: there's life in the old girl yet', pages 177–97 in J.W.D. Dougherty (ed.), *Directions in Cognitive Anthropology*, Urbana, IL: University of Illinois Press.

Boulton, A., Panizzon, D. and Prior, J. (2005) 'Explicit knowledge structures as a tool for overcoming obstacles to interdisciplinary research', *Conservation Biology* 19: 2026–9.

Bourque, N. (2001) 'Eating your words: communicating with food in Ecuadorian Andes', pages 85–100 in J. Hendry, and C. Watson, (eds), *An Anthropology of Indirect Communication*, ASA Monographs 37, London: Routledge.

Boycott, A.E. (1928) President's Address: The Transition from Live to Dead: the Nature of Filtrable Viruses. *Proceedings of the Royal Society of Medicine* 22(1): 55–69.

Brace, N., Kemp, R., and Snelgar, R. (2009) *SPSS for Psychologists*, 4th edn, Basingstoke: Palgrave Macmillan.

Bradshaw, C., Brook, B. and McMahon, C. (2007) 'Dangers of sensationalising conservation biology', *Conservation Biology* 21: 570–71.

Briggs, J. (1970) *Never in Anger: Portrait of an Eskimo Family*, Cambridge, MA: Harvard University Press.

Bryman, A. (2004) *Social Research Methods*, 2nd edn, Oxford: Oxford University Press.

Burton, D. (2000) *Research Training for Social Scientists: A Handbook for Postgraduate Scientists, Part 2: Ethical and Legal Issues in Social Science Research*, London: SAGE Publications.

Buscher, B. and Wolmer. W. (2007) 'Introduction: the politics of engagement between biodiversity conservation and the social sciences', *Conservation and Society* 5: 1–29.

Campbell, L. (2003) 'Challenges for interdisciplinary sea turtle research: perspectives of a social scientist', *Marine Turtle Newsletter* 100: 28–32.

—— (2005) 'Overcoming obstacles to interdisciplinary research', *Conservation Biology* 19: 574–7.

Caplan, P. (ed.) (2003) *The Ethics of Anthropology: Debates and Dilemmas*, London: Routledge.

Chambers, R. (1992) *Rural Appraisal: Rapid, Relaxed and Participatory*, Brighton: University of Sussex.

Chapin, M., Lamb, Z. and B. Threlkeld. (2005) 'Mapping indigenous lands', *Annual Review of Anthropology* 34(1): 619–38.

Clark, T. (2001) 'Developing policy-oriented curricula for conservation biology: professional and leadership education in the public interest', *Conservation Biology* 15: 31–9.

Clayton, S. and Brook. A. (2005) 'Can psychology help save the world? A model for conservation psychology', *Analyses of Social Issues and Public Policy* 5: 87–102.

Cooke, B. and Kothari, U. (eds) (2001) *Participation: The New Tyranny*? London: Zed Books.

Corbett, J. (2009) *Good Practices in Participatory Mapping: A Review Prepared for the International Fund for Agricultural Development (IFAD)*, Rome: IFAD. Online at <http://www.ifad.org/pub/map/PM_web.pdf>.

Cornwall, A. and Pratt, G. (eds) (2003) *Pathways to Participation: Reflections on PRA*, London: Institute of Development Studies (IDS).

Creswell, J.W. (2009) *Research Design: Qualitative, Quantitative and Mixed Methods Approaches*, 3rd edn, London: SAGE Publications.

Czaja, R. and Blair, J. (2005) *Designing Surveys: A Guide to Decisions and Procedures*, 2nd edn, London: SAGE Publications.

Czech, B. (2006) 'If Rome is burning, why are we fiddling?', *Conservation Biology* 20: 1563–5.

Dale, P. (1986) *The Myth of the Unique Japanese*, New York: St Martin's Press.

De Munck, V.C. and Sobo, E.J. (eds) (1998) *Using Methods in the Field: A Practical Introduction and Casebook*, London: Altamira Press.

De Vaus, D. (2002) *Surveys in Social Research*, 5th edn, London: Routledge.

DEFRA (Department for Environment, Farming and Rural Affairs) (2007) *'Survey of Public Attitudes and Behaviours toward the Environment, 2007'* [computer file], Environment Statistics and Indicators Division and BMRB. Social Research, Colchester, Essex: UK Data Archive [distributor], November 2007. SN: 5741.

Denscombe, M. (2007) *The Good Research Guide for Small-scale Social Research Projects*, 3rd edn, Maidenhead: Open University Press.

Dillon, P. (2008) 'A pedagogy of connection and boundary crossings: methodological and epistemological transactions in working across and between disciplines', *Innovations in Education and Teaching International* 45: 255–62.

Donovan, D.G. (1999) 'Strapped for Cash: Asians plunder their forests and endanger their future', *Asia Pacific Issues* 39, Honolulu, HI: East-West Center.

Dunlap, R., Van Liere, K., Mertig, A. and Jones, R. (2000) 'Measuring endorsement of the New Ecological Paradigm: a revised NEP scale', *Journal of Social Issues* 56: 425–42.

Eddleston, M. Davidson, R. Wilkinson, R. and Pierini, S. (2004) *The Oxford Handbook of Tropical Medicine*, Oxford: Oxford University Press.

Eghenter, C. (2000) *Mapping People's Forests: The Role of Mapping in Planning Community-based Management of Conservation Areas in Indonesia*, Washington, DC: Biodiversity Support Program.

Emmerson, D. (1976) *Indonesia's Elite: Political Culture and Cultural Politics*, Ithaca, NY: Cornell University Press.

Eriksson, L. (1999) 'Graduate conservation education', *Conservation Biology* 13: 955.

Everitt, B.S. and Dunn, G. (2000) *Applied Multivariate Analysis*, 2nd edn, London: Hodder Arnold.

Field, A. (2009) *Discovering Statistics Using SPSS*, 3rd edn, London: SAGE Publications.

Fowler, F. (1995) *Improving Survey Questions: Design and Evaluation*, Applied Social Research Methods Series 38, London: SAGE Publications.

Fowler, J., Cohen, L. and Jarvis, P. (1998) *Practical Statistics for Field Biology*, 2nd edn, Chichester: John Wiley & Sons.

Fox, H., Christian, C., Cully-Nordby, J., Pergams, O., Peterson, G. and Pyke, C. (2006) 'Perceived barriers to integrating social science and conservation', *Conservation Biology* 20: 1817–20.

Fox, J., Rindfuss, R., Walsh, S.J. and Mishra, V. (2003) *People and the Environment: Approaches for Linking Household and Community Surveys to Remote Sensing and GIS* Boston, MA: Kluwer Academic.

Frake, C. (1964) 'How to ask for a drink in Subanun', in J. Gumperz and D. Hymes (eds), The Ethnography of Communication, *American Anthropologist* 66: 123–32.

Freeman, D. (1983) *Margaret Mead and Samoa: The Making and Unmaking of an Anthropological Myth*, Cambridge, MA: Harvard University Press.

Geertz, C. (2001) *Thinking as a Moral Act: Ethical Dimensions of Anthropological Fieldwork in the New State in Available Light*, Princeton, NJ: Princeton University Press.

Gifford, R. (1997) *Environmental Psychology: Principles and Practice*, 2nd edn, Needham Heights, MA: Allyn and Bacon.

Gilchrist, T. (2008) 'Emerging issues in water recreation: an investigation into the conflict between kitesurfing and birds in the UK', unpublished MSc thesis, DICE, University of Kent, Canterbury.

Gill, R. (2001) 'Professionalism, advocacy and credibility: a futile cycle?', *Human Dimensions of Wildlife* 6: 21–32.

Govan, H., Aalbersberg, W., Tawake, A. and Parks, J. (2008) *Locally-managed Marine Areas: A Guide for Practitioners.* The Locally-Managed Marine Area Network. Available online at <www.LMMAnetwork.org>

Greenway, M. (1979) *'Essex Members of Friends of the Earth'*, [computer file], Colchester: UK Data Archive [distributor], 1981. SN: 1561.

Harper, D. (2002) 'Talking about pictures: a case for photo elicitation', *Visual Studies* 17(1): 13–26.

Harrison, S., Massey, D. and Richards, K. (2008) 'Conversations across the divide', *Geoforum* 39: 549–51.

Harvey, G. (1998) *Writing with Sources: A Guide for Students*, Indianapolis, IN: Hackett Publishing Company.

Hendry, J. (1999) *An Anthropologist in Japan: Glimpses of Life in the Field*, London: Routledge.

Hickey, S. and Mohan, G. (2004) *Participation – From Tyranny to Transformation? Exploring New Approaches to Participation in Development*, London: Zed Books.

Holland, J. and Campbell J. (eds) (2005) *Methods in Development Research: Combining Qualitative and Quantitative Approaches*, Swansea: Practical Action Publishing.

IAPAD (2008) Participatory Mapping Toolbox. Available online at http:/www.iapad.org/toolbox.htm) (accessed 16 October 2009).

IIRR (International Institute of Rural Reconstruction) (1996) *Recording and Using Indigenous Knowledge: A Manual*, compiled by R.M. Pastores and R. E. San Buenaventura, Silang, Philippines: IIRR. http://www.panasia.org.sg/iirr/ikmanual/layong.htm.

INEI (Instituto Nacional de Estadísticas e Informaci6n) (n.d.) *Gráfico: Ingreso de divisas por turismo, 1986–2000*, Available online at <http://www1.inei.gob.pe/biblioineipub/bancopub/Est/Lib0412/cap10.htm> (accessed 12 November 2009).

Inouye, D. and Dietz, J. (2000) 'Creating academically and practically trained graduate students', *Conservation Biology* 14: 595–6.

Jacobson, S.K. and McDuff, M.D. (1998) 'Training idiot savants: the lack of human dimensions in conservation biology', *Conservation Biology* 12(2): 263–7.

Jakobsen, C., Hels, T. and McLaughlin, W. (2004) 'Barriers and facilitators to integration among scientists in transdisciplinary landscape analyses: a cross-country comparison', *Forest Policy and Economics* 6: 15–31.

Johnson, K. (2008) 'Community orchards in England as indigenous and community conserved areas (ICCAs)', unpublished MSc thesis, DICE, University of Kent, Canterbury.

Joshi, P. (1979) 'Fieldwork experience: relived and reconsidered. Rural Uttar Pradesh', Pages 73–99 in M. Srinivas, A. Shah and E. Ramaswamy (eds), *The Fieldworker and the Field: Problems and Challenges in Sociological Investigation*, Delhi: Oxford University Press.

Kaplowitz, M. and Hoehn, J. (2001) 'Do focus groups and individual interviews reveal the same information for natural resource valuation?', *Ecological Economics* 36: 237–47.

Klein, J.T. (1990) *Interdisciplinarity: History, Theory and Practice*, Detroit, MI: Wayne State University Press.

Krech, S., McNeill, J.R. and Merchant, C. (2004) *Encyclopedia of World Environmental History*. New York, London: Routledge.

Kroll, A.J. (2005) 'Integrating professional skills in wildlife student education', *Journal of Wildlife Management* 71(1): 226–30.

Kumar, S. (2002) *Methods for Community Participation: A Complete Guide for Practitioners*, Bourton on Dunsmore: Practical Action Publishing.

Lassiter, L.E. (2005) *The Chicago Guide to Collaborative Ethnography* (2005). Chicago, IL: University of Chicago Press.

Lau, L. and Pasquini, M. (2004) 'Meeting grounds: perceiving and defining interdisciplinarity across the arts, social sciences and sciences', *Interdisciplinary Science Reviews* 29: 49–64.

Lidicker, W.Z. (1998) 'Revisiting the Human Dimension in Conservation Biology', *Conservation Biology*, 12(6): 1170–1.

Little, P.D. (1994) 'The link between local participation and improved conservation: a review of issues and experiences', pages 347–72 in D. Western and R.M. Wright (eds), *Natural Connections: Perspectives in Community-based Conservation*, Washington, DC: Island Press.

Lopez, R., Hays, K., Wagner, M., Locke, S., McLeery, R. and Silvy, N. (2006) 'Integrating land conservation planning in the classroom', *Wildlife Society Bulletin* 34: 223–8.

MacMillan, D., L. Hanley, P.N. and Alvarez-Farizo, B. (2002) 'Valuing the non-market benefits of wild goose conservation: a comparison of interview and group-based approaches', *Ecological Economics* 43: 49–59.

Malinowski, B. (1922) *Argonauts of the Western Pacific*, London: Routledge and Sons.
Martin, G.J. (1995) *Ethnobotany: A Methods Manual*, London: Chapman and Hall; reprinted in 2004 by Earthscan.
—— (2004) *Ethnobotany: A Methods Manual*, London: Earthscan.
Martinich, J., Solarz, S. and Lyons, J. (2006) 'Preparing students for conservation careers through project-based learning', *Conservation Biology* 20: 1579–83.
Marzano, M., Carss, D.N. and Bell, S. (2006) 'Working to make interdisciplinarity work: investing in communication and interpersonal relationships', *Journal of Agricultural Economics* 57(2): 185–97.
Mascia, M., Brosius, J., Dobson, T., Forbes, B., Horowitz, L., McKean, M. and Turner, N. (2003) 'Conservation and the social sciences', *Conservation Biology* 17: 649–50.
Meffe, G. (1998) 'Softening the boundaries', *Conservation Biology* 12: 259–60.
—— (2007) 'Conservation focus: policy advocacy and conservation science', *Conservation Biology* 21: 11.
Meine, C., Soule, M. and Noss, R. (2006) '"A mission-driven discipline": the growth of conservation biology', *Conservation Biology* 20: 631–51.
Menon, R.V.G. (2010) *An Introduction to the History and Philosophy of Science*. Noida, Dorling Kindersley India.
Metcalf, P. (2002) *They Lie, We Lie: Getting on with Anthropology*, London: Routledge.
Mikkelsen, B. (2005) *Methods for Development Work and Research: A New Guide for Practitioners*, 2nd edn, London: SAGE Publications.
Miles, M. and Huberman, A. (1994) *Qualitative Data Analysis: An Expanded Sourcebook*, 2nd edn, London: SAGE Publications.
Moeran, B. (1985) *Okubo Diary: Portrait of a Japanese Valley*, Stanford, CA: Stanford University Press.
Moran, J. (2002) *Interdisciplinarity*, Routledge, London.
Morgan, D. and Krueger, R. (1998) *The Focus Group Kit*, Thousand Oaks, CA: SAGE Publications.
Mosse, D. (2005) *Cultivating Development: An Ethnography of Aid Policy and Practice*, London: Pluto Press.
Mulder, M.B. and Coppolillo, P. (2005) *Conservation: Linking Ecology, Economics and Culture*, Princeton, NJ: Princetown University Press.
Mulvey, M. and Lydeard, C. (2000) 'Let's not abandon science for advocacy: reply to Berg and Berg', *Conservation Biology* 14: 1924–5.
National Centre for Social Research (2000) *British Social Attitudes Survey, 2000* [computer file]. Colchester, Essex: UK Data Archive [distributor], March 2002. SN: 4486.
Nelson, M.P. and Vucetich, J.A. (2009) 'On advocacy by environmental scientists: what, whether, why and how', *Conservation Biology* 23(5): 1090–101.
Newing, H. (2007) 'Social impacts of industrial logging concessions: effects on forest user rights', pages 58–64 in S. Counsell, C. Long and S. Wilson (eds), *Concessions to Poverty: The Environmental, Social and Economic Impacts of Industrial Logging Concessions in Africa's Rainforests*, London: Rainforest Foundation UK / Forests Monitor.
—— (2009) 'It's not good to stay in one place': mobility and conservation in the Peruvian Amazon. The case of Tamshiyacu Tahuayo Communal Reserve', pages 97–114 in M. Alexiades (ed.), *Ethnobiology of Mobility, Displacement and Migration in Indigenous Lowland South America*, Studies in Environmental Anthropology and Ethnobiology, Oxford: Berghahn.
—— (forthcoming) 'Bridging the gap: interdisciplinarity, biocultural diversity and conservation', in Pretty, J. and Pilgrim, S. (eds), *Nature and Culture: Revitalising the Connection*. London: Earthscan.
Nichols, P. (1991) *Social Survey Methods: A Field Guide for Development Workers*, Oxford: Oxfam.
Nordenstam, B. and Smardon, R. (2000) 'A perspective of educational needs in environmental science and policy for the next century', *Environmental Science and Policy* (3): 57–8.
Noss, R. (1997) 'The failure of universities to produce conservation biologists', *Conservation Biology* 11: 1267–9.

—— (1999) 'Is there a special conservation biology?', *Ecography* 22: 113–22.

—— (2000) 'Science on the bridge', *Conservation Biology* 14: 333–5.

Pallant, J. (2007) *SPSS Survival Manual: A Step by Step Guide to Data Analysis Using SPSS for Windows*, 3rd edn, Milton Keynes: Open University Press.

Pearce, D. (2002) 'An intellectual history of environmental economics', *Annual Review of Energy and the Environment* 27: 57–81.

Peluso, N.L. (1995) 'Whose woods are these? Counter-mapping forest territories in Kalimantan, Indonesia', *Antipode* 27(4): 383–406.

Penn, L. (2005) 'An exploration of zoo theatre's contribution to the directives of zoos: a case study from the Central Park Zoo in New York', unpublished thesis, University of Kent.

Perez, H. (2005) 'What students can do to improve graduate education in conservation biology?', *Conservation Biology* 19: 2033–5.

Pickett, S., Burch Jr., W. and Grove, J. (1999) 'Interdisciplinary research: maintaining the constructive impulse in a culture of criticism', *Ecosystems* 2: 302–7.

Pomeroy, R.S. and Rivera-Guieb, R. (2006) *Fishery Co management: A Practical Handbook*, Wallingford: CAB International/International Development Research Centre.

PPGIS.NET Open Forum on Participatory Geographic Information Systems and Technologies. Available online at http://ppgis.iapad.org (accessed on 10 January 2009).

Pratt, B. and Loizos, P. (1992) *Choosing Research Methods: Data Collection for Development Workers*, Oxford: Oxfam.

Pretty, J., Guijt, I. Thompson, J. and Scoones, I. (1995) *Participatory Learning and Action: A Trainer's Guide*, London: IIED.

Pride, J. and J. Holmes (1972) *Sociolinguistics: Selected Readings*, Harmondsworth: Penguin.

Puri, R.K. (2005) *Deadly Dances in the Bornean Rainforest: Hunting Knowledge of the Penan Benalui*, Leiden: KITLV Press.

Puri, R.K. and Maxwell, O. (2002) 'Forest use and natural resource management in two villages adjacent to Pu Luong Nature reserve and Cuc Phuong National Park, Vietnam', in U. Apel, O.C. Maxwell, T.N. Nguyen, M. Nurse, R.K.Puri and V.C. Trieu (eds), *Collaborative Management and Conservation: A Strategy for Community Based Natural Resource Management of Special Use Forest in Vietnam – Case Studies from Pu Luong Nature Reserve, Thanh Hoa Province*, Part II, pages 37–122, Cambridge: Fauna and Flora International/World Bank.

Puri, R.K. and Vogl, C. (2005) *A Methods Manual for Ethnobiological Research and Cultural Domain Analysis, with analysis using ANTHROPAC*, unpublished manuscript, Anthropology Department, University Kent, Canterbury, UK.

Rambaldi, G., Corbett, J., McCall, M., Olson, R., Muchemi, J., Kyem, P.K., Wiener, D. and Chambers, R. (2006) 'Mapping for change: practice, technologies and communication', Participatory Learning and Action No. 54, London: IIED.

Redford, K. and Sanjayan, M. (2003) 'Retiring Cassandra', *Conservation Biology* 17: 1473–4.

Robben, A.C.G.M. and Suka, J.A. (eds) (2007) *Ethnographic Fieldwork: An Anthropological Reader*, Oxford: Blackwell.

Roebuck, P. and Phifer, P. (1999) 'The persistence of positivism in conservation biology', *Conservation Biology* 13: 444–6.

Rosa, E. and Machlis, G. (2002) 'It's a bad thing to make one thing into two: disciplinary distinctions as trained incapacities', *Society and Natural Resources* 15: 251–61.

Rowntree, D. (2000) *Statistics Without Tears: An Introduction for Non-mathematicians*, London: Penguin.

Russell, D. and Harshbarger, C. (2003) *Groundwork for Community-based Conservation: Strategies for Social Research*, Walnut Creek, CA; Oxford: Altamira Press.

Ryan, G. and Weisner, T. (1998) 'Content analysis of words in brief descriptions: how fathers and mothers describe their children', pages 57–68 in V. de Munck, and E. Sobo (eds), *Using Methods in the Field: A Practical Introduction and Casebook*. Walnut Creek, CA: Altamira Press.

Saberwal, V.K. and Kothari, A. (1996) 'The human dimension in conservation biology curricula in developing countries', *Conservation Biology* 10: 1328–31.

Saunders, C. and Myers Jr., O. (2003) 'Exploring the Potential of Conservation Psychology', *Human Ecology Review* 10: 10(2): iii–v.

Scholte, P. (2003) 'Curriculum development at the African Regional Wildlife Colleges, with special reference to the Ecole de Faune, Cameroon', *Environmental Conservation* 30(3): 249–58.

Scott, J., Rachlow, J., Lackey, R., Pidgorna, A., Aycrigg, J., Feldman, G., Svancara, L., Rupp, D., Stanish, D. and Steinhorst, R. (2007) 'Policy advocacy in science: prevalence, perspectives and implications for conservation biologists', *Conservation Biology* 21: 29–35.

Selener, D., Endara, N. and Carvajal, J. (1999) *Participatory Rural Appraisal and Planning Workbook*, Quito, Ecuador: International Institute of Rural Reconstruction.

Sheil, D. and Liswanti, N. (2006) 'Scoring the importance of tropical forest landscapes with local people: patterns and insights', *Environmental Management* 38(1): 126–36.

Sheil, D., Puri, R.K., Basuki, I., van Heist, M., Rukmiyati, M.A., Sardjono, A., Samsoedin, I., Sidiyasa Chrisandini, K., Permana, E. Mangopo Angi E., Gatzweiler, F., Johnson, B. and Wijaya, A. (2003) *Exploring Biological Diversity, Environment and Local People's Perspectives in Forest Landscapes: Methods for a Multidisciplinary Landscape Assessment*, Bogor, Indonesia: Center for International Forestry Research.

Sheil, D., Puri, R.K., Wan, M., Basuki, I., van Heist, M., Liswanti Rukmiyati N., Rachmatika, I. and Samsoedin, I. (2006) 'Local people's priorities for biodiversity: examples from the forests of Indonesian Borneo', *Ambio* 15(1): 17–24.

Shore, C. (1999) 'Fictions of Fieldwork: depicting the 'self' in ethnographic writing', pp. 25–48 in C.W. Watson (ed.), *Being There*. London: Pluto Press.

Siebert, S. (2000) 'Creating academically and practically training graduate students', *Conservation Biology* 14: 595–96.

Sillitoe, P. (2004) 'Interdisciplinary experiences: working with indigenous knowledge in development', *Interdisciplinary Science Reviews* 29: 9–23.

Sirait, M.T., S. Prasodjo, N. Podger, A. Flavelle and J. Fox, (1994) 'Mapping customary land in East Kalimantan, Indonesia: a tool for forest management', *Ambio* 23: 411–17.

Slade, L. (2007) 'Community perceptions and stakeholder relationships in nature-based tourism in Ngezi-Vumawimbi area of Pemba, Zanzibar', unpublished MSc thesis, DICE, University of Kent, Canterbury.

Smith, J. J. (1993) 'Using ANTHROPAC 3.5 and a spreadsheet to compute a free-list salience index', *Cultural Anthropology Methods Journal* 5(3): 1–3.

Smith, R.J., Muir, R.D.J., Walpole, M.J., Balmford, A. and Leader-Williams, N. (2003) 'Governance and the loss of biodiversity', *Nature* 426: 67–70.

Soule, M. and B. Wilcox (1980) *Conservation Biology: An Evolutionary-ecological Perspective*, Sunderland, MA: Sinauer Associates.

Spindler. G.D. (ed.) (1970) *Being an Anthropologist. Fieldwork in Eleven Cultures*, New York: Holt, Rinehart and Winston.

Spradley, J. (1979) *The Ethnographic Interview*, New York, London: Rinehart & Winston.

Srinivas, M., Shah, A. and Ramaswamy, E. (1979) *The Fieldworker and the Field: Problems and Challenges in Sociological Investigation, Delhi*. New Delhi: Oxford University Press.

Stapp, W., Bennett, D., Bryan, W., Fulton, J., Havlick, S., MacGregor, J., Nowak, P., Swan, J. and Wall, R. (1969) 'The Concept of Environmental Education', *Journal of Environmental Education* 1: 30–1.

Stevens, S.S. (1946) 'On the theory of scales of measurement', *Science* 103: 677–80.

Stockdale, M.C. and B. Ambrose, (1996) 'Mapping and NTFP Inventory: Participatory Assessment Methods for Forest-Dwelling Communities in East Kalimantan, Indonesia', pp. 170–211 in J. Carter (ed.), *Recent Approaches to Participatory Forest Resource Assessment*. London: Overseas Development Institute.

Strang, V. (2009) 'Integrating the social and natural sciences in environmental research: a discussion paper', *Environment Development and Sustainability* 11: 1–18.

Strauss, A. and Corbin, J. (1990) *Basics of Qualitative Research: Grounded Theory Procedures and Techniques*. London: SAGE Publications.

Stross, B. (1969) 'Aspects of language acquisition by Tzeltal Children', unpublished thesis, University of California, Berkeley.

Swinscow-Hall, O.C. (2007) 'Processes of knowledge transmission among a displaced Batwa community in southwest Uganda', unpublished MSc thesis, University of Kent, Canterbury.

Takacs, D., Shapiro, D. and Head, W. (2006) 'From is to should: helping students translate conservation biology into conservation policy', *Conservation Biology* 20: 1342–8.

Tashakkori, A. and Teddlie, C. (1998) *Mixed Methodology: Combining Qualitative and Quantitative Approaches.* Applied Social Research Methods Series volume 46. London: SAGE Publications.

Touval, J. and Dietz, J. (1994) 'The problem of teaching conservation problem solving', *Conservation Biology* 8: 902–4.

Trudgill, P. (1995) *Sociolinguistics: An Introduction to Language and Society*, London: Penguin.

Turner, V. (1960) 'Muchona the hornet, interpreter of religion', pages 334–42 in J. Casagrande (ed.), *In the Company of Man*, New York: Harper and brothers.

Twain, M. (1902-1903) 'Mark Twain's Notebook', Available HTTP: <http://editorialengine.com/?p=808> (accessed 22 July 2010).

Twyman, C. (2000) 'Participatory conservation? Community-based natural resource management in Botswana', *The Geographical Journal* 166(4): 323–35.

Unni, K. (1979) 'On the tracks and tracts in my fieldwork. Rural Kerala' pp. 58–72 in M. Srinivas, A. Shah and E. Ramaswamy (eds), *The Fieldworker and the Field: Problems and Challenges in Sociological Investigation.* Delhi: Oxford University Press.

Van Leeuwen, T. and Jewitt, C. (eds) (2001) *Handbook of Visual Analysis*, London: SAGE Publications.

Vogl, C., Vogl-Lukasser, B. and Puri, R.K. (2004) 'Tools and methods for data collection in ethnobotanical studies of homegardens', *Field Methods* 16(3): 285–306.

Wade, Peter (ed.) (1995) *Advocacy in Anthropology: The GDAT Debate*, Manchester: Department of Anthropology, Manchester University.

Wallace, M. and Wray, A. (2006) *Critical Reading and Writing for Postgraduates*, SAGE Study Skills Series, London: SAGE Publications.

Weller, S.C., and Romney, K.A. (1988) *Systematic Data Collection*, Newbury Park, CA: SAGE Publications.

Werner, D., Thuman, C., Maxwell, J., Pearson, A. and Cary, F. (1994) *Where There is No Doctor: Village Health Care Handbook for Africa*, Oxford: MacMillan Education.

White, P.C.L., Vaughan Jennings, N., Renwick, A.R. and Barker, N.H.L. (2005) 'Questionnaires in Ecology: A Review of Past Use and Recommendations for Best Practice', *Journal of Applied Ecology* 42: 421–30.

White, R., Fleischner, T. and Trombulak, S. (2000) 'The status of undergraduate education in conservation biology', Available HTTP: <http://www.conbio.org/Resources/education/StatusUndergradEducation.pdf> (accessed 23 February 2009).

Whyte, W.F. (1943) *Street Corner Society*, Chicago, IL: University of Chicago Press.

Wilson, I. (2005) 'Some practical sampling procedures for development research', pp. 37–51 in J. Holland and J. Campbell (eds), *Methods in Development Research: Combining Qualitative and Quantitative Approaches*, Bourton on Dunsmore: Intermediate Technology Applications/Practical Action Publishing.

Winser, S. (ed.) (2004) *Royal Geographical Society Expedition Handbook*, London: Profile Books.

Wollenberg, E., with D. Edmunds and L. Buck (2000) *Anticipating Change: Scenarios as a Tool for Adaptive Management*, Bogor, Indonesia: Centre for International Forestry Research.

World Bank (2001) *World Development Report 2000/2001: Attacking Poverty*, New York: Oxford University Press.

Young, I. and Gherardin, T. (2008) *Africa: Healthy Travel Guide*, London: Lonely Planet.

Index

Figures in **Bold**; Tables in *Italics*